New Aspects of Cancer Stem Cell Biology

New Aspects of Cancer Stem Cell Biology: Implications for Innovative Therapies

Editors

Ugo Cavallaro
Marco Giordano

MDPI • Basel • Beijing • Wuhan • Barcelona • Belgrade • Manchester • Tokyo • Cluj • Tianjin

Editors
Ugo Cavallaro
Unit Gynecol Oncol Res
European Institute of Oncology IRCCS
Italy

Marco Giordano
Unit Gynecol Oncol Res
European Institute of Oncology
IRCCS Italy

Editorial Office
MDPI
St. Alban-Anlage 66
4052 Basel, Switzerland

This is a reprint of articles from the Special Issue published online in the open access journal *Journal of Clinical Medicine* (ISSN 2077-0383) (available at: https://www.mdpi.com/journal/jcm/special_issues/Cancer_Stem_Cell_Biology).

For citation purposes, cite each article independently as indicated on the article page online and as indicated below:

LastName, A.A.; LastName, B.B.; LastName, C.C. Article Title. *Journal Name* **Year**, *Article Number*, Page Range.

ISBN 978-3-03943-406-0 (Hbk)
ISBN 978-3-03943-407-7 (PDF)

Cover image courtesy of Ugo Cavallaro.
The cover image shows a human lung cancer section which has been stained for a putative cancer stem cell marker. A subset of marker-positive tumor cells are present at the invasive front of the tumor.

Contents

About the Editors

Ugo Cavallaro, PhD. Director of Unit of Gynecological Oncology Research. I have a broad background in cell biology, biochemistry, and molecular oncology. As a junior scientist, I focused on the biological mechanisms that underlie the role of cell adhesion molecules in cancer cells. Upon establishing my own group, I built on that knowledge to investigate the cellular and molecular events that drive ovarian cancer. The projects that I have conducted so far as PI or co-investigator provided not only the basis and the rationale for the ongoing research, but also a series of clinically relevant experimental models. Over the last years, my group collected a repository of viable tissue from ovarian cancer and healthy normal ovary and fallopian tubes. We also have a repository of primary cell cultures and patient-derived xenografts. Recent achievements of my lab have included profiling the proteome and the phosphoproteome of primary ovarian cancer cells vs. their normal counterpart, which resulted in the discovery of the CDK7/POLR2A pathway as a novel driver in ovarian cancer. We were also the first to report the novel role of CD73 in ovarian cancer stem cells and the possibility to design CD73-based cancer stem cell-targeted therapy. My group is also interested in tumor/microenvironment interactions, with a particular emphasis on the vasculature. We identified L1CAM as a master regulator of cancer vessel morphology and function, and have recently discovered a novel, soluble L1CAM isoform with angiogenic activity involved in ovarian cancer vascularization.

Marco Giordano, PhD. Postdoc researcher. I obtained my Master's degree magna cum laude in Biological Sciences at the University of Calabria in December 2009. Then, I started my PhD in the laboratory of Genetics at the same University, headed by Prof. Giuseppe Passarino, working in the pioneering field of mitochondrial DNA epigenetics. I received my PhD in Operative Research (bio-pathologic curriculum) in December 2012. My first postdoc experience started in April 2013 at the University Magna Graecia of Catanzaro in the laboratory of molecular hematopoiesis and stem cell biology, headed by the late Prof. Giovanni Morrone, studying the functional cross-talk between the zinc finger protein ZNF521 and the sonic hedgehog pathway effectors Gli1/2. In 2014, I spent one year at DKFZ in Heidelberg as a visiting postdoc at the Department of Molecular Genetics headed by Prof. Dr. Peter Lichter. There, I completed two projects that focused on the functional role of ZNF521 in the biology of glioblastoma stem-like cells and its interaction with the polycomb machinery to determine their stem-like phenotype. In 2015, I joined the group of Dr. Ugo Cavallaro at IEO, which made my research more translational working on patient-derived samples and taking advantage of the collaboration with IEO clinicians. Currently, I am studying the functional contribution of the adhesion molecule L1 to the pathophysiology of ovarian cancer stem cells, focusing on the related biological mechanism and its therapeutic implications.

Preface to "New Aspects of Cancer Stem Cell Biology: Implications for Innovative Therapies"

The cancer stem cell (CSC) paradigm represents one of the most prominent breakthroughs of the last decades in tumor biology. CSCs are broadly defined as a subpopulation of tumor cells that, besides sharing some functional features with normal stem cells (self-renewal, low cycling rate, and differentiation capacity), can drive tumor initiation, to rebuild the heterogeneity of the original tumor, to evade the immune system, and to escape conventional treatments such as chemo- and radiotherapy. In light of these properties, CSCs have been implicated in tumor metastasis and recurrence. Nevertheless, the biological characteristics of CSCs, their molecular profile, their contribution to cancer progression, and even their existence, remain matters of intense debate and controversy among tumor biologists. The difficulty in achieving a better definition of CSC biology may be explained by the plasticity of such a cell subpopulation. Indeed, the emerging view is that CSCs represent a dynamic "state" of tumor cells that can acquire stemness-related properties under specific circumstances, rather than referring to a well-defined group of cells.

Regardless of their origin, it is clear that designing novel anti-tumor treatments based on the eradication of CSCs will only be possible upon unraveling the biological mechanisms that underlie their pathogenic role in tumor progression and therapy resistance. The Special Issue on "New Aspects of Cancer Stem Cell Biology: Implications for Innovative Therapies" will highlight recent insights into CSC features that can make them an attractive target for novel therapeutic strategies.

Ugo Cavallaro, Marco Giordano
Editors

Journal of
Clinical Medicine

Article

Patient-Derived Colorectal Cancer Organoids Upregulate Revival Stem Cell Marker Genes Following Chemotherapeutic Treatment

Rebekah M. Engel [1,2,3,†], Wing Hei Chan [1,2,†], David Nickless [4], Sara Hlavca [1,2], Elizabeth Richards [1,2], Genevieve Kerr [1,2,5], Karen Oliva [3], Paul J. McMurrick [3], Thierry Jardé [1,2,5,6,*] and Helen E. Abud [1,2,5,*]

[1]   Department of Anatomy and Developmental Biology, Monash University, Clayton Victoria 3800, Australia; rengel@cabrini.com.au (R.M.E.); horace.chan@monash.edu (W.H.C.); Sara.Hlavca1@monash.edu (S.H.); elizabeth.a.richards@monash.edu (E.R.); genevieve.kerr@monash.edu (G.K.)
[2]   Stem Cells and Development Program, Monash Biomedicine Discovery Institute, Monash University, Clayton, Victoria 3800, Australia
[3]   Cabrini Monash University Department of Surgery, Cabrini Hospital, Malvern Victoria 3144, Australia; koliva@cabrini.com.au (K.O.); pjm@colorectal.com.au (P.J.M.)
[4]   Anatomical Pathology Department, Cabrini Pathology, Cabrini Hospital, Malvern, Victoria 3144, Australia; David.Nickless@mps.com.au
[5]   Monash BDI Organoid Program, Monash Biomedicine Discovery Institute, Monash University, Wellington Road, Clayton, Victoria 3800, Australia
[6]   Centre for Cancer Research, Hudson Institute of Medical Research, Clayton, Victoria 3168, Australia
*   Correspondence: thierry.jarde@monash.edu (T.J.); helen.abud@monash.edu (H.E.A.)
†   These authors contributed equally to this work.

Received: 24 November 2019; Accepted: 31 December 2019; Published: 2 January 2020

Abstract: Colorectal cancer stem cells have been proposed to drive disease progression, tumour recurrence and chemoresistance. However, studies ablating leucine rich repeat containing G protein-coupled receptor 5 (LGR5)-positive stem cells have shown that they are rapidly replenished in primary tumours. Following injury in normal tissue, LGR5+ stem cells are replaced by a newly defined, transient population of revival stem cells. We investigated whether markers of the revival stem cell population are present in colorectal tumours and how this signature relates to chemoresistance. We examined the expression of different stem cell markers in a cohort of patient-derived colorectal cancer organoids and correlated expression with sensitivity to 5-fluorouracil (5-FU) treatment. Our findings revealed that there was inter-tumour variability in the expression of stem cell markers. Clusterin (*CLU*), a marker of the revival stem cell population, was significantly enriched following 5-FU treatment and expression correlated with the level of drug resistance. Patient outcome data revealed that *CLU* expression is associated with both lower patient survival and an increase in disease recurrence. This suggests that *CLU* is a marker of drug resistance and may identify cells that drive colorectal cancer progression.

Keywords: bowel cancer; organoid; tumoroid; colorectal; colon; stem cell; chemotherapy resistance

---

1. Introduction

Colorectal cancer (CRC) is the most frequently diagnosed cancer of the digestive tract and a principal cause of cancer-related deaths worldwide [1,2]. The majority of deaths from CRC can be attributed to cancer recurrence after initial treatment, which presents as distant metastases in secondary sites such as the liver or lung. If left untreated, the 5 year survival rate of patients with metastatic CRC can be as low as 5% [3,4]. Combined chemotherapy treatment, which includes the commonly used drug 5-fluorouracil (5-FU), can help to increase this survival rate [5,6]. However, development of more targeted and personalised treatments are necessary to decrease overall CRC mortality [7]. Understanding the mechanisms underlying resistance to common treatments incorporating 5-FU is complex and requires further elucidation in models that recapitulate the diversity of primary tumours arising [8].

Cellular heterogeneity is a feature of CRC tumours, where only a subset of cells display tumour-initiating activity [9,10]. The cancer stem cell (CSC) hypothesis supports a model where a small population of stem cells drive tumour growth and metastasis and may even predict disease relapse [11]. Furthermore, CSCs may enter a quiescent state, rendering them inherently resistant to anti-proliferative drugs. They may also stimulate tumour recurrence following therapy. [12]. CSC markers are promising prognostic biomarkers and therapeutic targets. However, several studies have now demonstrated the considerable plasticity of stem cell populations within tumours which creates further complexity when targeting these cell populations [13,14]. Characterising the dynamics of these cell populations in response to chemotherapeutic challenge is therefore of critical importance.

Normal intestinal crypt-based columnar (CBC) stem cell markers, including leucine rich repeat containing G protein-coupled receptor 5 (*LGR5*) and EPH receptor B2 (*EPHB2*), have been shown to be over-expressed in colorectal adenocarcinomas, and colorectal tumours that display a stem cell signature correlate with a decrease in disease-free survival of patients and an increase in relapse [15,16]. However, specific targeted ablation of LGR5+ cells in cancerous tissues revealed that the LGR5+ stem cell population is dispensable for primary tumour maintenance [13,14,17]. This is also the case when LGR5+ cells are deleted within the intestinal epithelium during normal homeostasis [17,18]. In normal tissues and tumours, it appears that cellular plasticity within intestinal cell populations allows LGR5- progenitors to revert back into a stem-like state and reconstitute the ablated LGR5+ stem cells [13,14,17,18]. In normal tissue, a quiescent reserve population of stem cells marked by BMI1 polycomb ring finger proto-oncogene (*BMI1*) has been postulated to be activated upon damage and regenerate lost *Lgr5*+ cells [18,19]. Whether a population of cells with these characteristics are present in tumours is unclear. However, colon tumours with elevated levels of *BMI1* have been associated with reduced overall patient survival [20].

Recently, a novel unique stem cell population, marked by clusterin (*CLU*) and annexin A1 (*ANXA1*) expression, has been identified by single cell RNA sequencing in normal mouse small intestinal tissue. This stem cell pool is activated at the onset of tissue injury and repopulates the damaged small intestinal crypts, including the LGR5+ stem cell population [21]. The exact role of this revival stem cell population in a cancer context remains yet to be determined.

In this study, we aim to characterise the expression profile of the revival, CBC and quiescent stem cell markers in colorectal cancer and examine how this expression is modified upon chemotherapeutic treatment with 5-FU. To achieve this objective, we make use of patient-derived colorectal tumour organoids that were generated from treatment-naïve patients. This enables analysis of expression of stem cell markers in a cohort of organoid lines derived directly from patients. Subsequently, we investigate how the profile of different stem cell markers correlates with resistance to therapy and patient survival.

2. Experimental Section

2.1. Ethics and Consent

This study was conducted in accordance with the Declaration of Helsinki, and the protocol was approved by the Cabrini Human Research Ethics Committee (CHREC 04-19-01-15) and the Monash Human Research Ethics Committee (MHREC ID 2518). Patient recruitment was led by the colorectal surgeons in the Cabrini Monash University Department of Surgery. Tissue was obtained from treatment naïve patients diagnosed with colorectal cancer undergoing surgical resection at the Cabrini Hospital, Malvern, Australia. All subjects provided written informed consent.

2.2. Patient Data

Patient information including clinical characteristics, treatment regimen and outcome data (Table S1) was obtained from the prospectively maintained, clinician-led Cabrini Monash University Department of Surgery colorectal neoplasia database (CMCND) [22]. This dataset has been adopted in a minimum dataset format as the Binational Colorectal Cancer Audit of the Colorectal Surgical Society of Australia and New Zealand (https://cssanz.org/bcca-database/).

2.3. Establishing Colorectal Cancer Organoids

CRC tissue specimens were cut into 5 mm pieces and washed eight times with 1× PBS supplemented with antibiotics. Tissue fragments were digested with 0.125 mg/mL dispase type II (Sigma, St Louis, MO, USA) and 1 mg/mL collagenase A (Roche Diagnostics, Mannheim, Germany) at 37 °C for 30 min and then mechanically dissociated by repetitive pipetting in cold PBS. Cancer tissue fragments were allowed to settle by gravity before supernatant was collected and pelleted by centrifugation at 240× g for 5 min at 4 °C. The isolated cells/fragments were passed through a 70 μm cell strainer (Corning, NY, USA), centrifuged and resuspended in Matrigel (Corning).

Matrigel containing cancer cell clusters were seeded into 24-well tissue culture plates (Thermo Scientific Nunc, Foster City, CA, USA) and allowed to polymerize for 10 min at 37 °C. The cancer cells were overlaid with 500 μL of culture medium composed of advanced Dulbecco's modified Eagle medium/F12 supplemented with 1X B27, Glutamax, 10 mM HEPES (all from Gibco, Waltham, MA, USA), 100 μg/mL Primocin (InvivoGen, San Diego, CA, USA), 50 ng/mL recombinant human EGF (Peprotech, Rochy Hill, NJ, USA), 10 nM Gastrin (Sigma), 500 nM A83-01 (Tocris Bioscience, Bristol, UK), 1.25 mM N-acetylcysteine (Sigma), 10 mM nicotinamide (Sigma) and 100 ng/mL recombinant human Noggin (Peprotech) or 10% Noggin conditioned media, 20% R-spondin1 conditioned media. Following initial seeding of the cultures, 10 μM Y-27632 dihydrochloride kinase inhibitor (Tocris Bioscience) was also added to the media for 2–3 days.

2.4. Organoid Drug Sensitivity Testing

After the establishment of cancer-derived organoids, organoids were dissociated using TrypLE Express enzyme (Gibco) and re-seeded in Matrigel into a 48-well plate in triplicate. Organoids were cultured in complete media until small organoids were formed. Reference viability values were determined at day 0 by adding 200 μL of 1X Presto Blue reagent (Invitrogen, Carlsbad, CA, USA) diluted in culture medium to each well. Organoids were incubated for 45 min at 37 °C before the Presto Blue solution was removed into a black microplate and the fluorescence was measured (excitation of 560 nm and an emission of 590 nm) on the PHERAstar FS (BMG Labtech, Ortenberg, Germany). Complete media supplemented with 0, 0.1, 1, 10, 20 and 50 μm 5-fluorouracil (5-FU) (Sigma) was replaced onto the organoids at day 0 and day 2. Cell viability was measured at day 5, as for day 0.

2.5. Histological Sections

Primary tissue samples were fixed in 4% paraformaldehyde (PFA) and embedded in paraffin blocks. Mature organoids were fixed in 4% PFA before being dissociated from the Matrigel. Organoids were collected into a tube and gently centrifuged before being embedded into low melting agarose (2% diluted in PBS). The agarose blocks were processed before being embedded into paraffin. Sections (4 μm) of both primary tissue and patint-matched organoids were subjected to routine haematoxylin and eosin (H&E) staining.

2.6. Immunohistochemistry

The immunohistochemical procedure was conducted as previously described [23]. Briefly, slides were deparaffinized in histosol and rehydrated in graded alcohols. Antigen retrieval was performed by heating the slides for 10 min in a pressure cooker in 10 mM citrate buffer (pH = 6). Slides were blocked with CAS block (Invitrogen) for 1 h at room temperature. Sections were incubated overnight at 4 °C with the primary antibody diluted in PBS containing 1% bovine serum albumin. The following antibodies were used: anti-cleaved caspase 3 (Cell Signaling Technology, Danvers, MA, USA), anti-Cytokeratin 20 (Roche Ventana, Oro Valley, AZ, USA), anti-caudal type homeobox 2 (CDX2) (Abcam, Cambridge, UK) and anti-LGR5 ([24] gift from Dr Melissa R. Junttila, GenenTech, South San Francisco, CA, USA) (Table 1). For the detection of primary antibodies, sections were exposed to anti-goat or anti-rabbit horseradish peroxidase coupled antibodies (Life technologies, Carlsbad, CA, USA) in PBS with 1% bovine serum albumin for 1 h at room temperature. Peroxidase activity was detected with the 3, 3'-diaminobenzidine (DAB) liquid kit (Dako, Burlingame, CA, USA). Sections were counterstained with haematoxylin, dehydrated and mounted. Imaging was carried out using a Zeiss Axio Imager running ZEN digital imaging software (Carl Zeiss, Oberkochen, Germany).

Table 1. List of antibodies used for immunohistochemistry.

Antibodies/Dye	Host	Dilution	Supplier	Cat/Lot number
Primary				
Cleaved Caspase 3 (Asp175)	Rabbit	1:250	Cell Signaling	9661S
Cytokeratin 20 (CK20) (SP33)	Rabbit	1:1	Roche Ventana	790-4431
Caudal type homeobox 2 (CDX2)	Rabbit	1:1000	Abcam	Ab76541
Leucine rich repeat containing G protein-coupled receptor 5 (LGR5)	Rabbit	1:200	Genentech	n/a
Secondary				
Anti-rabbit horseradish peroxidase	Goat	1:200	Life Technologies	G21234

2.7. Quantitative RT–PCR Analysis

Total RNA was isolated from tumour-derived organoids using the RNeasy Micro kit (Qiagen, Hilden, Germany). cDNA was synthesized from 400 ng of total RNA using QuantiTect Reverse Transcription Kit (Qiagen). Quantitative PCR was performed in technical triplicates on a LightCycler 480 II (Roche) using QuantiNova SYBR® Green PCR Kit with thermal cycle as follows: 95 °C for 10 min, 45 cycles of 95 °C for 20 s, 60 °C for 30 s followed by 72 °C for 40 s. The average expression levels ($2^{-\Delta Ct}$) for each gene was calculated relative to beta-2-microglobulin and β-actin expression levels. The primers used in the current study are listed in Table 2.

Table 2. List of primers used for qPCR analysis.

Gene	Forward Primer Sequences	Reverse Primer Sequences	Product Length (bp)
ACTB	CTGGCACCACACCTTCTACAATG	GGTCTCAAACATGATCTGGGTC	124
ANXA1	TTTGCAAGAAGGTAGAGATAAAGAC	GGATGACTTCACAGTTTGAACAT	121
B2M	GTGCTCGCGCTACTCTCTC	GTCAACTTCAATGTCGGAT	142
BMI1	GGTACTTCATTGATGCCACAACC	CTGGTCTTGTGAACTTGGACATC	104
CLU	CAGGCCATGGACATCCACTT	GTCATCGTCGCCTTCTCGTA	78
EPHB2	TTGGGCTCTCACGCTTTCTA	AGGTGAACTTCCGGTACTGG	120
Ki67	CAGCACCTGCTTGTTTGGAAG	TAATATTGCCTCCTGCTCATGGAT	109
KRT20	CTGAGGTTCAACTAACGGAGCTG	AACAGCGACTGGAGGTTGGCTA	129
LGR5	CCTTCCAACCTCAGCGTCTT	AGGGATTGAAGGCTTCGCAA	250

2.8. Survival Analysis

The survival analysis of The Cancer Genome Atlas (TCGA) data were performed using online tools OncoLnc (http://www.oncolnc.org/) [25] and SurvExpress (http://bioinformatica.mty.itesm.mx/SurvExpress) [26]. In OncoLnc, the correlation between *CLU* expression and prognosis of patient were analysed with the upper and lower 25 percentile of *CLU* expression in colon adenocarcinoma (COAD) patients (n = 220). Survival rate is represented by Kaplan–Meier plot and analysed using Log-rank test. In SurvExpress, Cox survival analysis with *CLU* expression was performed in Colon GSE41258 database (n = 244) and censored with both survival and recurrence data. A value of $p < 0.05$ was considered to be significant.

3. Results

3.1. Patient-Derived Colorectal Cancer Organoids Recapitulate the Histopathological Characteristics of Their Primary Tumours and Display Inter-Tumoural Heterogeneity in Stem Cell Signatures

In order to determine whether CRC patient-derived colorectal cancer organoids (PDCOs) are robust models for examining the expression pattern of stem cell markers, we first compared the biological characteristics of primary tumours and their matched PDCOs. Histological analysis of 10 primary tumours and their respective PDCOs (Figure 1A) was conducted by a trained pathologist. The histological profile of the primary tumour was generally well maintained in PDCOs, with organoids typically having a more cuboidal appearance, but present otherwise similar cellular morphology to the primary specimen. The primary tumour in patient 30T is a moderately differentiated colonic adenocarcinoma with the PDCO displaying a more cuboidal appearance with less nuclear pleomorphism evident. In tissue from 38T, the primary tumour is a well differentiated colonic adenocarcinoma and matched PDCOs appear more cuboidal and show less nuclear stratification, but are otherwise similar (Figure 1A). The primary tumour in 53T represents an intramucosal adenocarcinoma with luminal necrosis. Similar cell morphology and luminal necrosis is evident in the PDCO section (Figure 1A). In 63T, the primary tumour is a moderately differentiated adenocarcinoma displaying cuboidal cell morphology with PDCOs closely resembling this morphology (Figure 1A).

Commonly used for differential diagnosis of colorectal cancer, caudal type homeobox 2 (CDX2) is a transcription factor critical for intestinal development that is highly expressed in normal and neoplastic intestinal epithelium [27,28]. To confirm whether intestinal identity of epithelial tissue is maintained under culture conditions, we compared the expression of CDX2 in primary tumour and PDCOs. Strong and diffuse nuclear staining of CDX2+ was observed in all 10 paired primary tumour and PDCO sections, confirming the intestinal origin of these adenocarcinomas is maintained in culture. Representative images of CDX2+ staining are provided in Figure 1B.

The ability of PDCOs to recapitulate the profile of stem and differentiated cell populations of the primary tumour was examined by immunohistochemical analysis. The number of LGR5+ cells as well as the cytoplasmic staining intensity were variable across the primary tumour specimens. However, the levels of expression between the primary tumour and matched PDCO were comparable. Similarly, positive immunostaining for cytokeratin 20 (CK20), the protein encoded by the Keratin 20 (*KRT20*) gene, was observed in the cytoplasm and/or cell membrane of the PDCOs and the abundance of CK20+ cells closely resembled that of the patient-matched primary tumour. Representative images of LGR5 and CK20 staining are illustrated in Figure 1C.

To more broadly assess the expression profiles of stem cell markers in CRC, we performed qRT-PCR analysis on 10 PDCO lines, analysing basal expression levels of CBC stem cell markers *LGR5* and *EPHB2*, a quiescent stem cell marker *BMI1*, recently identified revival stem cell markers *CLU* and *ANXA1*, and a marker of differentiated intestinal cells, Keratin 20 (*KRT20*) (Figure 1D). These results were consistent with the immunostaining for LGR5 and CK20 (Figure 1C) and revealed unique expression profiles for each of the 10 PDCO lines. Given that it has previously been reported that aggressive CRCs are enriched in intestinal stem cell marker expression and this is predictive of relapse in CRC patients [16], we were interested in investigating whether the differential expression profiles of the PDCOs were predictive of chemotherapeutic drug response.

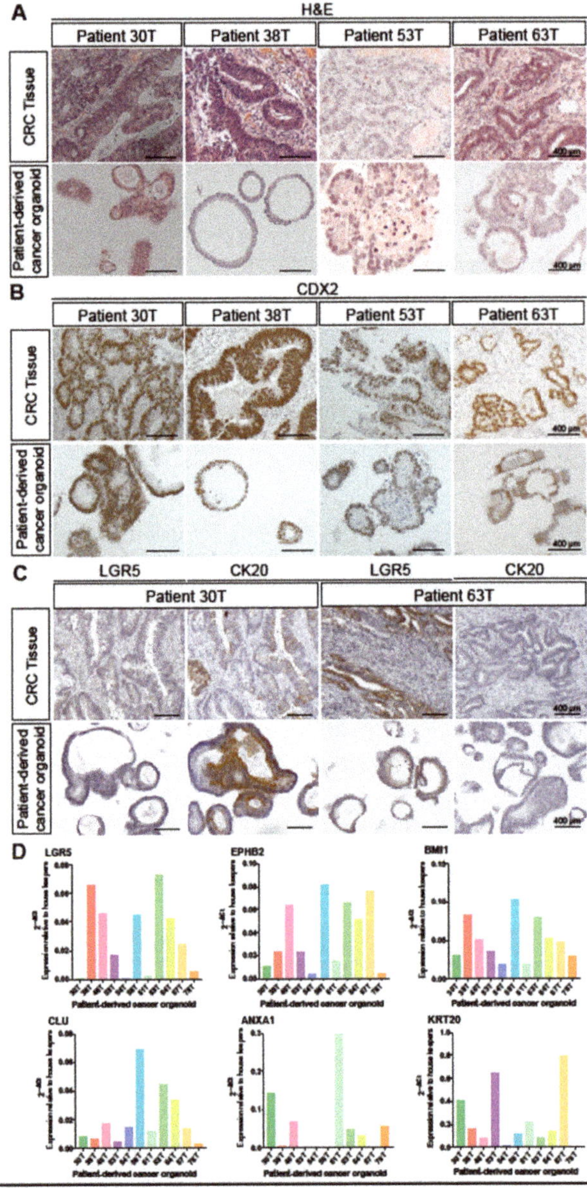

Figure 1. Patient-derived cancer organoids recapitulate the histopathological characteristics of their primary tumours and inter-tumoural heterogeneity in stem cell signatures. (**A**): Haematoxylin and eosin (H&E) staining of sectioned tissue from primary colorectal adenocarcinoma and patient-derived cancer organoids derived from the same tumour. Scale bar, 400 μm. (**B**): Immunohistochemical detection of CDX2 (marker of adenocarcinomas of intestinal origin) in primary colorectal adenocarcinoma compared to patient-derived colorectal cancer organoids (PDCOs). Scale bar, 400 μm. (**C**): Immunohistochemical detection of LGR5 (CBC stem cell marker) and CK20 (intestinal epithelial marker) in the colorectal adenocarcinomas and PDCOs. Scale bar, 400 μm. (**D**): The expression levels ($2^{-\Delta Ct}$) for *LGR5, EPHB2, BMI1, CLU, ANXA1* and *KRT20* were calculated relative to beta-2-microglobulin and β-actin by qRT-PCR in individual PDCO lines.

3.2. Elevated Expression of a Subset of Stem Cell Markers Correlates with Resistance to Chemotherapy

PDCOs from six treatment-naïve CRC patients with stage III (present in local lymph nodes) and IV (spread to distant organs) disease were treated with an increasing dose of common chemotherapeutic, 5-fluorouracil (5-FU). The 5-FU serves as the main backbone of adjuvant chemotherapy regimens for the treatment of CRC. Cell viability was assessed five days following initial treatment and was analysed by the slope of the dose-response curve (Figure 2A,B), and area under the curve (AUC) analysis (Figure 2C). The PDCOs display differential responses to 5-FU treatment including distinct phenotypic changes in organoids that are responsive to drug treatment (ORG54T) and those that are less sensitive to treatment (ORG64T) (Representative images; Figure 2A). The presence of cleaved caspase-3-positive apoptotic cells in 5FU-treated PDCO 54T further illustrates sensitivity to treatment compared with the absence of staining in treatment resistant PDCO 64T (Figure 2A, right panel).

To determine whether basal expression levels of stem cell markers in PDCOs are predictive of resistance to chemotherapeutic drug treatment, gene expression analysis (Figure 1D) was correlated with AUC (Figure 2C). We identified an inverse correlation with marker of intestinal differentiation, *KRT20* ($p = 0.008$). High *KRT20* expression was significantly associated with lower AUC values, conferring sensitivity to treatment. There is also a strong positive correlation with revival stem cell gene *CLU*, when *CLU* expression in PDCO 54T was eliminated as an outlier from analysis which fall away from the 95% prediction interval of the best-fit line. High expression of *CLU* was significantly correlated with high AUC values ($R^2 = 0.885$; $p = 0.004$).

Following treatment with chemotherapeutic drug 5-FU, we analysed PDCO expression profiles for stem cell markers from the same six patient lines (Figure 2E). We performed qRT-PCR analysis of PDCOs five days following initial treatment with 0, 1 and 10 μM 5-FU. We observed a modest decrease in LGR5 expression at 1 μM (0.88-fold decrease; $p = 0.0115$). There was upregulation of expression at 10 μM for both *BMI1* (1.5-fold increase; $p < 0.0001$) and *KRT20* (2.4-fold increase; $p < 0.0001$). However, it was revival stem cell-associated genes *CLU* and *ANXA1* that were the most significantly upregulated, with expression robustly increasing up to 5-fold between the untreated (0 μM) and 10 μM treated PDCOs.

We further explored the prognostic relevance of *CLU* expression in CRC using the OncoLnc online system to analyse colon adenocarcinoma patient data from TCGA. Kaplan–Meier survival analysis revealed increased expression of *CLU* was associated with poorer survival outcomes ($p = 0.0286$) (Figure 3A). Using the SurvExpress online tool, we evaluated differences in overall survival and recurrence-free survival between the predicted high-risk and low-risk groups in the Sheffer series (GSE41258). The high-risk group that had decreased overall survival and increased disease recurrence was associated with significantly higher expression of *CLU* ($p = 0.032$ and 0.024 respectively) (Figure 3B,C), which is in line with previous observations [29,30].

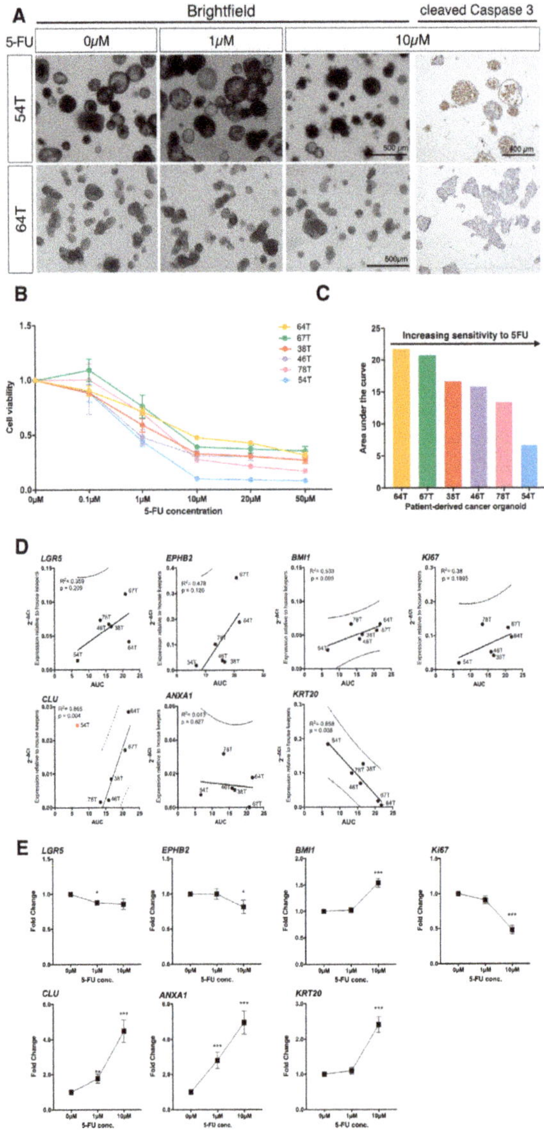

Figure 2. Elevated expression of a subset of stem cell markers correlates with resistance to chemotherapy. (**A**): Representative brightfield images of patient-derived cancer organoids (PDCOs) in response to chemotherapeutic treatment with 5-fluorouracil (5-FU) and the immunohistochemical detection of the active cleaved form of caspase 3 (apoptotic marker) in 10 μM treated PDCOs. Scale bar, 500 μm. (**B**): Dose-response curve of PDCOs in response to chemotherapeutic treatment with 5-FU ($n = 6$, mean ± SEM). (**C**): Area under the curve (AUC) analysis of 5-FU sensitivity in six patient-derived tumoroids. (**D**): Linear regression correlation analysis between the expression of stem cell marker genes and AUC showing the best-fit line (solid line) and 95% prediction interval (dash-line). (**E**): Expression of stem cell marker genes in PDCOs in response to an increasing dose of 5-FU. Fold change is calculated by the average gene expression levels ($2^{-\Delta Ct}$) relative to the vehicle control ($n = 6$, mean ± SEM). In each graph, asterisks indicate pairs of means (compared to vehicle control) that were significantly different using Mann–Whitney test (*, $p < 0.05$; **, $p < 0.01$; ***, $p < 0.001$).

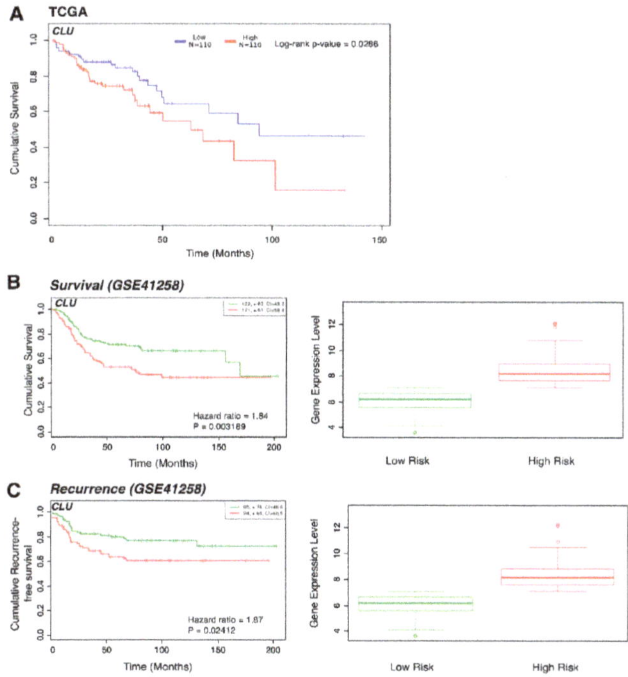

Figure 3. *CLU* expression in colon adenocarcinoma patients is associated with decreased overall survival. (**A**): Kaplan–Meier survival plot comparing The Cancer Genome Atlas (TCGA) colon adenocarcinoma patients with high (n = 110) and low (n = 110) expression of *CLU* using the OncoLnc tool. The associated log-rank p-value is 0.0286. (**B**): Kaplan–Meier survival plot for high- (red, n = 121) and low- (green, n = 122) risk groups in GSE41258 database by SurvExpress tool shows cumulative survival against time (months) and the box plot shows the corresponding *CLU* expression across groups. The number of individuals, the number of censored, and the CI of each risk group are shown in the top-right insets. Censoring samples are shown as "+" marks. (**C**): Kaplan–Meier survival plot for high- (red, n = 95) and low- (green, n = 94) risk groups in GSE41258 database by SurvExpress tool shows cumulative recurrence-free survival against time (months) and the box plot shows the corresponding *CLU* expression across groups. The number of individuals, the number of censored, and the CI of each risk group are shown in the top-right insets. Censoring samples are shown as "+" marks.

4. Discussion

The initial objective of this study was to profile the expression of stem cell markers in a cohort of colorectal cancer-derived organoid cultures from different patients. We first determined how faithfully the organoids replicated the histological features of the primary tumour from which they were derived. The overall morphological features and level of differentiation was maintained in the PDCOs with some tumours displaying a moderate degree of differentiation and others with a more differentiated phenotype. This is consistent with previous studies conducted on PDCOs which suggest most epithelial features of the primary tumours are present in vitro [31–34]. All of the PDCOs were positive for the definitive intestinal marker CDX2 [28,35] and the expression of the CBC stem cell marker, LGR5 [36,37] and the cytokeratin marker CK20 was similar in the primary tissue and organoids. Some organoids displayed more widespread expression and this was reflected in the corresponding primary tissue, while others showed little expression of LGR5. This indicated that the PDCOs were closely modelling the epithelial features of the tumours from which they were derived. Subsequently, we examined individual PDCOs for the expression of stem cell markers. The results revealed considerable inter-tumour variability of expression of the different markers. Overall, this suggests marked heterogeneity between individual PDCOs. All of the cultures utilised in our study were derived from patient tumours naïve to treatment, so these initial results represent cellular expression without the selective pressure of chemotherapeutic treatment.

Using knowledge of the behaviour of different stem cell populations in normal tissue, we predicted that there may be a difference in the expression of the repertoire of markers following exposure to 5-FU. We treated each of the PDCO lines with 5-FU and measured cell viability to determine relative sensitivity and correlated this with the stem cell expression signatures. LGR5 clearly did not correlate with resistance to 5-FU. LGR5+ cells are extremely susceptible to damage induced by chemotherapy or radiation in normal mouse tissue and rapidly undergo apoptosis [17–19]. This is also observed in LGR5+ cells in primary intestinal tumours in mice and human colorectal tumours [13,14]. This suggests the susceptibility to treatment observed here and in other studies is a conserved feature of LGR5+ cells in both normal tissue and tumours. In contrast, markers of both quiescent reserve cells and the newly described revival stem cells were enriched upon exposure to 5-FU in PDCOs. In mouse models, BMI1+ cells appear to be more resistant to injury and lineage tracing experiments have revealed these cells are activated to replenish lost LGR5+ cells [18]. Revival stem cells cannot be detected in normal tissue and are only identified in a mouse injury model, where damage-induced revival stem cells are also capable of reconstituting LGR5+ stem cells and regenerating the intestinal epithelium [21]. Our study shows for the first time that a revival stem cell signature strongly correlates with chemoresistance in PDCOs and suggests that the process of re-population of LGR5+ cells observed in normal tissue, may also operate with a similar mechanism in human cancer. It remains to be elucidated whether a comparable hierarchical relationship between LGR5+ cells and CLU+ cells is conserved in human intestinal cancer tissues.

Patient-derived models of cancer, including PDCOs show great promise in trialling therapies before they reach the patient and evidence supporting how accurately these ex vivo models can predict patient response is beginning to emerge [32,38–42]. A recent report showed convincing data from ten patients that a viability assay could be used to predict patient response to irinotecan-based therapies using CRC-derived PDCOs [39]. However, the same testing protocol was not predictive for patient outcomes following treatment with 5-FU plus oxaliplatin. How predictive PDCO testing is for a range of therapies and what the most effective testing regime to implement is still being investigated. Here, we used PDCO testing to look at how markers of different stem cell populations behave following 5-FU treatment and identified *CLU*, a marker of the revival stem cell population, as being enriched. Interestingly, both the patients from whom PDCO cultures 64T and 67T were derived, which were most resistant in our assay of sensitivity to 5-FU, have progressive disease. These PDCOs also exhibited the highest levels of *CLU* expression following treatment. Although the sample size is limited, this provides the basis for future studies on the role of revival stem cells in progressive disease in colorectal cancer patients. This cell type may be a potential therapeutic target and a marker of drug resistance.

5. Conclusions

Our study demonstrates that PDCOs are relevant in vitro models for studying the heterogeneity of stem cell populations represented in the primary tumours. We show that there is considerable variation between individual PDCOs in the repertoire of stem cell markers present. We have identified that *CLU*, a marker of the revival stem cell population, is enriched in PDCOs resistant to 5-FU and this is consistent with overall patient data showing that *CLU* correlates with lower survival and an increase in disease recurrence.

Supplementary Materials: The following are available online at http://www.mdpi.com/2077-0383/9/1/128/s1. Table S1: Patient clinical information.

Author Contributions: Conceptualization, H.E.A., T.J., R.M.E., W.H.C. and P.J.M.; Methodology, H.E.A., T.J., R.M.E., W.H.C., S.H., K.O., D.N., E.R., G.K. and P.J.M.; Data Analysis, H.E.A., T.J., R.M.E., W.H.C., S.H., K.O., D.N., E.R., G.K. and P.J.M.; Writing—Original Draft Preparation, R.M.E., W.H.C., S.H., T.J., H.E.A.; Writing—Review and Editing, R.M.E., W.H.C., S.H., T.J. and H.E.A.; Supervision, H.E.A., T.J., P.J.M., W.H.C. and R.M.E.; Project

Administration, H.E.A., G.K., K.O. and R.M.E.; Funding Acquisition, H.E.A., P.J.M., T.J. and R.M.E. All authors have read and agreed to the published version of the manuscript.

Funding: This work was supported by grants from the Monash Strategic scheme SGS15-0156, Victorian Cancer Agency TRP15021, Cancer Australia 1145907, NH&MRC project 1100531 and the Collie Foundation. Work was also supported in part by "Let's Beat Bowel Cancer" (www.letsbeatbowelcancer.com), a benevolent fundraising and public awareness foundation that has had no part in the design, conduct, outcomes or the drafting of this manuscript.

Acknowledgments: The authors acknowledge the colorectal surgeons in the Cabrini Monash University Department of Surgery for their contributions to specimen collections as well as the collation and interpretation of patient data in the CMCND. The authors also acknowledge the resources, facilities, and scientific and technical assistance of the Monash BDI Organoid Program, and the Monash Histology Platform, Department of Anatomy and Developmental Biology, Monash Biomedicine Discovery Institute, Victoria, Australia.

Conflicts of Interest: The authors declare no conflict of interest.

References

1. Favoriti, P.; Carbone, G.; Greco, M.; Pirozzi, F.; Pirozzi, R.E.; Corcione, F. Worldwide burden of colorectal cancer: A review. *Updates Surg.* **2016**, *68*, 7–11. [CrossRef] [PubMed]

2. Meyer, B.; Are, C. Current Status and Future Directions in Colorectal Cancer. *Indian J. Surg Oncol.* **2017**, *8*, 455–456. [CrossRef] [PubMed]

3. Lintoiu-Ursut, B.; Tulin, A.; Constantinoiu, S. Recurrence after hepatic resection in colorectal cancer liver metastasis—Review article. *J. Med. Life* **2015**, *8*, 12–14. [PubMed]

4. Regnard, J.-F.; Grunenwald, D.; Spaggiari, L.; Girard, P.; Elias, D.; Ducreux, M.; Baldeyrou, P.; Levasseur, P. Surgical treatment of hepatic and pulmonary metastases from colorectal cancers. *Ann. Thorac. Surg.* **1998**, *66*, 214–218. [CrossRef]

5. Grothey, A.; Sargent, D.; Goldberg, R.M.; Schmoll, H.J. Survival of patients with advanced colorectal cancer improves with the availability of fluorouracil-leucovorin, irinotecan, and oxaliplatin in the course of treatment. *J. Clin. Oncol.* **2004**, *22*, 1209–1214. [CrossRef]

6. Tournigand, C.; Andre, T.; Achille, E.; Lledo, G.; Flesh, M.; Mery-Mignard, D.; Quinaux, E.; Couteau, C.; Buyse, M.; Ganem, G.; et al. FOLFIRI followed by FOLFOX6 or the reverse sequence in advanced colorectal cancer: A randomized GERCOR study. *J. Clin. Oncol.* **2004**, *22*, 229–237. [CrossRef]

7. Grossman, J.G.; Nywening, T.M.; Belt, B.A.; Panni, R.Z.; Krasnick, B.A.; DeNardo, D.G.; Hawkins, W.G.; Goedegebuure, S.P.; Linehan, D.C.; Fields, R.C. Recruitment of CCR2(+) tumor associated macrophage to sites of liver metastasis confers a poor prognosis in human colorectal cancer. *Oncoimmunology* **2018**, *7*, e1470729. [CrossRef]

8. Holohan, C.; Van Schaeybroeck, S.; Longley, D.B.; Johnston, P.G. Cancer drug resistance: An evolving paradigm. *Nat. Rev. Cancer* **2013**, *13*, 714–726. [CrossRef]

9. O'Brien, C.A.; Pollett, A.; Gallinger, S.; Dick, J.E. A human colon cancer cell capable of initiating tumour growth in immunodeficient mice. *Nature* **2007**, *445*, 106–110. [CrossRef]

10. Ricci-Vitiani, L.; Lombardi, D.G.; Pilozzi, E.; Biffoni, M.; Todaro, M.; Peschle, C.; De Maria, R. Identification and expansion of human colon-cancer-initiating cells. *Nature* **2007**, *445*, 111–115. [CrossRef]

11. Batlle, E.; Clevers, H. Cancer stem cells revisited. *Nat. Med.* **2017**, *23*, 1124–1134. [CrossRef] [PubMed]

12. Touil, Y.; Igoudjil, W.; Corvaisier, M.; Dessein, A.F.; Vandomme, J.; Monte, D.; Stechly, L.; Skrypek, N.; Langlois, C.; Grard, G.; et al. Colon cancer cells escape 5FU chemotherapy-induced cell death by entering stemness and quiescence associated with the c-Yes/YAP axis. *Clin. Cancer Res.* **2014**, *20*, 837–846. [CrossRef] [PubMed]

13. de Sousa e Melo, F.; Kurtova, A.V.; Harnoss, J.M.; Kljavin, N.; Hoeck, J.D.; Hung, J.; Anderson, J.E.; Storm, E.E.; Modrusan, Z.; Koeppen, H.; et al. A distinct role for Lgr5(+) stem cells in primary and metastatic colon cancer. *Nature* **2017**, *543*, 676–680. [CrossRef] [PubMed]

14. Shimokawa, M.; Ohta, Y.; Nishikori, S.; Matano, M.; Takano, A.; Fujii, M.; Date, S.; Sugimoto, S.; Kanai, T.; Sato, T. Visualization and targeting of LGR5(+) human colon cancer stem cells. *Nature* **2017**, *545*, 187–192. [CrossRef]

15. Jardé, T.; Kass, L.; Staples, M.; Lescesen, H.; Carne, P.; Oliva, K.; McMurrick, P.J.; Abud, H.E. ERBB3 Positively Correlates with Intestinal Stem Cell Markers but Marks a Distinct Non Proliferative Cell Population in Colorectal Cancer. *PLoS ONE* **2015**, *10*, e0138336. [CrossRef]

16. Merlos-Suárez, A.; Barriga, F.M.; Jung, P.; Iglesias, M.; Céspedes, M.V.; Rossell, D.; Sevillano, M.; Hernando-Momblona, X.; da Silva-Diz, V.; Muñoz, P.; et al. The Intestinal Stem Cell Signature Identifies Colorectal Cancer Stem Cells and Predicts Disease Relapse. *Cell Stem Cell* **2011**, *8*, 511–524. [CrossRef]

17. Asfaha, S.; Hayakawa, Y.; Muley, A.; Stokes, S.; Graham, T.A.; Ericksen, R.E.; Westphalen, C.B.; Von Burstin, J.; Mastracci, T.L.; Worthley, D.L.; et al. Krt19+/Lgr5− Cells Are Radioresistant Cancer-Initiating Stem Cells in the Colon and Intestine. *Cell Stem Cell* **2015**, *16*, 627–638. [CrossRef]

18. Tian, H.; Biehs, B.; Warming, S.; Leong, K.G.; Rangell, L.; Klein, O.D.; de Sauvage, F.J. A reserve stem cell population in small intestine renders Lgr5-positive cells dispensable. *Nature* **2011**, *478*, 255–259. [CrossRef]

19. Metcalfe, C.; Kljavin, N.M.; Ybarra, R.; de Sauvage, F.J. Lgr5+ stem cells are indispensable for radiation-induced intestinal regeneration. *Cell Stem Cell* **2014**, *14*, 149–159. [CrossRef]

20. Li, D.W.; Tang, H.M.; Fan, J.W.; Yan, D.W.; Zhou, C.Z.; Li, S.X.; Wang, X.L.; Peng, Z.H. Expression level of Bmi-1 oncoprotein is associated with progression and prognosis in colon cancer. *J. Cancer Res. Clin. Oncol.* **2010**, *136*, 997–1006. [CrossRef]

21. Ayyaz, A.; Kumar, S.; Sangiorgi, B.; Ghoshal, B.; Gosio, J.; Ouladan, S.; Fink, M.; Barutcu, S.; Trcka, D.; Shen, J.; et al. Single-cell transcriptomes of the regenerating intestine reveal a revival stem cell. *Nature* **2019**, *569*, 121–125. [CrossRef] [PubMed]

22. McMurrick, P.J.; Oliva, K.; Carne, P.; Reid, C.; Polglase, A.; Bell, S.; Farmer, K.C.; Ranchod, P. The first 1000 patients on an internet-based colorectal neoplasia database across private and public medicine in Australia: Development of a binational model for the Colorectal Surgical Society of Australia and New Zealand. *Dis. Colon Rectum* **2014**, *57*, 167–173. [CrossRef] [PubMed]

23. Horvay, K.; Jardé, T.; Casagranda, F.; Perreau, V.M.; Haigh, K.; Nefzger, C.M.; Akhtar, R.; Gridley, T.; Berx, G.; Haigh, J.J.; et al. Snai1 regulates cell lineage allocation and stem cell maintenance in the mouse intestinal epithelium. *EMBO J.* **2015**, *34*, 1319–1335. [CrossRef] [PubMed]

24. Junttila, M.R.; Mao, W.; Wang, X.; Wang, B.E.; Pham, T.; Flygare, J.; Yu, S.F.; Yee, S.; Goldenberg, D.; Fields, C.; et al. Targeting LGR5+ cells with an antibody-drug conjugate for the treatment of colon cancer. *Sci. Transl. Med.* **2015**, *7*, 314ra186. [CrossRef] [PubMed]

25. Anaya, J. OncoLnc: Linking TCGA survival data to mRNAs, miRNAs, and lncRNAs. *PeerJ Comput. Sci.* **2016**, *2*, e67. [CrossRef]

26. Aguirre-Gamboa, R.; Gomez-Rueda, H.; Martínez-Ledesma, E.; Martínez-Torteya, A.; Chacolla-Huaringa, R.; Rodriguez-Barrientos, A.; Tamez-Peña, J.G.; Treviño, V. SurvExpress: An online biomarker validation tool and database for cancer gene expression data using survival analysis. *PLoS ONE* **2013**, *8*, e74250. [CrossRef]

27. Moskaluk, C.A.; Zhang, H.; Powell, S.M.; Cerilli, L.A.; Hampton, G.M.; Frierson, H.F., Jr. Cdx2 protein expression in normal and malignant human tissues: An immunohistochemical survey using tissue microarrays. *Mod. Pathol.* **2003**, *16*, 913–919. [CrossRef]

28. Werling, R.W.; Yaziji, H.; Bacchi, C.E.; Gown, A.M. CDX2, a highly sensitive and specific marker of adenocarcinomas of intestinal origin: An immunohistochemical survey of 476 primary and metastatic carcinomas. *Am. J. Surg. Pathol.* **2003**, *27*, 303–310. [CrossRef]

29. Redondo, M.; Rodrigo, I.; Alcaide, J.; Tellez, T.; Roldan, M.J.; Funez, R.; Diaz-Martin, A.; Rueda, A.; Jimenez, E. Clusterin expression is associated with decreased disease-free survival of patients with colorectal carcinomas. *Histopathology* **2010**, *56*, 932–936. [CrossRef]

30. Kevans, D.; Foley, J.; Tenniswood, M.; Sheahan, K.; Hyland, J.; O'Donoghue, D.; Mulcahy, H.; O'Sullivan, J. High clusterin expression correlates with a poor outcome in stage II colorectal cancers. *Cancer Epidemiol. Biomark. Prev.* **2009**, *18*, 393–399. [CrossRef]

31. Sato, T.; Stange, D.E.; Ferrante, M.; Vries, R.G.; Van Es, J.H.; Van Den Brink, S.; Van Houdt, W.J.; Pronk, A.; Van Gorp, J.; Siersema, P.D.; et al. Long-term expansion of epithelial organoids from human colon, adenoma, adenocarcinoma, and Barrett's epithelium. *Gastroenterology* **2011**, *141*, 1762–1772. [CrossRef] [PubMed]

32. Vlachogiannis, G.; Hedayat, S.; Vatsiou, A.; Jamin, Y.; Fernández-Mateos, J.; Khan, K.; Lampis, A.; Eason, K.; Huntingford, I.; Burke, R.; et al. Patient-derived organoids model treatment response of metastatic gastrointestinal cancers. *Science* **2018**, *359*, 920–926. [CrossRef] [PubMed]

33. Weeber, F.; van de Wetering, M.; Hoogstraat, M.; Dijkstra, K.K.; Krijgsman, O.; Kuilman, T.; Gadellaa-van Hooijdonk, C.G.; van der Velden, D.L.; Peeper, D.S.; Cuppen, E.P.; et al. Preserved genetic diversity in organoids cultured from biopsies of human colorectal cancer metastases. *Proc. Natl. Acad. Sci. USA* **2015**, *112*, 13308–13311. [CrossRef] [PubMed]
34. Van de Wetering, M.; Francies, H.E.; Francis, J.M.; Bounova, G.; Iorio, F.; Pronk, A.; van Houdt, W.; van Gorp, J.; Taylor-Weiner, A.; Kester, L.; et al. Prospective derivation of a living organoid biobank of colorectal cancer patients. *Cell* **2015**, *161*, 933–945. [CrossRef] [PubMed]
35. Saad, R.S.; Ghorab, Z.; Khalifa, M.A.; Xu, M. CDX2 as a marker for intestinal differentiation: Its utility and limitations. *World J. Gastrointest. Surg.* **2011**, *3*, 159–166. [CrossRef] [PubMed]
36. Barker, N.; Van Es, J.H.; Kuipers, J.; Kujala, P.; Van Den Born, M.; Cozijnsen, M.; Haegebarth, A.; Korving, J.; Begthel, H.; Peters, P.J.; et al. Identification of stem cells in small intestine and colon by marker gene Lgr5. *Nature* **2007**, *449*, 1003–1007. [CrossRef]
37. Dame, M.K.; Attili, D.; McClintock, S.D.; Dedhia, P.H.; Ouillette, P.; Hardt, O.; Chin, A.M.; Xue, X.; Laliberte, J.; Katz, E.L.; et al. Identification, isolation and characterization of human LGR5-positive colon adenoma cells. *Development* **2018**, *145*, dev153049. [CrossRef]
38. Lee, S.H.; Hu, W.; Matulay, J.T.; Silva, M.V.; Owczarek, T.B.; Kim, K.; Chua, C.W.; Barlow, L.J.; Kandoth, C.; Williams, A.B.; et al. Tumor Evolution and Drug Response in Patient-Derived Organoid Models of Bladder Cancer. *Cell* **2018**, *173*, 515–528. [CrossRef]
39. Ooft, S.N.; Weeber, F.; Dijkstra, K.K.; McLean, C.M.; Kaing, S.; van Werkhoven, E.; Schipper, L.; Hoes, L.; Vis, D.J.; van de Haar, J.; et al. Patient-derived organoids can predict response to chemotherapy in metastatic colorectal cancer patients. *Sci. Transl. Med.* **2019**, *11*, eaay2574. [CrossRef]
40. Sachs, N.; de Ligt, J.; Kopper, O.; Gogola, E.; Bounova, G.; Weeber, F.; Balgobind, A.V.; Wind, K.; Gracanin, A.; Begthel, H.; et al. A Living Biobank of Breast Cancer Organoids Captures Disease Heterogeneity. *Cell* **2018**, *172*, 373–386. [CrossRef]
41. Tiriac, H.; Belleau, P.; Engle, D.D.; Plenker, D.; Deschênes, A.; Somerville, T.D.; Froeling, F.E.; Burkhart, R.A.; Denroche, R.E.; Jang, G.H.; et al. Organoid Profiling Identifies Common Responders to Chemotherapy in Pancreatic Cancer. *Cancer Discov.* **2018**, *8*, 1112–1129. [CrossRef] [PubMed]
42. Du Puch, C.B.; Nouaille, M.; Giraud, S.; Labrunie, A.; Luce, S.; Preux, P.M.; Labrousse, F.; Gainant, A.; Tubiana-Mathieu, N.; Le Brun-Ly, V.; et al. Chemotherapy outcome predictive effectiveness by the Oncogramme: Pilot trial on stage-IV colorectal cancer. *J. Transl. Med.* **2016**, *14*, 10. [CrossRef] [PubMed]

 Journal of
Clinical Medicine

Article

B4GALT1 Is a New Candidate to Maintain the Stemness of Lung Cancer Stem Cells

Claudia De Vitis [1,†], Giacomo Corleone [2,†], Valentina Salvati [3], Francesca Ascenzi [4], Matteo Pallocca [2], Francesca De Nicola [2], Maurizio Fanciulli [2], Simona di Martino [5], Sara Bruschini [6], Christian Napoli [7], Alberto Ricci [8], Massimiliano Bassi [9], Federico Venuta [9], Erino Angelo Rendina [10], Gennaro Ciliberto [11,*] and Rita Mancini [1]

[1] Department of Clinical and Molecular Medicine, Sant'Andrea Hospital, "Sapienza" University of Rome, 00161 Rome, Italy; claudia.devitis@uniroma1.it (C.D.V.); rita.mancini@uniroma1.it (R.M.)

[2] SAFU Laboratory, Department of Research, Advanced Diagnostic, and Technological Innovation, IRCCS "Regina Elena" National Cancer Institute, 00144 Rome, Italy; giacomo.corleone@ifo.gov.it (G.C.); matteo.pallocca@ifo.gov.it (M.P.); francesca.denicola@ifo.gov.it (F.D.N.); maurizio.fanciulli@ifo.gov.it (M.F.)

[3] Preclinical Models and New Therapeutic Agents Unit, IRCCS-Regina Elena National Cancer Institute, 00144 Rome, Italy; salvati.sv@gmail.com

[4] Tumor Immunology and Immunotherapy Unit, Department of Research, Advanced Diagnostic and Technological Innovation, IRCCS Regina Elena National Cancer Institute, 00144 Rome, Italy; francesca.ascenzi@gmail.com

[5] Pathology Unit, IRCSS "Regina Elena" National Cancer Institute, 00144 Rome, Italy; simona.dimartino@ifo.gov.it

[6] Department of Experimental and Clinical Medicine, Magna Graecia University of Catanzaro, 88100 Catanzaro, Italy; sarabruschini@hotmail.it

[7] Department of Medical Surgical Sciences and Translational Medicine, Sant'Andrea Hospital, "Sapienza" University of Rome, 00189 Rome, Italy; christian.napoli@uniroma1.it

[8] Department of Clinical and Molecular Medicine, Division of Pneumology, Sapienza University of Rome, Sant'Andrea Hospital, 00189 Rome, Italy; alberto.ricci@uniroma1.it

[9] Department of Thoracic Surgery, University of Rome Sapienza, 00161 Rome, Italy; massimiliano.bassi@uniroma1.it (M.B.); federico.venuta@uniroma1.it (F.V.)

[10] Department of Thoracic Surgery, Sant'Andrea Hospital, "Sapienza" University of Rome, 00189 Rome, Italy; erinoangelo.rendina@uniroma1.it

[11] Scientific Direction, IRCCS "Regina Elena" National Cancer Institute, 00144 Rome, Italy

* Correspondence: gennaro.ciliberto@ifo.gov.it; Tel.: +39-06-5266-2728/2726

† These authors contributed equally to this work.

Received: 16 September 2019; Accepted: 5 November 2019; Published: 9 November 2019

Abstract: Background: According to the cancer stem cells (CSCs) hypothesis, a population of cancer cells with stem cell properties is responsible for tumor propagation, drug resistance, and disease recurrence. Study of the mechanisms responsible for lung CSCs propagation is expected to provide better understanding of cancer biology and new opportunities for therapy. Methods: The Lung Adenocarcinoma (LUAD) NCI-H460 cell line was grown either as 2D or as 3D cultures. Transcriptomic and genome-wide chromatin accessibility studies of 2D vs. 3D cultures were carried out using RNA-sequencing and Assay for Transposase Accessible Chromatin with high-throughput sequencing (ATAC-seq), respectively. Reverse transcription polymerase chain reaction (RT-PCR) was also carried out on RNA extracted from primary cultures derived from malignant pleural effusions to validate RNA-seq results. Results: RNA-seq and ATAC-seq data disentangled transcriptional and genome accessibility variability of 3D vs. 2D cultures in NCI-H460 cells. The examination of genomic landscape of genes upregulated in 3D vs. 2D cultures led to the identification of 2D cultures led to the identification of Beta-1,4-galactosyltranferase 1 (B4GALT1) as the top candidate. B4GALT1 as the top candidate. B4GALT1 was validated as a stemness factor, since its silencing caused strong inhibition of 3D spheroid formation. Conclusion: Combined transcriptomic and chromatin accessibility study of 3D vs. 2D LUAD cultures led to the identification of B4GALT1 as a new factor involved in the propagation and maintenance of LUAD CSCs.

Keywords: cancer stem cells; genome-wide; transcriptome; lung cancer; ATAC-seq; RNA-seq; CSCs; NSCLC; B4GALT1; LUAD

---

1. Introduction

Lung cancer is the leading cause of cancer mortality worldwide [1]. Lung cancer can be divided into two major types, non-small cell lung cancer (NSCLC), which accounts for 80% of cases, and small cell lung cancer (SCLC). There are two main subtypes of NSCLC: Adenocarcinoma (LUAD) and squamous cell carcinoma (LUSC), of which LUAD is the most common [2,3]. The five-year survival rate of patients with advanced lung cancer remains very low, in spite of the introduction of targeted therapies for patients with specific genetic alterations such as EGFR mutations, or ALK and ROS translocations, and, more recently, of immunotherapy with anti-PD1 or anti-PD-L1 antibodies as first or second line therapy [4–7]. Unfortunately, failure of immunotherapy in NSCLC still occurs in a high proportion of cases because several factors, both tumor intrinsic and tumor extrinsic, are responsible for drug resistance [8,9]. In this scenario, lung cancer, and in particular its most frequent form LUAD, remains a highly unmet medical need in search of a deeper understanding of mechanisms of drug resistance and of the identification of new targets for more efficient therapy.

A widely accepted concept is that tumor cells are organized as a hierarchical population sustained by a subset of cells called Cancer Stem Cells (CSCs) capable of both symmetrical and asymmetrical divisions [10]. CSCs are resistant to conventional drugs and are responsible for tumor relapse after therapy. For these reasons, CSCs have been the object of intense study over the last two decades in the attempt to identify mechanisms responsible for their propagation and of ways to inhibit their growth.

Our group previously described a protocol able to propagate LUAD as 3D spheroids obtained either from malignant pleural effusions (MPEs) or from stable cell lines [11]. We showed that 3D spheroids are highly enriched in cells with stemness markers, including upregulation of Nanog, Oct4, SOX2, and ALDH1A1 activity [12,13]. Enrichment of CSCs was confirmed in transplantation studies in immune-deficient mice showing high rates of tumor engraftment [11,13]. Spheroids showed upregulation of the expression of a key protein in lipid metabolism, namely, Stearoyl–CoA–desaturase 1 (SCD1), the key factor in the biogenesis of monounsaturated fatty acids as precursors of phospholipids [14,15]. SCD1 pharmacological inhibition determined a decreased efficiency of 3D spheroid formation accompanied by a negative impact on the architecture of 3D spheroids. Moreover, SCD1 depletion induced a reduction in ALDH1A1 activity, a marker of cancer stem cells, determining apoptosis specifically in ALDH1A1-positive cells. Silencing of SCD1 impaired in vivo tumorigenicity of 3D lung cancer stem cells [11,16]. We also observed that LUAD 3D spheroids treated with an SCD1 inhibitor reverted resistance to cisplatin. Finally, SCD1 expression correlated with poor survival and worse clinical outcome [17]. Mechanistically, SCD1 inhibition induced the inactivation of β-catenin and YAP/TAZ signalling pathways [18]. Based on these observations and on those of several other laboratories, we consider the 3D spheroid assay a valuable approach to identify factors responsible for the maintenance and propagation of lung CSCs [19]. An intrinsic feature of CSCs is considered to be their plasticity, i.e., their ability to change their physiological properties, which has been linked to epigenetic reprogramming [20–22].

In this context, the study of epigenetic alterations in CSCs acquires particular relevance [23], also because epigenetic alterations might deregulate signalling pathways controlling self-renewal and differentiation, including Wnt, Myc, Notch, and Hedgehog pathways [24,25]. Many epigenetic studies on lung cancer focused on the methylation level of DNA [26,27]. Likewise, post-translational modifications of histone proteins are known to influence chromatin accessibility to transcription factors and complex gene expression machinery [28–30]. In recent years, ATAC-seq has been developed as an emerging technology able to provide a fast and sensitive analysis of the epigenome. ATAC-seq allows to obtain a chromatin accessibility map of the entire genome in order to identify functional gene-regulatory regions [31,32]. Hence, the intersection of transcriptomic analysis and of chromatin accessibility through ATAC-seq can provide a unique opportunity to better identify CSC-specific changes in gene regulatory networks.

In the present study, we have applied this approach to identify regulatory principles responsible for CSCs maintenance. For this purpose, we used the 3D spheroid assay on the MPEs derived LUAD cell line NCI-H460 and intersected transcriptomic with epigenetic signatures of CSCs-enriched (3D cultures) vs. differentiated cells (2D cultures) by RNA-seq and ATAC-seq, respectively. Interestingly, we observed that 3D cultures of NCI-H460 cells are characterized by the activation of pathways enriched of genes responsible for Epithelial to Mesenchymal Transition (EMT). Furthermore, we singled out and validated the role of the B4GALT1 as an emerging gene in CSCs propagation and as a target for their inhibition.

2. Materials and Methods

2.1. Cell Cultures

Established human NSCLC cell NCI-H460 was obtained from the American Type Culture Collection (ATCC, Manassasn, VA, USA) and was cultured according to the manufacturer's instructions. BBIRE T-238 and BBIRE T-248 primary cultures were isolates from Malignant Pleural Effusion (MPEs) of patients with LUAD, as previously described [11,13]. The study was approved by the Ethics Committee (protocol number 1032/17). Cells in adherent condition were cultured in RPMI-1640 supplemented with 10% FBS, 1% L–Glutammine and 1% Penicillin/Streptomycin (Sigma, St. Louis, MO, USA), while 3D cultures in suspension were obtained as previously described [11,13,18]. Briefly, 20,000 cells/mL were resuspended in an appropriate amount of Spheroid Medium and seeded onto Ultralow Attachment plates (Corning Costar, MA, USA) to form 3D structures. After 4 days, the number of 3D spheres in single wells of a 96 well low attachment culture plate were counted, and their volume was estimated using Image J Software v1.50i. Cells were routinely checked for mycoplasma contamination by PCR.

2.2. RNA-Sequencing and Bioinformatic Analysis

Total RNA was extracted from NCI-H460 2D and 3D cells cultures, using Qiazol (Qiagen, Hilden, Germany), purified from DNA contamination through a DNase I (Qiagen) digestion step, and further enriched by Qiagen RNeasy columns profiling (Qiagen). Quantity and integrity of the extracted RNA were assessed by Nanodrop Spectrophotometer (Nanodrop Technologies LCC, Thermofisher, Waltham, MA, USA) and by Agilent 2100 Bioanalyzer (Agilent Technologies, Santa Clara, CA, USA), respectively. All RNA used for subsequent library preparation had an RNA integrity number greater than9.0. RNA libraries for sequencing were generated in triplicate using the same amount of RNA for each sample according to the standard Illumina TrueSeq Stranded Total RNA kit with an initial ribosomal depletion step using Ribo Zero Gold (Illumina, San Diego, CA, USA). The libraries were quantified by qPCR and sequenced in paired-end mode (2 × 75 bp) with NextSeq 500 (Illumina). Each sample was generated with the Illumina platform. A pre-step for quality control was performed to assess sequence data quality and to discard low quality reads [33]. Paired 75 bp long reads were quality checked and processed using Kallisto v0.46.0 [34] with parameters: "quant -t 8 -b 30" to the genome Ensemble GRCh38.96 [35] downloaded from https://github.com/pachterlab/kallisto-transcriptome-indices/releases/download/ensembl-96/homo_sapiens.tar.gz. Differential expression data were obtained running the R package Sleuth v0.27.3 [36] following the pipeline suggested in the Sleuth manual available https://pachterlab.github.io/sleuth_walkthroughs/trapnell/analysis.html using both available Likely ratio [37] and Wald tests [38]. Differentially expressed genes were considered those with FDR < 0.001 in both tests and B value over 0.3 for upregulation and less than −0.3 for downregulation. Volcano plots were drawn with an in-house R script (R v3.6.1). Gene set enrichment analysis was performed, submitting the top 100 significant genes to the hallmark.MSigDB [39] available in ShinyGo v0.60 [40] and plotted with an in-house R script.

2.3. Transcriptome Analyses of Public Datasets

TCGA cohort coupled with GTEx cohort was used to run a pan-cancer, LUAD, and LUSC specific analysis of the gene B4GALT1 expression. Box-plots and overall survival analysis were obtained using GEPIA v1 [41] (http://gepia.cancer-pku.cn/index.html) selecting default parameters. The Kaplan–Meier overall survival analysis of pooled B4GALT1 and SCD1 genes expression data was performed by re-analysis of the TCGA dataset available in Kaplan–Meier plotter [42,43] of 513 (LUAD) and 501 (LUSC) patients, respectively, with default parameters. Gene expression data were scaled in $\log_2(TPM+1)$.

2.4. Assay for Transposase Accessible Chromatin with High-Throughput Sequencing (ATAC) and Bioinformatic Analysis

To profile open chromatin, we used the ATAC-seq protocol developed by Buenrostro et al., with minor modifications [44]. 50,000 NCI–H460 2D and 3D cells were washed once with 1x PBS and centrifuged at 500× g for 5 min at 4 °C. The cell pellet was lysed in ice-cold lysis buffer (10 mM Tris-HCl pH 7.4, 10 mM NaCl, 3 mM $MgCl_2$, 0.1% IGEPAL CA–630) to isolate the nuclei. The nuclei were centrifuged at 500× g for 5 min at 4 °C and subsequently resuspended on ice in 50 μL transposase reaction buffer containing 2.5 μL of Tn5 transposase and 25 μL of 2xTD buffer (Nextera DNA Sample preparation kit from Illumina). After incubation at 37 °C for 30 min, the samples were purified with MiniElute PCR Purification Kit (Qiagen), eluting in 10 μL elution buffer (10 mM Tris–HCl pH 8). To amplify transposed DNA fragments, we used NEBNext High-Fidelity 2x PCR Master Mix (New England Labs, Ipswich, MA, USA) and the Customized Nextera PCR Primers. Libraries were purified by adding Agencourt Ampure XP (Beckman, Brea, CA, USA) magnetic beads (1:1 ratio) to remove remaining adapters (left side selection) and double purified (1:0.5 and 1:1.15 ratio) for right side selection. Libraries were controlled using a High Sensitivity DNA Kit on a Bioanalyzer. Each library was then paired-end sequenced (2 × 75 bp) on a NextSeq 500 instrument (Illumina). Paired-end 75 bp long reads were quality checked using FASTQC and aligned to Hg38 using Bowtie v 2.3.4.2 setting: "mode = local and p = 6". The aligned reads were processed by Samtools v1.9 to be converted in BAM format then sorted and indexed. Peaks were called by MACS2 v2.1.2 with parameters "nomodel–shift–100–extsize 200–B–SPMR–call–summits". Peaks with a –log10(q-value) lower than 2.0 were discarded. Peaks in bdg format were converted in bw format using bedGraphToBigWig v4 tool available on https://www.encodeproject.org/software/bedgraphtobigwig/. All peaks matching blacklisted regions were discarded from further processing. After peak generation, an in-house pipeline based on BEDTOOLS v2.25.0 and custom BASH scripts were developed to build a master list of accessible sites by pooling the significant peaks of all the sample dataset. The master list of accessible sites was produced with a multistep procedure: 1. To identify the common overlapping signal amongst all the samples, promoter peaks were intersected using BEDTOOLS multiinter. 2. The book-ended regions from the core signal file were merged using BEDTOOLS merge, then intersected with the original peak calls and sorted.

2.5. Comparative Analysis of DNA Accessibility

Differential analysis of the two groups (3D vs. 2D signals) was obtained using an in-house script which deploys edgeR suite (v3.28.0) [45]. Data were processed and normalized with the TMM method [46]. The differential comparison was performed using an in-house script which relies on the edgeR "exactTest" function (v3.28.0). Data were further adjusted for Benjamini Hochberg correction. The sites showing FDR < 0.05 were considered significant. Boxplot of differential enriched sites enrichment was performed with an in-house script. To show normalized read count differences observed between 3D and 2D culture, we developed an R script to build a heatmap showing differences between our groups. We centered the data to the row mean and fixed the color palette from the lowest to the highest value. Centered data were hierarchically clustered (Pearson distance, average linkage) using the hclust package. Results were visualized using heatmap.2 available in the gplot R package (v3.0.1.1).

2.6. Reverse Transcription Polymerase Chain Reaction (RT-PCR)

Total RNA was isolated by Trizol (Thermofisher) following the manufacturer's instructions. First-strand cDNA was synthesized with PrimeScript RT reagent Kit, genomic DNA contamination is eliminated with gDNA Eraser (Takara Bio Inc, Kusatsu, Prefettura di Shiga, Japan). The cDNA was used for RT-PCR experiments carried out in a 7500 StepOnePlus (Applied Biosystems, Foster City, CA, USA) as previously described [47,48]. Primers for B4GALT1: 5′-CTATATCTCGCCCAAATGCTG-3′ (forward) and 5′-GTGCAATTCGGTCAAACCTC-3′ (reverse), and other primers are previously described [18]. The relative amount of all mRNAs was calculated using the comparative method ($2^{-\Delta\Delta Ct}$) after normalization to H3. Three independent experimental replicates were performed.

2.7. siRNA Transfection

NCI-H460 cells were transfected with siRNA of B4GALT (s5726; Thermofisher) or control siRNA (Santa Cruz, CA, USA). DNA transfections were done with Lipofectamine RNAiMAX Reagents (Invitrogen Co, Carlsad, CA, USA) according to the manufacturer's instructions.

2.8. Aldehyde Dehydrogenase (ALDH) Activity Assays

ALDH activity was performed with Activity kit (ab155893, Abcam, Cambridge, UK), in 3D transfected cells. Following the manufacturer's recommendation, NADH standard and sample were added into a 96 well plate in duplicate. Subsequently, 50 µL of the Reaction mix, containing Acetaldehyde and ALDH substrate, were added to each well. Finally, it was incubated for 20–60 min at room temperature and measured at OD 450 nm. The activity was calculated according to the datasheet.

2.9. Statistical Analysis

All experiments were performed in triplicate and the statistical was carried out to GraphPad Prism v6.0 software. Group comparison were performed by Student's test and values were expressed as mean ± Standard Error of Measurement (SEM).

3. Results

3.1. A Combinatorial Approach to Identify Gene Expression Features of Cancer Stem Cells (CSC)-Enriched Lung Adenocarcinoma (LUAD) Cell Cultures

In order to characterize regulatory cues responsible for orchestrating gene expression in CSCs, we used the approach described in Figure 1. The stable LUAD cell line NCI–H460 derived from MPEs was grown either in standard 2D attachment conditions or as 3D cultures enriched of CSCs markers [16,47] (Figure 1A). For transcriptomic analysis, RNA was extracted from both types of cultures. In parallel, in order to achieve genome-wide chromatin accessibility maps and to identify active gene regulatory regions, nuclei were isolated from both 2D and 3D cultures, chromatin was extracted and subjected to ATAC-seq (see materials and methods) (Figure 1B).

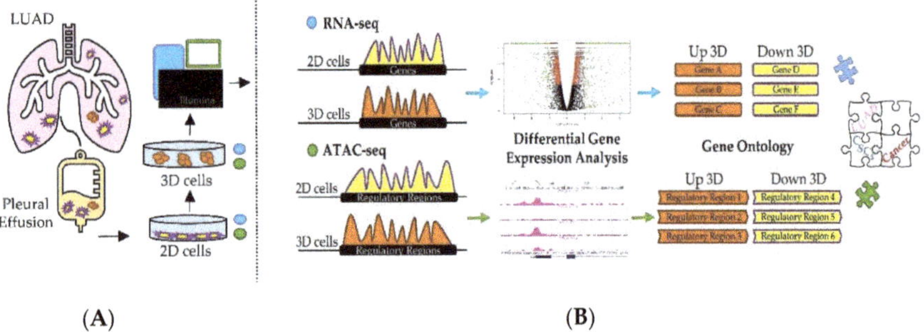

(A)   (B)

Figure 1. Graphical overview of the investigation. (**A**) 2D and 3D cell cultures were obtained from the stable NCI–H460 cell line obtained from the malignant pleural effusion of a lung adenocarcinoma (LUAD) patient. Total RNA was extracted and subjected to RNA-seq. Nuclei were isolated and processed to perform ATAC-seq. (**B**) Computational workflow developed to identify differentially expressed genes, pathways and biological processes in 3D compared to 2D cell cultures.

Transcriptomics technologies and chromatin accessibility assays are powerful tools to infer active cell regulatory states by analyzing the relative number of transcripts (Figure 1) and cis-regulatory activity, respectively. Although there are several difficulties in linearly correlating the activity of cis-regulatory regions with the relative production of gene transcripts, numerous studies [49–51] revealed that the large majority of regulatory elements are strongly linked to RNA production and occur in the proximity (within 50 kb) to the closest Transcriptional Start Sites (TSS). Thus, we took advantage of these observations and assayed the differences of RNA abundance and active chromatin accessibility sites in 3D vs. 2D cultures obtained from stable NCI–H460 cell line in a unique computational workflow. Each set of data was analyzed first individually then jointly to better identify gene regulatory elements capable of simultaneously controlling key cellular pathways specifically activated in CSC-enriched cell cultures.

3.2. Integrative RNA-seq and ATAC-seq Analysis Reveals Overarching Transcriptomic Features of Cancer Stem Cells (CSC)-Enriched LUAD Cell Cultures

RNA sequencing analysis revealed a global transcriptional rewiring from 2D to 3D cultures. The differential analysis revealed 3095 genes significantly enriched (FDR < 10^{-3}) (detailed in Figure 2A, Figure S1, Table S1) of which 1854 were downregulated and 1241 upregulated in 3D spheroids. To gain insights into the functional role of the differentially expressed transcripts, we computationally predicted the impact in their relative abundance to the cell physiological state (Figure 2B). Not unexpectedly, gene set enrichment analysis of the differentially regulated genes in the two systems revealed the emergence of a vastly different regulatory landscape. In particular, the 3D spheroids cell population exhibited a significant enrichment of gene pathways involved in oncogenesis and cancer progression, including EMT (FDR < 10^{-13}), Hypoxia (FDR < 10^{-10}), Cholesterol homeostasis (FDR < 10^{-5}), and Apoptosis (FDR < 10^{-4}) (Figure 2B).

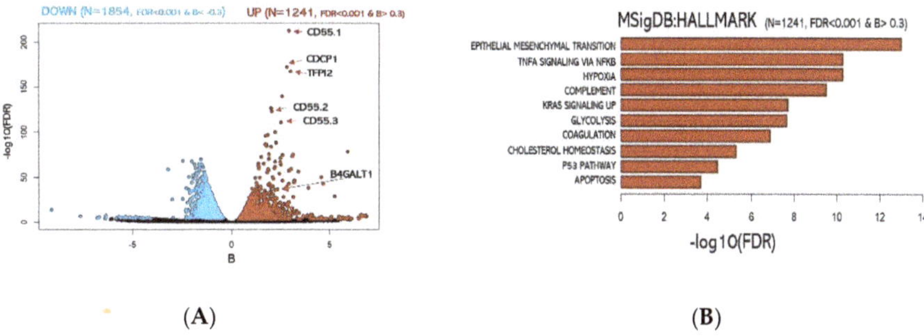

(**A**)   (**B**)

Figure 2. Analysis of 2D vs. 3D culture transcriptomes. (**A**) Volcano plot of over-expressed or under-expressed genes in 3D transcriptome vs. 2D. (blue: significantly under-expressed transcripts; red: significantly over-expressed transcripts). (**B**) Functional enrichment for upregulated genes in 3D cultures.

Among the most significant upregulated genes, we identified 3 different isoforms of CD55 a well-known protein coding gene frequently linked to cancer aggressiveness in many carcinomas. Surprisingly, among the top 10% genes exhibiting the most significant transcript abundance shift (Table S1), we found the gene B4GALT1. Notably, B4GALT1 protein has been previously associated with multi-drug-resistant cell phenotype in human leukemia [52] and other haematological malignancies [53,54]. On the other side, genes significantly downregulated were associated with MYC and E2F target genes (Figure S2). Importantly, our transcriptomic data further support the model of cancer aggressiveness recently proposed by Padmanaban et al. [55] in which significant loss of E–cadherin transcription (CDH1 gene) guides the upregulation of the transforming growth factor-β (TGFβ) expression (see Table S1) as a requirement for metastatic invasion.

Then, ATAC-seq data were utilized to determine genome-wide chromatin accessibility of 3D vs. 2D NCI–H460 cells. DNA nucleosome-free regions store regulatory information in the form of active regulatory elements, including enhancers and promoters. These elements are highly plastic and act as essential players in guiding cell phenotypic states. The analysis revealed only 404 genomic loci (FDR < 0.05) significantly different between the two groups (see Figure 3A,B) of which 236 were closed and 168 open in 3D cultures, respectively.

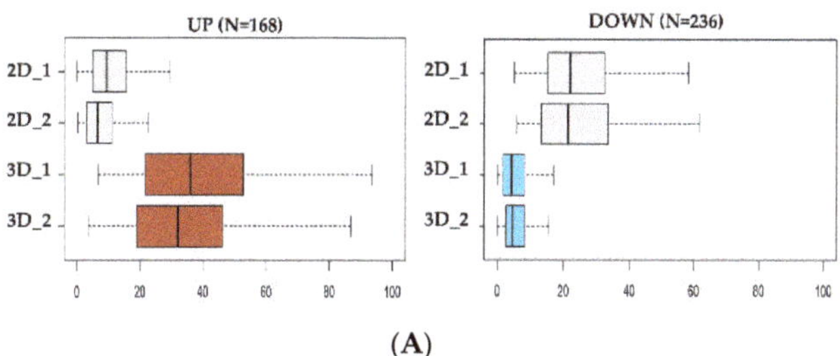

(**A**)

Figure 3. *Cont.*

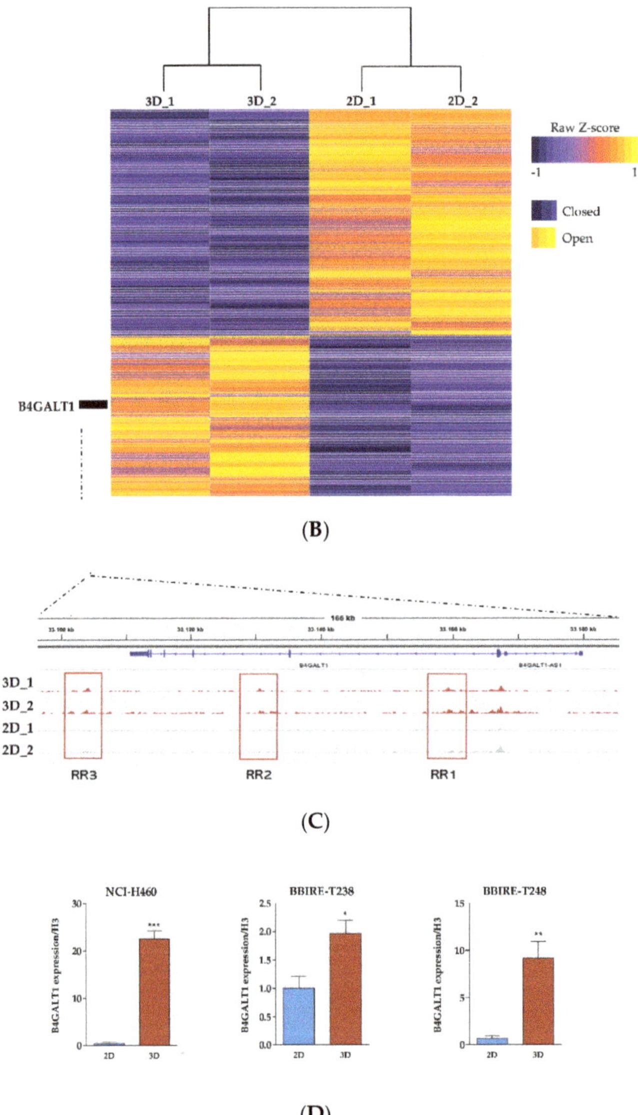

Figure 3. Analysis of DNA accessibility in 3D vs. 2D cultures. (**A**) ATAC-seq signal enrichment of normalized read counts at differential accessible sites in 3D vs. 2D (red = open; blue = closed). (**B**) Heatmap of differential chromatin accessibility sites (N = 404) showing the cell condition specificity of 2D and 3D ATAC-seq peaks. (**C**) Regulatory landscape of B4GALT1 locus. Peaks represent Assay for Transposase Accessible Chromatin with high-throughput sequencing (ATAC-seq) raw signals. (**D**) Reverse transcription polymerase chain reaction (RT-PCR) validation level of B4GALT1 upregulated in 3D cells compared to 2D culture in NCI-H460 (stable lung cell line) and BBIRE T-238, BBIRE T248 (primary lung cell lines). H3 reference gene have been used for normalization. Bars represent the mean of three biological replicates and technical replicates with their corresponding standard error of the mean (SEM). *** $p < 0.0002$; * $p < 0.02$; ** $p < 0.0085$.

Recent observations [56,57] confirmed the notion that the signal intensity of ATAC-seq is linearly dependent on the number of cells contributing to the signal. On the basis of this, we observed that the normalized signal of the open sites in 3D cells was consistently doubling the signal of sites only accessible in 2D cells, thus implying that 3D cultures are composed of a more homogeneous cell population compared to the 2D culture signal (Figure S3, Table S2). Finally, we tested the relationship between gene expression and chromatin openness among each gene-specific neighborhood (|50 kb| to the TSS) to identify regulatory patterns exhibiting both significant changes in transcriptomic abundance and chromatin plasticity (Figure S4). By integrating RNA-seq and ATAC-seq data, we found that only a small proportion of differential ATAC-seq signal was occurring within |50 kb| from the TSSs of differentially regulated genes (Table S3). However, the largest proportion of promoters associated with expressed genes were consistently accessible to regulatory factors in both groups, suggesting active transcription (Figure S5). These data imply that the differential abundance of transcripts in 3D vs. 2D cultures does not depend only on chromatin shifts in the vast majority of cases, but may be favored by other mechanisms whose further investigation could shed light on their identity (Table S2). Examining the genomic landscape of the genes upregulated in 3D, we identified B4GALT1 as the top candidate. Interestingly, the B4GALT1 locus presents three regulatory sites (Figure 3C) differentially opened in 3D vs. 2D cell cultures. These regulatory sites may represent active enhancers; however, no previously published data annotate these genomic sites as regulatory regions, and further investigations need to be conducted to confirm this hypothesis.

In order to confirm B4GALT1 upregulation in LUAD 3D spheroids we carried out qRT-PCR assays on RNA extracted from 2D vs. 3D cultures of NCI–H460 cells and from two additional primary LUAD cell lines isolated from malignant pleural effusions in our laboratory (BBIRE–T238 and BB–IRE T–248). The results (Figure 3D) show that B4GALT1 is strongly upregulated in 3D cultures in all cases analyzed.

3.3. Validation of B4GALT1 as Novel Factor Responsible for the Propagation of Lung Adenocarcinoma (LUAD) Cancer Stem Cells (CSCs)

As shown in the previous paragraph, the combined RNA-seq/ATAC-seq analysis of 3D vs. 2D cultures of NCI–H460 cells highlighted B4GALT1 as one of the top genes overexpressed in CSCs enriched 3D cultures with a corresponding highly opened chromatin structure of its gene regulatory domains. In order to validate B4GALT1 as a critical factor for CSC propagation and a potential target for intervention, we carried out a series of analyses. We first performed an explorative analysis of B4GALT1 expression by interrogating TCGA and GTEx dataset. Remarkably, 84% of cancer types exhibit upregulation of B4GALT1 expression compared to healthy tissues, including both lung cancers LUAD and LUSC (see Figure 4A, Figure S6).

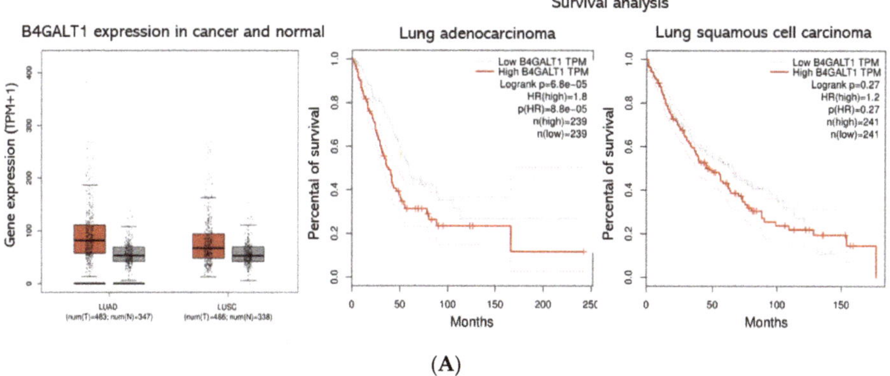

(A)

Figure 4. *Cont.*

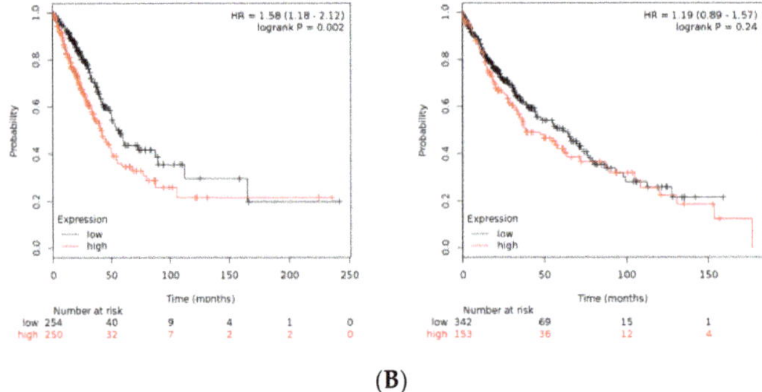

(B)

Figure 4. B4GALT1 gene expression and survival analysis in lung adenocarcinoma (LUAD) and lung squamous cell carcinoma (LUSC) patients. (**A**) On the left panel is shown a Box Plot of B4GALT1 expression in lung adenocarcinoma (LUAD) and lung squamous cell carcinoma (LUSC). Each dot represents a patient. (red: cancer tissues; grey: healthy tissues). On the right panel is shown a survival curves depict the B4GALT1 prognostic value in LUAD (N = 239 high expression tissues +239 low expression tissues) and LUSC cohort (N = 241 high expression +241 low expression). Comparison of survival curves was performed using a log-rank (Mantel–Cox) test. HR = Hazard ration. Dotted lines represent the 95% of Confidence Interval. (**B**) Kaplan–Meier curves depict the cumulative prognostic value of B4GALT1 and SCD1 gene expressions in LUAD (N = 250 high expression tissues + 254 low expression tissues) and LUSC (N = 153 high expression tissues + 342 low expression tissues) Abbreviations: LUAD, lung adenocarcinoma; LUSC lung squamous cell carcinoma; num, number; T, tumor; N, normal.

These data were consistent with Immunohistochemistry of normal and pathologic tissues available on the Human Protein Atlas data portal [58]. Consistent with these predictions, a meta-analysis of the TCGA datasets re-computed from raw RNA-seq by XENA project revealed that higher expression of B4GALT1 mRNA is linked with poor survival in many cancers (data not shown). In particular, LUAD patients with higher expression of B4GALT1 at diagnosis have a significantly worse outcome; conversely, this is not occurring in LUSC patients (Figure 4A). In agreement, a meta-analysis of pooled B4GALT1 and SCD1 mRNA expression data from TCGA conducted to similar results (Figure 4B), thus suggesting a potential co-regulatory activity in tumor progression of SCD1 and B4GALT1 genes.

Furthermore, in order to assess the role of B4GALT1 in the generation of 3D spheroids we transfected a B4GALT1 siRNA in NCI–H460 lung cancer cells. Transient knockdown of B4GALT1 strongly impaired 3D spheroids formation, both as to number and size (Figure 5A,B). The role of B4GALT1 in the propagation of lung CSCs enriched cell cultures was confirmed by the demonstration that its depletion reduced the expression level of other stemness markers such as Oct4, Sox2, and Nanog compared to scramble siRNA (Figure 5C). Moreover, the silencing of B4GALT1 affected SCD1 mRNA levels in 3D cells. To further confirm the role of B4GALT1 in CSCs, ALDH activity was also evaluated. As shown in Figure 5D, inhibition of B4GALT1 resulted in a significant reduction of ALDH activity in 3D cells, as well as a decrease of Nanog protein expression compared to control scrambled siRNA (data not shown). These results taken together suggest that B4GALT1 plays an important role in the propagation and maintenance of CSC-enriched LUAD cell cultures.

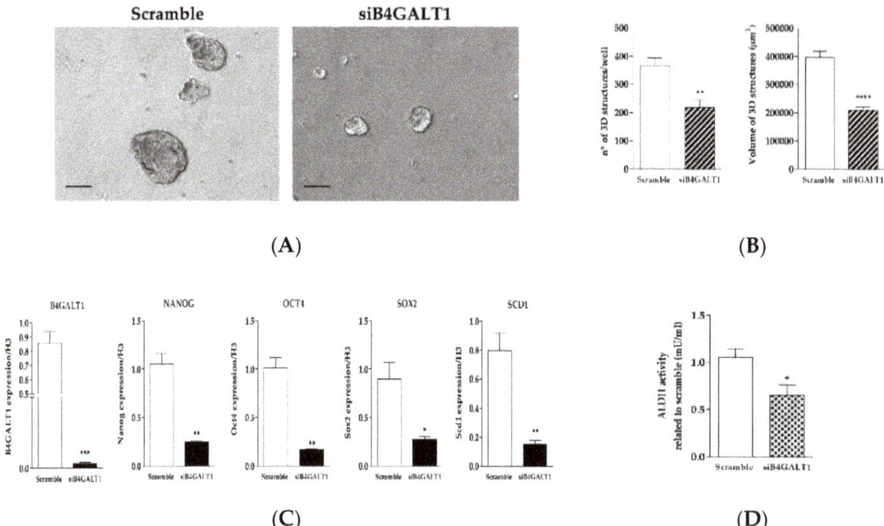

Figure 5. Knockdown of B4GALT1 RNA decreases the 3D structure formation and the expression of stemness markers. (**A**) Representative images of silencing of B4GALT1 reduction of 3D structure formation potential compared to the scramble of H460 cells. Scale bar = 100 μm. (**B**) Graphs show that silencing of B4GALT1 in 3D induces a decrease of volume and number of 3D spheroids. Number and volume of the 3D cells counted in each well after four days of culture. (**C**) Validation of B4GALT1 silencing in 3D cells, the results show a strongly decreases of stemness markers mRNA levels, such as Oct4, Sox2, Nanog, and SCD1. Expression of each gene was normalized to that of H3. (**D**) ALDH activity decrease substantially in 3D siB4GALT1 vs. Scramble cells. Experiments were performed in triplicate, and the background interference and the blank value were subtracted from the absorbance of the samples. In the bar plots, the mean ± standard error of mean (SEM) was shown from at least three independent experiments * $p < 0.05$, ** $p < 0.005$, **** $p < 0.0001$ (vs. scramble).

4. Discussion

In this study, in order to better identify transcriptional cues active in lung adenocarcinoma stem cells, we sought to link genes undergoing expression changes to their regulatory elements by leveraging RNA-seq and ATAC-seq data obtained from 2D and CSC-enriched 3D cell cultures. Indeed, the parallel profiling of gene expression and chromatin accessibility within the same cell bulk is a well-described approach to reveal causal regulatory relationships [59]. In this respect, our data revealed a global shift of gene expression which was accompanied by discrete changes in chromatin accessibility. Using a rigorous and conservative computational strategy, we identified more than 3000 differentially regulated transcripts and approximately 400 cis-regulatory regions affected in the passage from 2D to 3D cultures. In this regard, modifications of chromatin openness are linked to the activation and inhibition of regulatory pathways able to confer a selective growth advantage to cancer cells. Regulatory regions such as enhancers are key distal cis-regulatory elements that elevate the expression of nearby genes, independently from the distance to the target gene or orientation [60,61]. Our data predict that significant chromatin openness changes might collectively affect several hundreds of genes. Interestingly, both sets of differentially opened regulatory regions and RNA transcribed enrich for "cell migration" and "EMT pathways", which are both distinctive features of cancer stem cells [62,63], and which we know from previous studies to be enriched in 3D vs. 2D cultures [64–68].

One of the most exciting findings of our combined transcriptomic and epigenomic analysis was the identification of B4GALT1 as one of the top candidates. Indeed, besides being one of the most highly transcriptionally upregulated genes in 3D cultures, the B4GALT1 locus shows three genomic regions located downstream the transcriptional start site whose chromatin is accessible only in 3D cultures as compared with 2D cultures.

We believe these are regulatory regions which may act as enhancers. Further studies are needed to confirm this hypothesis. An apparent limitation of our study is that we have applied our approach so far only to one LUAD cell line. However, we are confident that our observation about the involvement of B4GALT1 is of more general significance because of various reasons. Firstly, B4GALT1 transcriptional upregulation in 3D LUAD spheroids was confirmed by RT-PCR, not only in NCI–H460 but also in other LUAD primary cells, generated in our laboratory from malignant pleural effusions of LUAD patients. Secondly, our meta-analysis of TCGA datasets revealed that overexpression of B4GALT1 is linked to poorer survival in LUAD patients.

B4GALT1 has been reported before to facilitate cancer cell proliferation, invasiveness, and metastasis in several cancer types [69,70]. However, we provide the first evidence that this gene may be involved in the propagation of cancer stem cells, namely in lung cancer. In this respect the previous observation in hematologic cancers that B4GALT1 is responsible for drug resistance by regulating the expression of P-gp and MDR- associated protein [52,69] acquires particular relevance given the known ability of CSCs to be drug resistant [71,72].

We have observed that B4GALT1 silencing potently inhibits the 3D spheroid formation and impairs the expression of a set of stem cell markers. Hence, we believe that B4GALT1 is needed for lung CSCs propagation. At the moment the mechanism by which B4GALT1 activity facilitates lung CSCs propagation can only be the object of speculations. The enzyme catalyzes the transfer of galactose from UDP–Galactose to N-linked sugar chains of glycoproteins and is therefore important for the biosynthesis of glycoconjugates which may be required to facilitate the formation and maintenance of cell–cell contact and resistance to anoikis during the formation of 3D structures in non-adherent conditions. Further studies are required to investigate about this possibility.

Finally, B4GALT1 expression is somehow linked in LUAD to the expression of SCD1, which has been the object of intense studies by our group and other laboratories in recent years. TCGA data clearly show that co-overexpression of B4GALT1 and SCD1 is a negative prognostic factor. Furthermore, our silencing data demonstrate that inhibiting the expression of B4GALT1 strongly reduces SCD1 expression in LUAD 3D spheroids. We tend to believe that this is not a direct effect, but rather an indirect consequence of decreased survival of lung CSCs linked to the absent expression of B4GALT1.

5. Conclusions

In conclusion, our study proposes for the first time the involvement of B4GALT1 in lung CSCs maintenance and propagation. Therefore, this enzyme can become a new potential target of intervention. It will be important in the future to assess the effect of in vivo inhibition of B4GALT1 in LUAD tumor growth either alone or combined with inhibitors of other CSC-enriched targets, such as SCD1.

Supplementary Materials: The following are available online at http://www.mdpi.com/2077-0383/8/11/1928/s1, Figure S1: quality Control of RNA-seq; Figure S2: MSigDB: HALLMARK enrichment of the downregulated genes in 3D vs. 2D cultures; Figure S3: Quality Control of ATAC-seq; Figure S4: Differential significant ATAC-seq sites in relationship with the relative genomic neighbourhood; Figure S5: Enrichment analysis of top 100 up-regulated sites in 3D vs. 2D ATAC-seq; Figure S6: PanCancer analysis of B4GALT1 expression; Table S1: Upregulated genes in 3D vs. 2D associated with most significant MSigDB: HALLMARK; Table S2: output_ATACseq_3Dvs2D, Table S3: ATAC_RNAseq_data. Data availability: Raw ATAC-seq and RNA-seq data used in the study have been deposited at the ENA (http://www.ebi.ac.uk/ena) under "project no.PRJEB35287".

Author Contributions: Conceptualization, C.D.V., G.C. (Giacomo Corleone), methodology, F.D.N., F.A., V.S., S.d.M., S.B., validation, V.S., formal analysis, G.C. (Giacomo Corleone), M.P., M.F., writing-original draft preparation, G.C. (Giacomo Corleone), C.D.V., supervision, G.C. (Gennaro Ciliberto), R.M., manuscript revision and editing, G.C. (Gennaro Ciliberto), R.M., visualization, M.B., F.V., E.A.R., A.R., C.N.

Funding: This work was supported by Italian Association for Cancer Research (AIRC) (grants IG17007 to R. Mancini and IG15216 to G. Ciliberto), by the LazioInnova (grant 2018 n. 85-2017-13750 to R. Mancini) and by Fondo di Ateneo 2018, grant/Award number: B86C19001510005 to R. Mancini.

Acknowledgments: The author are grateful Luigi Fattore for his advice and comments and Tommaso Mancuso for technical support. The author, however, bears full responsibility for the paper.

Conflicts of Interest: The authors declare no conflict of interest. The funders of this study had no role in the design of the study; in the collection, analyses or interpretation of data; in the writing of the manuscript, or in the decision to publish the results.

References

1. Bray, F.; Ferlay, J.; Soerjomataram, I.; Siegel, R.L.; Torre, L.A.; Jemal, A. Global Cancer Statistics 2018: GLof Incidence and Mortality World in 185 Countries. *CA Cancer J. Clin. Anticancer. Res.* **2018**, *55*, 78–108.
2. Nasim, F.; Sabath, B.F.; Eapen, G.A. Lung Cancer. *Med. Clin. N. Am.* **2019**, *3*, 463–473. [CrossRef] [PubMed]
3. Testa, U.; Castelli, G.; Pelosi, E. Lung Cancers: Molecular Characterization, Clonal Heterogeneity and Evolution, and Cancer Stem Cells. *Cancers* **2018**, *10*, 248. [CrossRef] [PubMed]
4. Moon, E. Immunotherapy: Beyond Anti–PD-1 and Anti–PD-L1 Therapies. *Am. Soc. Clin. Oncol. Educ. Book* **2016**, *36*, e450–e458.
5. Darvin, P.; Toor, S.M.; Nair, V.S.; Elkord, E. Immune Checkpoint Inhibitors: Recent Progress and Potential Biomarkers. *Exp. Mol. Med.* **2018**, *50*, 165. [CrossRef] [PubMed]
6. Roscilli, G.; De Vitis, C.; Ferrara, F.F.; Noto, A.; Cherubini, E.; Ricci, A.; Mariotta, S.; Giarnieri, E.; Giovagnoli, M.R.; Torrisi, M.R.; et al. Human Lung Adenocarcinoma Cell Cultures Derived from Malignant Pleural Effusions as Model System to Predict Patients Chemosensitivity. *J. Transl. Med.* **2016**, *14*, 61. [CrossRef]
7. Roscilli, G.; Cappelletti, M.; De Vitis, C.; Ciliberto, G.; Di Napoli, A.; Ruco, L.; Mancini, R.; Aurisicchio, L. Circulating MMP11 and Specific Antibody Immune Response in Breast and Prostate Cancer Patients. *J. Transl. Med.* **2014**, *12*, 54. [CrossRef] [PubMed]
8. Chae, Y.K.; Pan, A.; Davis, A.A.; Raparia, K.; Mohindra, N.A.; Matsangou, M.; Giles, F.J. Biomarkers for PD-1/PD-L1 Blockade Therapy in Non–Small-cell Lung Cancer: Is PD-L1 Expression a Good Marker for Patient Selection? *Clin. Lung Cancer* **2016**, *17*, 350–361. [CrossRef]
9. Wang, Q.; Wu, X. Primary and Acquired Resistance to PD-1/PD-L1 Blockade in Cancer Treatment. *Int. Immunopharmacol.* **2017**, *46*, 210–219. [CrossRef]
10. Lathia, J.D.; Liu, H. Overview of Cancer Stem Cells and Stemness for Community Oncologists. *Target. Oncol.* **2017**, *12*, 387–399. [CrossRef]
11. Mancini, R.; Giarnieri, E.; De Vitis, C.; Malanga, D.; Roscilli, G.; Noto, A.; Marra, E.; Laudanna, C.; Zoppoli, P.; De Luca, P.; et al. Spheres Derived from Lung Adenocarcinoma Pleural Effusions: Molecular Characterization and Tumor Engraftment. *PLoS ONE* **2011**, *6*, e21320. [CrossRef] [PubMed]
12. Noto, A.; Raffa, S.; De Vitis, C.; Roscilli, G.; Malpicci, D.; Coluccia, P.; Di Napoli, A.; Ricci, A.; Giovagnoli, M.R.; Aurisicchio, L.; et al. Stearoyl-CoA Desaturase-1 is a Key Factor for Lung Cancer-Initiating Cells. *Cell Death Dis.* **2013**, *4*, e947. [CrossRef] [PubMed]
13. Bruschini, S.; Di Martino, S.; Pisanu, M.E.; Fattore, L.; De Vitis, C.; Laquintana, V.; Buglioni, S.; Tabbi, E.; Cerri, A.; Visca, P.; et al. CytoMatrix for a Reliable and Simple Characterization of Lung Cancer Stem Cells from Malignant Pleural Effusions. *J. Cell. Physiol.* **2019**, 1–11. [CrossRef] [PubMed]
14. Ariel Igal, R. Roles of stearoylcoa Desaturase-1 in the Regulation of Cancer Cell Growth, Survival and Tumorigenesis. *Cancers* **2011**, *3*, 2462–2477. [CrossRef] [PubMed]
15. Aljohani, A.M.; Syed, D.N.; Ntambi, J.M. Insights into Stearoyl-CoA Desaturase-1 Regulation of Systemic Metabolism. *Trends Endocrinol. Metab.* **2017**, *28*, 831–842. [CrossRef] [PubMed]
16. Pinkham, K.; Park, D.J.; Hashemiaghdam, A.; Kirov, A.B.; Adam, I.; Rosiak, K.; da Hora, C.C.; Teng, J.; Cheah, P.S.; Carvalho, L.; et al. Stearoyl CoA Desaturase Is Essential for Regulation of Endoplasmic Reticulum Homeostasis and Tumor Growth in Glioblastoma Cancer Stem Cells. *Stem Cell Rep.* **2019**, *12*, 712–727. [CrossRef] [PubMed]
17. Pisanu, M.E.; Noto, A.; De Vitis, C.; Morrone, S.; Scognamiglio, G.; Botti, G.; Venuta, F.; Diso, D.; Jakopin, Z.; Padula, F.; et al. Blockade of Stearoyl-CoA-Desaturase 1 Activity Reverts Resistance to Cisplatin in Lung Cancer Stem Cells. *Cancer Lett.* **2017**, *406*, 93–104. [CrossRef]

18. Noto, A.; De Vitis, C.; Pisanu, M.E.; Roscilli, G.; Ricci, G.; Catizone, A.; Sorrentino, G.; Chianese, G.; Taglialatela-Scafati, O.; Trisciuoglio, D.; et al. Stearoyl-CoA-Desaturase 1 Regulates Lung Cancer Stemness Via Stabilization and Nuclear Localization of YAP/TAZ. *Oncogene* **2017**, *36*, 4671–4672. [CrossRef]

19. Mancini, R.; Noto, A.; Pisanu, M.E.; De Vitis, C.; Maugeri-Sacca, M.; Ciliberto, G. Metabolic Features of Cancer Stem Cells: The Emerging Role of Lipid Metabolism. *Oncogene* **2018**, *37*, 2367–2378. [CrossRef]

20. Poli, V.; Fagnocchi, L.; Zippo, A. Tumorigenic Cell Reprogramming and Cancer Plasticity: Interplay Between Signaling, Microenvironment, and Epigenetics. *Stem Cells Int.* **2018**, *2018*, 4598195. [CrossRef]

21. Li, W.; Bai, H.; Liu, S.; Cao, D.; Wu, H.; Shen, K.; Tai, Y.; Yang, J. Targeting Stearoyl-CoA Desaturase 1 to Repress Endometrial Cancer Progression. *Oncotarget* **2018**, *9*, 12064–12078. [CrossRef] [PubMed]

22. Poeta, M.L.; Massi, E.; Parrella, P.; Pellegrini, P.; De Robertis, M.; Copetti, M.; Rabitti, C.; Perrone, G.; Muda, A.O.; Molinari, F.; et al. Aberrant Promoter Methylation of Beta-1,4 Galactosyltransferase 1 as Potential Cancer-Specific Biomarker of Colorectal Tumors. *Genes. Chromosom. Cancer* **2012**, *51*, 1133–1143. [CrossRef] [PubMed]

23. Zhang, Q.; Thakur, C.; Shi, J.; Sun, J.; Fu, Y.; Stemmer, P.; Chen, F. New Discoveries of Mdig in the Epigenetic Regulation of Cancers. *Semin. Cancer Biol.* **2019**, *57*, 27–35. [CrossRef] [PubMed]

24. Prasetyanti, P.R.; Medema, J.P. Intra-Tumor Heterogeneity from a Cancer Stem Cell Perspective. *Mol. Cancer* **2017**, *16*, 41. [CrossRef]

25. Pattabiraman, D.R.; Weinberg, R.A. Tackling the Cancer Stem Cells—What Challenges Do They Pose? *Nat. Rev. Drug Discov.* **2014**, *13*, 497–512. [CrossRef]

26. Bruno, P.; Gentile, G.; Mancini, R.; De Vitis, C.; Esposito, M.C.; Scozzi, D.; Mastrangelo, M.; Ricci, A.; Mohsen, I.; Ciliberto, G.; et al. WT1 CpG Islands Methylation in Human Lung Cancer: A Pilot Study. *Biochem. Biophys. Res. Commun.* **2012**, *426*, 306–309. [CrossRef]

27. Mehta, A.; Dobersch, S.; Romero-Olmedo, A.J.; Barreto, G. Epigenetics in Lung Cancer Diagnosis and Therapy. *Cancer Metastasis Rev.* **2015**, *34*, 229–241. [CrossRef]

28. Chapman-Rothe, N.; Curry, E.; Zeller, C.; Liber, D.; Stronach, E.; Gabra, H.; Ghaem-Maghami, S.; Brown, R. Chromatin H3K27me3/H3K4me3 Histone Marks Define Gene Sets in High-Grade Serous Ovarian Cancer that Distinguish Malignant, Tumour-Sustaining and Chemo-Resistant Ovarian Tumour Cells. *Oncogene* **2013**, *32*, 4586–4592. [CrossRef]

29. Ngollo, M.; Lebert, A.; Daures, M.; Judes, G.; Rifai, K.; Dubois, L.; Kemeny, J.L.; Penault Llorca, F.; Bignon, Y.J.; Guy, L.; et al. Global Analysis of H3K27me3 as an Epigenetic Marker in Prostate Cancer Progression. *BMC Cancer* **2017**, *17*, 261. [CrossRef]

30. Hunt, C.R.; Ramnarain, D.; Horikoshi, N.; Iyengar, P.; Pandita, R.K.; Shay, J.W.; Pandita, T.K. Histone Modifications and DNA Double-Strand Break Repair after Exposure to Ionizing Radiations. *Radiat. Res.* **2013**, *179*, 383–392. [CrossRef]

31. Thurman, R.E.; Rynes, E.; Humbert, R.; Vierstra, J.; Maurano, M.T.; Haugen, E.; Sheffield, N.C.; Stergachis, A.B.; Wang, H.; Vernot, B.; et al. The Accessible Chromatin Landscape of the Human Genome. *Nature* **2012**, *489*, 75–82. [CrossRef] [PubMed]

32. Tsompana, M.; Buck, M.J. Chromatin Accessibility: A Window into the Genome. *Epigenetics Chromatin* **2014**, *7*, 33. [CrossRef]

33. Folgiero, V.; Sorino, C.; Pallocca, M.; De Nicola, F.; Goeman, F.; Bertaina, V.; Strocchio, L.; Romania, P.; Pitisci, A.; Iezzi, S.; et al. Che-1 is Targeted by c-Myc to Sustain Proliferation in Pre-B-Cell Acute Lymphoblastic Leukemia. *EMBO Rep.* **2018**, *19*, e44871. [CrossRef] [PubMed]

34. Srivastava, A.; Sarkar, H.; Gupta, N.; Patro, R. RapMap: A Rapid, Sensitive and Accurate Tool for Mapping RNA-Seq Reads to Transcriptomes. *Bioinformatics* **2016**, *32*, i192–i200. [CrossRef] [PubMed]

35. Manzoni, C.; Kia, D.A.; Vandrovcova, J.; Hardy, J.; Wood, N.W.; Lewis, P.A.; Ferrari, R. Genome, Transcriptome and Proteome: The Rise of Omics Data and Their Integration in Biomedical Sciences. *Brief. Bioinform.* **2018**, *19*, 286–302. [CrossRef]

36. Yi, L.; Pimentel, H.; Bray, N.L.; Pachter, L. Gene-Level Differential Analysis at Transcript-Level Resolution. *Genome Boil.* **2018**, *19*, 53. [CrossRef]

37. Yalamanchili, H.K.; Wan, Y.W.; Liu, Z. Data Analysis Pipeline for RNA-Seq Experiments: From Differential Expression to Cryptic Splicing. *Curr. Protoc. In Bioinform.* **2017**, *59*, 11–15.

38. Hwang, K.B.; Lee, I.H.; Li, H.; Won, D.G.; Hernandez Ferrer, C.; Negron, J.A.; Kong, S.W. Comparative Analysis of Whole-Genome Sequencing Pipelines to Minimize False Negative Findings. *Sci. Rep.* **2019**, *9*, 3219. [CrossRef]

39. Liberzon, A.; Birger, C.; Thorvaldsdottir, H.; Ghandi, M.; Mesirov, J.P.; Tamayo, P. The Molecular Signatures Database Hallmark Gene Set Collection. *Cell Syst.* **2015**, *1*, 417–425. [CrossRef]

40. Ge, S.X.; Jung, D. ShinyGO: A Graphical Enrichment Tool for Ani-Mals and Plants. *Biorxiv* **2018**, 315150. [CrossRef]

41. Tang, Z.; Li, C.; Kang, B.; Gao, G.; Li, C.; Zhang, Z. GEPIA: A Web Server for Cancer and Normal Gene Expression Profiling and Interactive Analyses. *Nucleic Acids Res.* **2017**, *45*, W98–W102. [CrossRef] [PubMed]

42. Fattore, L.; Sacconi, A.; Mancini, R.; Ciliberto, G. MicroRNA-Driven Deregulation of Cytokine Expression Helps Development of Drug Resistance in Metastatic Melanoma. *Cytokine Growth Factor Rev.* **2017**, *36*, 39–48. [CrossRef] [PubMed]

43. Fattore, L.; Ruggiero, C.F.; Pisanu, M.E.; Liguoro, D.; Cerri, A.; Costantini, S.; Capone, F.; Acunzo, M.; Romano, G.; Nigita, G.; et al. Reprogramming miRNAs Global Expression Orchestrates Development of Drug Resistance in BRAF Mutated Melanoma. *Cell Death Differ.* **2019**, *26*, 1276–1282. [CrossRef] [PubMed]

44. Buenrostro, J.D.; Giresi, P.G.; Zaba, L.C.; Chang, H.Y.; Greenleaf, W.J. Transposition of Native Chromatin for Fast and Sensitive Epigenomic Profiling of Open Chromatin, DNA-Binding Proteins and Nucleosome Position. *Nat. Methods* **2013**, *10*, 1213–1218. [CrossRef]

45. Robinson, M.D.; McCarthy, D.J.; Smyth, G.K. EdgeR: A Bioconductor Package for Differential Expression Analysis of Digital Gene Expression Data. *Bioinformatics* **2009**, *26*, 139–140. [CrossRef]

46. Pisanu, M.E.; Noto, A.; De Vitis, C.; Masiello, M.G.; Coluccia, P.; Proietti, S.; Giovagnoli, M.R.; Ricci, A.; Giarnieri, E.; Cucina, A.; et al. Lung Cancer Stem Cell Lose Their Stemness Default State after Exposure to Microgravity. *BioMed Res. Int.* **2014**, *2014*, 470253. [CrossRef]

47. Costanzo, P.; Santini, A.; Fattore, L.; Novellino, E.; Ritieni, A. Toxicity of Aflatoxin B1 Towards the Vitamin D Receptor (VDR). *Food Chem. Toxicol.* **2015**, *76*, 77–79. [CrossRef]

48. Dunham, I.; Kundaje, A.; Aldred, S.F.; Collins, P.J.; Davis, C.A.; Doyle, F.; Epstein, C.B.; Frietze, S.; Harrow, J.; Kaul, R.; et al. An Integrated Encyclopedia of DNA Elements in the Human Genome. *Nature* **2012**, *489*, 57–74.

49. Visel, A.; Rubin, E.M.; Pennacchio, L.A. Genomic Views of Distant-Acting Enhancers. *Nature* **2009**, *461*, 199–205. [CrossRef]

50. Weintraub, A.S.; Li, C.H.; Zamudio, A.V.; Sigova, A.A.; Hannett, N.M.; Day, D.S.; Abraham, B.J.; Cohen, M.A.; Nabet, B.; Buckley, D.L.; et al. YY1 Is a Structural Regulator of Enhancer-Promoter Loops. *Cell* **2017**, *171*, 1573–1588. [CrossRef]

51. Zhou, H.; Ma, H.; Wei, W.; Ji, D.; Song, X.; Sun, J.; Zhang, J.; Jia, L. B4GALT Family Mediates the Multidrug Resistance of Human Leukemia Cells by Regulating the Hedgehog Pathway and the Expression of P-Glycoprotein and Multidrug Resistance-Associated Protein 1. *Cell Death Dis.* **2013**, *4*, e654. [CrossRef] [PubMed]

52. Lauc, G.; Huffman, J.E.; Pucic, M.; Zgaga, L.; Adamczyk, B.; Muzinic, A.; Novokmet, M.; Polasek, O.; Gornik, O.; Kristic, J.; et al. Loci Associated with N-Glycosylation of Human Immunoglobulin G Show Pleiotropy with Autoimmune Diseases and Haematological Cancers. *PLoS Genet* **2013**, *9*, e1003225. [CrossRef] [PubMed]

53. Pang, X.; Li, H.; Guan, F.; Li, X. Multiple Roles of Glycans in Hematological Malignancies. *Front. Oncol.* **2018**, *8*, 364. [CrossRef] [PubMed]

54. Padmanaban, V.; Krol, I.; Suhail, Y.; Szczerba, B.M.; Aceto, N.; Bader, J.S.; Ewald, A.J. E-Cadherin is Required for Metastasis in Multiple Models of Breast Cancer. *Nature* **2019**, *573*, 439–444. [CrossRef]

55. Patten, D.K.; Corleone, G.; Gyorffy, B.; Perone, Y.; Slaven, N.; Barozzi, I.; Erdos, E.; Saiakhova, A.; Goddard, K.; Vingiani, A.; et al. Enhancer Mapping Uncovers Phenotypic Heterogeneity and Evolution in Patients with Luminal Breast Cancer. *Nat. Med.* **2018**, *24*, 1469–1480. [CrossRef]

56. Buenrostro, J.D.; Wu, B.; Litzenburger, U.M.; Ruff, D.; Gonzales, M.L.; Snyder, M.P.; Chang, H.Y.; Greenleaf, W.J. Single-Cell Chromatin Accessibility Reveals Principles of Regulatory Variation. *Nature* **2015**, *523*, 486–490. [CrossRef]

J. Clin. Med. **2019**, *8*, 1928

57. Satpathy, A.T.; Saligrama, N.; Buenrostro, J.D.; Wei, Y.; Wu, B.; Rubin, A.J.; Granja, J.M.; Lareau, C.A.; Li, R.; Qi, Y.; et al. Transcript-Indexed ATAC-Seq for Precision Immune Profiling. *Nat. Med.* **2018**, *24*, 580–590. [CrossRef]

58. Jordan, D.M.; Frangakis, S.G.; Golzio, C.; Cassa, C.A.; Kurtzberg, J.; Davis, E.E.; Sunyaev, S.R.; Katsanis, N. Identification of Cis-Suppression of Human Disease Mutations by Comparative Genomics. *Nature* **2015**, *524*, 225–229. [CrossRef]

59. Lenhard, B.; Sandelin, A.; Mendoza, L.; Engstrom, P.; Jareborg, N.; Wasserman, W.W. Identification of Conserved Regulatory Elements by Comparative Genome Analysis. *J. Boil.* **2003**, *2*, 13. [CrossRef]

60. Rogerson, C.; Britton, E.; Withey, S.; Hanley, N.; Ang, Y.S.; Sharrocks, A.D. Identification of a Primitive Intestinal Transcription Factor Network Shared Between Esophageal Adenocarcinoma and its Precancerous Precursor State. *Genome Res.* **2019**, *29*, 723–736. [CrossRef]

61. Tieche, C.C.; Gao, Y.; Buhrer, E.D.; Hobi, N.; Berezowska, S.A.; Wyler, K.; Froment, L.; Weis, S.; Peng, R.W.; Bruggmann, R.; et al. Tumor Initiation Capacity and Therapy Resistance Are Differential Features of EMT-Related Subpopulations in the NSCLC Cell Line A549. *Neoplasia* **2019**, *21*, 185–196. [CrossRef] [PubMed]

62. Giarnieri, E.; De Vitis, C.; Noto, A.; Roscilli, G.; Salerno, G.; Mariotta, S.; Ricci, A.; Bruno, P.; Russo, G.; Laurenzi, A.; et al. EMT Markers in Lung Adenocarcinoma Pleural Effusion Spheroid Cells. *J. Cell. Physiol.* **2013**, *228*, 1720–1726. [CrossRef]

63. Ricci, A.; De Vitis, C.; Noto, A.; Fattore, L.; Mariotta, S.; Cherubini, E.; Roscilli, G.; Liguori, G.; Scognamiglio, G.; Rocco, G.; et al. TrkB is Responsible for EMT Transition in Malignant Pleural Effusions Derived Cultures from Adenocarcinoma of the Lung. *Cell Cycle* **2013**, *12*, 1696–1703. [CrossRef] [PubMed]

64. Song, J.; Wang, W.; Wang, Y.; Qin, Y.; Wang, Y.; Zhou, J.; Wang, X.; Zhang, Y.; Wang, Q. Epithelial-Mesenchymal Transition Markers Screened in a Cell-Based Model and Validated in Lung Adenocarcinoma. *BMC Cancer* **2019**, *19*, 680. [CrossRef]

65. Ishiwata, T. Cancer Stem Cells and Epithelial-Mesenchymal Transition: Novel Therapeutic Targets for Cancer. *Pathol. Int.* **2016**, *66*, 601–608. [CrossRef] [PubMed]

66. Picardo, F.; Romanelli, A.; Muinelo-Romay, L.; Mazza, T.; Fusilli, C.; Parrella, P.; Barbazan, J.; Lopez-Lopez, R.; Barbano, R.; De Robertis, M.; et al. Diagnostic and Prognostic Value of B4GALT1 Hypermethylation and Its Clinical Significance as a Novel Circulating Cell-Free DNA Biomarker in Colorectal Cancer. *Cancers* **2019**, *11*, 1598. [CrossRef]

67. Zhou, H.; Zhang, Z.; Liu, C.; Jin, C.; Zhang, J.; Miao, X.; Jia, L. B4GALT1 Gene Knockdown Inhibits the Hedgehog Pathway and Reverses Multidrug Resistance in the Human Leukemia K562/Adriamycin-Resistant Cell Line. *IUBMB Life* **2012**, *64*, 889–900. [CrossRef]

68. Xie, H.; Zhu, Y.; Zhang, J.; Liu, Z.; Fu, H.; Cao, Y.; Li, G.; Shen, Y.; Dai, B.; Xu, J.; et al. B4GALT1 Expression Predicts Prognosis and Adjuvant Chemotherapy Benefits in Muscle-Invasive Bladder Cancer Patients. *BMC Cancer* **2018**, *18*, 590. [CrossRef]

69. Pisanu, M.E.; Maugeri Sacca, M.; Fattore, L.; Bruschini, S.; De Vitis, C.; Tabbi, E.; Bellei, B.; Migliano, E.; Kovacs, D.; Camera, E.; et al. Inhibition of Stearoyl-CoA Desaturase 1 Reverts BRAF and MEK Inhibition-Induced Selection of Cancer Stem Cells In BRAF-Mutated Melanoma. *J. Exp. Clin. Cancer Res.* **2018**, *37*, 318. [CrossRef]

70. Codony-Servat, J.; Verlicchi, A.; Rosell, R. Cancer Stem Cells in Small Cell Lung Cancer. *Transl. Lung Cancer Res.* **2016**, *5*, 16–25.

71. Prieto Vila, M.; Takahashi, R.U.; Usuba, W.; Kohama, I.; Ochiya, T. Drug Resistance Driven by Cancer Stem Cells and Their Niche. *Int. J. Mol. Sci.* **2017**, *18*, 2574. [CrossRef] [PubMed]

72. Herreros Pomares, A.; De Maya Girones, J.D.; Calabuig Farinas, S.; Lucas, R.; Martinez, A.; Pardo-Sanchez, J.M.; Alonso, S.; Blasco, A.; Guijarro, R.; Martorell, M.; et al. Lung Tumorspheres Reveal Cancer Stem Cell-Like Properties and a Score with Prognostic Impact in Resected Non-Small-Cell Lung Cancer. *Cell Death Dis.* **2019**, *10*, 660. [CrossRef] [PubMed]

Journal of
Clinical Medicine

Article

SOX2 Expression Is an Independent Predictor of Oral Cancer Progression

Juan C. de Vicente [1,2,3,*], Paula Donate-Pérez del Molino [1,2], Juan P. Rodrigo [2,3,4,5],
Eva Allonca [3,5], Francisco Hermida-Prado [3,5], Rocío Granda-Díaz [3,5],
Tania Rodríguez Santamarta [1,2,3] and Juana M. García-Pedrero [3,4,5,*]

[1]   Department of Oral and Maxillofacial Surgery, Hospital Universitario Central de Asturias (HUCA).
      C/Carretera de Rubín, s/n, 33011 Oviedo, Asturias, Spain; pauladonatepdm@gmail.com (P.D.-P.d.M.);
      taniasantamarta@gmail.com (T.R.S.)
[2]   Department of Surgery, University of Oviedo. Avda. Julián Clavería, s/n, 33006 Oviedo, Asturias, Spain;
      jprodrigot@telefonica.net
[3]   Instituto de Investigación Sanitaria del Principado de Asturias (ISPA), Instituto Universitario de Oncología
      del Principado de Asturias (IUOPA), Universidad de Oviedo. C/Carretera de Rubín, s/n, 33011 Oviedo,
      Asturias, Spain; ynkc1@hotmail.com (E.A.); franjhermida@gmail.com (F.H.-P.);
      rocigd281@gmail.com (R.G.-D.)
[4]   Department of Otolaryngology, Hospital Universitario Central de Asturias (HUCA). C/Carretera de Rubín,
      s/n, 33011 Oviedo, Asturias, Spain
[5]   Ciber de Cáncer (CIBERONC), Instituto de Salud Carlos III, Av. Monforte de Lemos,
      3-5. 28029 Madrid, Spain
[*]   Correspondence: jvicente@uniovi.es (J.C.d.V.); juanagp.finba@gmail.com (J.M.G.-P.);
      Tel.: +34-85-103638 (J.C.d.V.); +34-985-107937 (J.M.G.-P.)

Received: 4 September 2019; Accepted: 16 October 2019; Published: 21 October 2019

Abstract: Potentially malignant oral lesions, mainly leukoplakia, are common. Malignant transformation varies widely, even in the absence of histological features such as dysplasia. Hence, there is a need for novel biomarker-based systems to more accurately predict the risk of cancer progression. The pluripotency transcription factor SOX2 is frequently overexpressed in cancers, including oral squamous cell carcinoma (OSCC), thereby providing a link between malignancy and stemness. This study investigates the clinical relevance of SOX2 protein expression in early stages of oral carcinogenesis as a cancer risk biomarker, and also its impact on prognosis and disease outcome at late stages of OSCC progression. SOX2 expression was evaluated by immunohistochemistry in 55 patients with oral epithelial dysplasia, and in 125 patients with OSCC, and correlated with clinicopathological data and outcomes. Nuclear SOX2 expression was detected in four (7%) cases of oral epithelial dysplasia, using a cut-off of 10% stained nuclei, and in 16 (29%) cases when any positive nuclei was evaluated. Univariate analysis showed that SOX2 expression and histopathological grading were significantly associated with oral cancer risk; and both were found to be significant independent predictors in the multivariate analysis. Nuclear SOX2 expression was also found in 49 (39%) OSCC cases, was more frequent in early tumor stages and N0 cases, and was associated with a better survival. In conclusion, SOX2 expression emerges as an independent predictor of oral cancer risk in patients with oral leukoplakia. These findings underscore the relevant role of SOX2 in early oral tumorigenesis rather than in tumor progression.

Keywords: oral cancer risk; oral epithelial dysplasia; SOX2; immunohistochemistry; oral squamous cell carcinoma

---

1. Introduction

Squamous cell carcinoma (SCC) of the oral cavity (OSCC) afflicts about 300,400 new cases and causes 145,400 deaths worldwide each year [1], with a predilection for South Asian and Southeast Asian populations [2]. According to the World Health Organization (WHO) mortality projections, there is an estimate of 679,941 mouth and oropharynx cancer-related deaths by 2060 [3]. OSCC shows an aggressive growth pattern with a high degree of local invasiveness and a propensity to metastasize to the cervical lymph nodes, even in early stages. In fact, metastases to neck lymph nodes occur in 40% of cases, which remain the main factor associated with poor prognosis [4]. Additionally, between 26% and 80% of patients with early-stage OSCC develop locoregional recurrence or distant metastasis [4]. Despite aggressive treatment, the prognosis of this disease remains dismal, with a five-year survival rate at around 55%–60% [2].

OSCC may develop from an apparently normal oral mucosa or from oral potentially malignant disorders (OPMDs), mainly oral leukoplakia (OLK), which shows a malignant transformation rate of 0.1% to 36% [5]. All premalignant lesions, such as OLK, erythroplakia, submucosal fibrosis, or oral lichen planus, may harbor many genetic alterations present in OSCC [6].

Oral cancer exhibits cellular heterogeneity and is composed by three different types of cells including highly differentiated bulk tumor cells, transit cells with maximum proliferation capacity, and a small subpopulation of cells with elevated self-renewal capacity and plasticity called cancer stem cells (CSCs) [7]. The CSCs are capable of long-term self-renewal and generation of the phenotypically diverse tumor cell population [8], and may be responsible for the genesis, anchorage-independent growth, cellular migration, and metastatic spread of the tumor [9,10]. Meantime, investigations by Takahashi and Yamanaka [11] revealed that the expression of four transcription factors (SOX2, Oct4, c-Myc, and Klf4) was sufficient to reprogram differentiated cells into induced pluripotent stem cells (iPSCs). One of these genes, the so-called sex-determining region Y (SRY)-related high-mobility-group (HMG)-box 2 (SOX2), located on chromosome 3q26, is implicated in the maintenance of embryonic stem cell pluripotency [12]. Numerous evidences indicate that SOX2 is involved in tumorigenesis, thereby acting as a link between malignancy and stemness [13]. Moreover, the proliferation of CSCs in head and neck SCC was inhibited both in vitro and in vivo when SOX2 was suppressed by all-trans-retinoic acid [14]. The degree of similarity to OSCC found in premalignant lesions depend upon the presence of atypia; however, individual lesions exhibit molecular genetic traits in common with OSCC, even in the absence of histologically-defined dysplasia [15]. Hence, better biomarker-based detection systems need to be developed to more accurately predict the risk of cancer progression in potentially malignant oral disorders.

It has been described that OLK lesions show higher expression of SOX2 than normal oral mucosa, suggesting its contribution to the pathogenesis of OSCC [16]. Consequently, SOX2 could be a potentially useful predictor of cancer risk in the oral cavity. SOX2 is mainly expressed in CSCs [13], and is one of the amplified genes in OSCC, where its expression has been closely associated with lymph node metastasis [17] and poor prognosis [18]. In marked contrast, several studies have demonstrated that increased levels of SOX2 were significantly associated with better prognosis in patients with OSCC, and also in squamous cell lung cancer [19,20]. Therefore, the role of SOX2 expression in OSCC prognosis remains controversial.

This prompted us to perform a thorough study to investigate the clinical significance of SOX2 in the development and progression of OSCC. To accomplish this, the expression pattern of SOX2 was evaluated at different stages of oral tumorigenesis, from potentially malignant oral disorders (i.e., oral epithelial dysplastic lesions) to invasive carcinomas, to ascertain its contribution to tumor initiation and malignant transformation, and also late stages of disease progression.

2. Materials and Methods

2.1. Patients and Tissue Specimens

Surgical tissue specimens from 55 patients who were diagnosed with oral mucosa dysplasia at the Hospital Universitario Central de Asturias between 2000 and 2005 were retrospectively collected. All selected patients met the following inclusion criteria: (i) pathological diagnosis of oral epithelial dysplasia; (ii) lesions of the oral mucosa (leukoplakia); (iii) no previous history of head and neck cancer, (iv) complete excisional biopsy of the lesion; and (v) a minimum follow-up of five years (or until progression). Patients were followed up as previously described [21].

In addition, surgical tissue specimens from 125 patients with histologically-confirmed OSCC surgically treated at the Hospital Universitario Central de Asturias between 1996 and 2007 were retrospectively collected, in accordance to approved institutional review board guidelines. All experimental procedures were conducted in accordance to the Declaration of Helsinki and approved by the Institutional Ethics Committee of the Hospital Universitario Central de Asturias and by the Regional (CEIC) from Principado de Asturias (date of approval 5th of May 2016; approval number: 70/16) for the project PI16/00280. Informed consent was obtained from all patients.

Tissue samples and data from donors were provided by the Principado de Asturias BioBank (PT17/0015/0023), integrated in the Spanish National Biobanks Network, and processed following standard operating procedures with the appropriate approval of the Ethical and Scientific Committees. Representative tissue samples were obtained from archival, formalin-fixed, paraffin-embedded blocks and the histological diagnosis was confirmed by an experienced pathologist.

2.2. Tissue Microarray (TMA) Construction

As previously described [21], three representative tissue cores (1 mm diameter) were taken from each tumor block to construct OSCC TMAs. Each TMA block also included three cores of normal epithelium as an internal control. These samples were obtained from non-oncological patients undergoing oral surgery.

2.3. Immunohistochemistry (IHC)

The TMAs were cut into 3 μm sections and dried on Flex IHC microscope slides (DakoCytomation, Glostrup, Denmark). Antigen retrieval was performed by heating the sections with Envision Flex Target Retrieval solution, high pH (Dako, Glostrup, Denmark). Staining was done at room temperature on an automatic staining workstation (Dako Autostainer Plus, Glostrup, Denmark) with anti-SOX2 rabbit polyclonal antibody (AB5603, Merck Millipore, Darmstadt, Germany) at 1:1000 dilution using the Dako EnVision Flex + Visualization System (Dako Autostainer, Glostrup, Denmark) and diaminobenzidine chromogen as substrate. Counterstaining with hematoxylin was the final step.

The IHC results were independently evaluated by two observers (JPR, and JMG-P), blinded to clinical data. SOX2 staining was evaluated as the percentage of cells with nuclear staining in the dysplastic epithelium or in the tumor tissue. SOX2 staining scores were classified as negative or positive staining on the basis of values below or above the median value of 10%. Since CSC-like subpopulations are usually limited to a very small percentage of cells, SOX2 staining in the dysplastic areas was also scored considering any positive nuclei.

2.4. Statistical Analysis

Bivariate analysis by χ^2 and Fisher's exact tests were used for comparison between SOX2 expression and clinicopathological categorical variables. Disease-specific survival (DSS) was determined from the date of treatment completion to the death of the tumor. For time-to-event analysis, survival curves were estimated using the Kaplan–Meier method. The log-rank test was used to compare the survival curves. Hazard ratios (HRs) with their 95% confidence intervals (CIs) for clinicopathological variables were calculated using univariate and multivariate Cox proportional hazards model. All tests were two-sided and *P*-values less than 0.05 were considered statistically significant. All statistical analyses were performed using SPSS version 18 (IBM Co., Armonk, NY, USA).

3. Results

3.1. Clinicopathological Features and Follow-Up in Patients with Oral Epithelial Dysplasia

Twenty-six patients (47%) were men and the remaining 29 were women (53%), with a mean age of 62.61 years (SD 12.56, range 39 to 83 years). Ten patients (18%) were smokers and four (7%) were habitual alcohol drinkers. Forty-two of 55 premalignant lesions (76%) were classified as mild dysplasia, six (11%) as moderate dysplasia, and the remaining seven (13%) as severe dysplasia, according to the WHO classification [22]. During the follow-up period (mean: 85.47, SD: 44.41, median: 75, range: four to 252 months), 12 (22%) of 55 patients developed an invasive OSCC. The most relevant clinical and pathological characteristics are summarized in Table S1.

3.2. SOX2 Protein Expression in Oral Epithelial Dysplasia

Nuclear SOX2 expression was detected in four (7%) cases when a cut-off of 10% stained nuclei was used (SOX2 > 10), and in 16 (29%) when any positive nuclei was applied (SOX2any). Normal adjacent epithelia showed negative SOX2 expression (Figure 1A–C). SOX2 protein expression was found to significantly increase with the grade of dysplasia (Table 1). In addition to the WHO three-tier classification, the binary WHO grading system (low-grade vs. high-grade) was used and correlated with SOX2 expression.

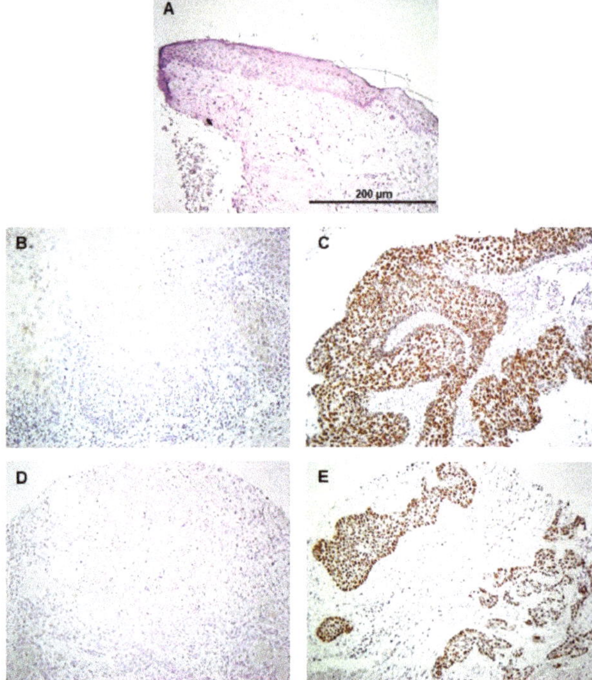

Figure 1. Immunohistochemical analysis of SOX2 expression in oral epithelial dysplasia. Normal adjacent epithelia showed negative staining (**A**). Representative examples of oral dysplasia showing negative (**B**) and positive nuclear SOX2 staining (**C**), and oral squamous cell carcinomas with negative (**D**) and positive SOX2 staining (**E**). Magnification 20 ×; scale bar 200 μm.

Table 1. Associations between SOX2 expression and patient characteristics.

Characteristics	SOX2 > 10% Positive Nuclei Negative Positive		P	SOX2 any Positive Nuclei Negative Positive		P
Age (years), Mean (SD)	62.93 (12.69)	60.50 (13.20)	0.72	61.00 (12.69)	65.55 (12.35)	0.34
Gender, number (%)						
• Female	27 (93)	2 (7)	1.00	22 (76)	7 (24)	0.39
• Male	24 (92)	2 (8)		17 (65)	9 (35)	
Smoking, number (%)						
• Yes	9 (90)	1 (10)	1.00	6 (60)	4 (40)	0.71
• No	18 (86)	3 (14)		14 (67)	7 (33)	
Ethanol intake, number (%)						
• Yes	3 (75)	1 (25)	0.44	2 (50)	2 (50)	0.60
• No	24 (89)	3 (11)		18 (67)	9 (33)	
Epithelial dysplasia						
• Mild	42 (100)	0 (0)	0.001	33 (79)	9 (21)	0.055
• Moderate	5 (83)	1 (17)		3 (50)	3 (50)	
• Severe	4 (57)	3 (43)		3 (43)	4 (57)	
Epithelial dysplasia						
• Low-grade	42 (100)	0 (0)	0.002	33 (79)	9 (21)	0.02
• High-grade	9 (69)	4 (31)		6 (46)	7 (54)	

There was a statistically significant correlation between the histopathological grade (both the WHO histological classification and the binary dysplasia grading) and the risk of progression to oral cancer in this cohort (log-rank test, $P < 0.001$; Figure 2A,B) (Table 2). In addition, positive SOX2 expression also significantly predicted oral cancer risk either considering SOX2 > 10 (log-rank test, $P = 0.02$; Figure 2C) or SOX2any (log-rank test, $P = 0.01$; Figure 2D) as cut-off points. Univariate Kaplan–Meier and Cox analysis showed that the SOX2 expression and histological grading were significantly associated with oral cancer risk (Table 3). When these factors were simultaneously analyzed using a multivariate Cox analysis, only SOX2 expression calculated using SOX2any as the cut-off point and the dysplasia grading were significant independent predictors of OSCC development (Table 4).

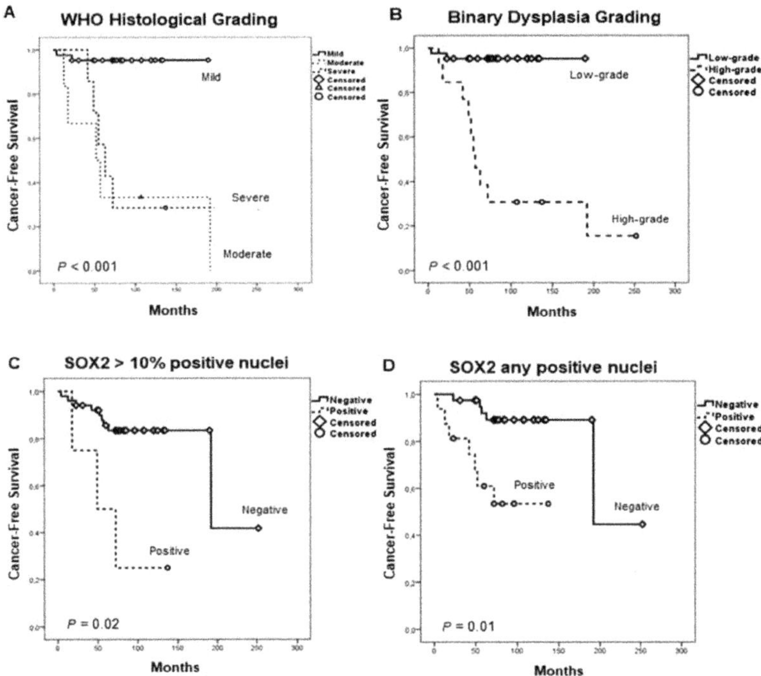

Figure 2. Kaplan–Meier cancer-free survival curves in the cohort of 55 patients with oral epithelial dysplasia categorized by the World Health Organization (WHO) histological grading (**A**), the binary dysplasia grading (**B**), and SOX2 protein expression dichotomized using the cut-off values of SOX2 staining > 10% positive nuclei (**C**) or SOX2 staining any positive nuclei (**D**).

Table 2. Evolution of the premalignant lesions in relation to histopathological diagnosis and SOX2 expression.

Variable	Number of Cases (%)	Progression to Carcinoma (%)	*P* *
Histopathological diagnosis			<0.001
• Low-grade dysplasia	42 (76)	2 (5)	
• High-grade dysplasia	13 (24)	10 (77)	
SOX2 > 10% positive nuclei			0.02
• Negative	51 (93)	9 (18)	
• Positive	4 (7)	3 (75)	
SOX2 any positive nuclei			0.01
• Negative	39 (71)	5 (13)	
• Positive	16 (29)	7 (44)	

* Fisher exact test.

Table 3. *Cont.*

Variable	No.	Censored Patients (%)	Mean Cancer-Free Survival Time (95% CI)	P	Hazard Ratio	95% CI
Epithelial Dysplasia				< 0.001	19.08	4.09–89.01
• High-grade	13	3 (23)	100.69 (54.14–147.24)			
• Low-grade	42	40 (95)	181.59 (170.21–192.98)			
SOX2 > 10% positive nuclei				0.002	6.13	1.62–23.27
• Positive	4	1 (25)	69.00 (26.18–111.81)			
• Negative	51	42 (82)	191.80 (152.14–231.47)			
SOX2 any positive nuclei				0.002	5.75	1.68–19.74
• Positive	16	9 (56)	90.40 (64.14–116.66)			
• Negative	39	34 (87)	203.22 (162.30–244.15)			

P-values were estimated using the log-rank test.

Table 4. Multivariate Cox proportional hazards model to estimate oral cancer risk.

Variable	P	Hazard Ratio	95% CI
Histology (high-grade vs. low-grade)	< 0.0001	21.88	4.13–116.07
SOX2 > 10% (positive vs. negative)	0.196	3.0	0.57–15.89
SOX2 any (positive vs. negative)	0.021	5.83	1.31–26.01

Similarly, we have recently reported a novel role for another pluripotency factor NANOG as a cancer risk marker using the same subset of 55 patients with oral epithelial dysplasia [21]. Since SOX2 and NANOG are functionally-related proteins, this prompted us to assess the impact of combined expression of SOX2 and NANOG in regards to malignization risk. Interestingly, results consistently showed that patients harboring positive expression of SOX2 (either SOX2 > 10% or SOX2any) and NANOG (either cytoplasmic or nuclear expression) significantly exhibited a much higher risk of developing oral cancer, compared to patients with positive expression of either SOX2 or NANOG or those patients with negative expression (Table 5).

Table 5. Univariate Cox cancer-free survival analysis in 55 patients with oral dysplasia categorized by both SOX2 and NANOG expression.

Variable	No.	Censored Patients (%)	Cancer-Free Survival Time (95% CI)	P	Hazard Ratio	95% CI
SOX2 > 10% positive nuclei and nuclear NANOG						
• Both negative	51	42 (84)	191.80 (152.13–231.46)		Ref	
• One positive	2	1 (50)	93.00 (32.01–153.98)	0.003	3.72	0.46–29.98
• Both positive	2	0 (0)	45.00 (8.00–97.92)		9.06	1.91–43.00
SOX2 > 10% positive nuclei and cytoplasmic NANOG						
• Both negative	45	38 (84)	171.10 (154.24–187.95)		Ref	
• One positive	7	5 (71)	182.28 (101.53–263.03)	< 0.0005	1.63	0.30–8.71
• Both positive	3	0 (0)	46.33 (15.66–76.99)		10.89	2.74–43.22

Table 5. *Cont.*

Variable	No.	Censored Patients (%)	Cancer-Free Survival Time (95% CI)	*P*	Hazard Ratio	95% CI
SOX2 any positive nuclei and cytoplasmic NANOG				< 0.0005		
• Both negative	36	31 (86)	175.49 (158.40–192.57)		Ref	
• One positive	13	11 (85)	215.99 (170.17–261.81)		1.091	0.20–5.85
• Both positive	6	1 (17)	44.50 (21.60–67.40)		11.36	3.18–40.60
SOX2 any positive nuclei and nuclear NANOG				< 0.0005		
• Both negative	39	34 (87)	203.22 (162.30–244.15)		Ref	
• One positive	14	9 (64)	97.76 (69.96–125.55)		4.62	1.23–17.29
• Both positive	2	0 (0)	45.00 (8.00–97.92)		14.82	2.69–81.56

3.3. Clinicopathological Features and Follow-Up in the Cohort of OSCC Patients

The mean and median follow-up times were 71.82 (SD: 57.55) and 61.0, respectively. Neck node metastases were present in 49 (39%) cases, and local recurrences were found in 54 (43%) cases. No patient had distant metastasis at the time of diagnosis. The five- and 10-year disease-specific survival rates were 60% and 44%, respectively. The mean and median survival times were 132.74 months (95% CI: 113.25 to 152.22 months) and 141 months (95% CI: 102.40 to 179.59 months), respectively. The remaining relevant clinical and pathological characteristics are summarized in Table S2.

3.4. SOX2 Protein Expression and Its Relation with Clinicopathological Variables and Follow-Up

Positive SOX2 staining was found in 49 (39%) cases located in the nucleus of tumor cells, whereas stromal cells and normal epithelium showed negligible expression (Figure 1A,D,E). SOX2 expression did not show any significant association with the clinicopathological variables studied (Table S3). In the survival analysis, tumor size and local extension (pT), neck node status (pN), and stage were significantly correlated to survival (Table S4). Positive SOX2 expression was associated with a longer disease-specific survival, although differences did not reach statistical significance (log-rank test, $P = 0.07$; Figure 3).

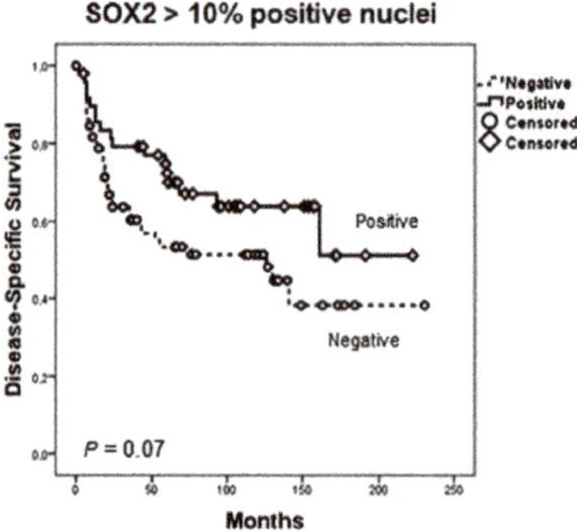

Figure 3. Kaplan–Meier disease-specific survival curves in the cohort of 125 patients with oral squamous cell carcinoma dichotomized according to SOX2 staining (positive versus negative). *P*-values were estimated using the log-rank test.

3.5. In Silico Analysis of SOX2 mRNA Expression and Copy Number Alterations using the Cancer Genome Atlas (TCGA) Data

The role of SOX2 mRNA expression and copy number alterations in OSCC was investigated by analyzing a subset of 172 OSCC patients from the TCGA Head and Neck Squamous Cell Carcinoma

(HNSCC) cohort [23] using the platform cBioPortal (http://cbioportal.org/) [24]. As shown in Figure 4A, SOX2 gene alterations were present in a total of 38 (22%) of 172 OSCC patients. Twenty-two (13%) cases harbored SOX2 mRNA up-regulation as previously reported [21], and 23 (13%) cases harbored *SOX2* gene amplification. SOX2 mRNA levels were associated to copy number alterations (Figure 4B). Overall, SOX2-amplified tumors showed higher mRNA levels; however, SOX2 amplification was only concomitantly accompanied by gene expression up-regulation in eight cases (5%). The impact of SOX2 mRNA expression on OSCC patient survival was also assessed (Figure 4C). The median survival times for patients with high (above the median) and low SOX2 mRNA levels (below the median) were 26.41 and 19.19 months, respectively, although differences did not reach statistical significance ($P = 0.495$, log-rank test).

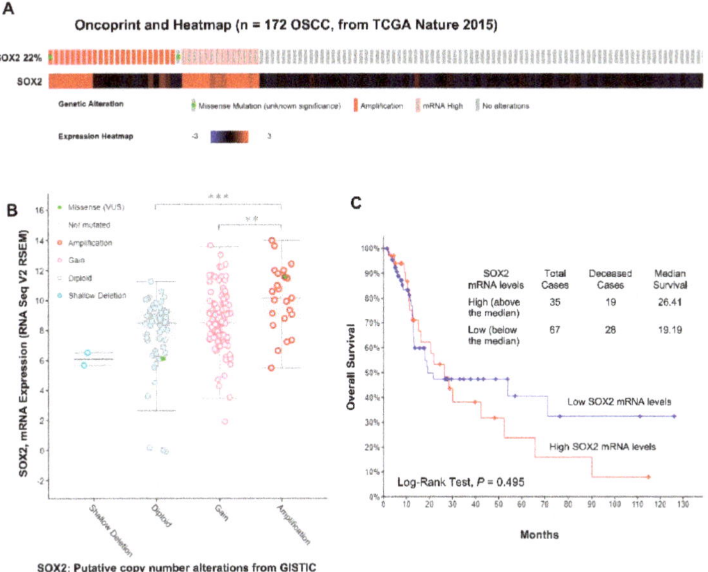

Figure 4. In silico analysis of mRNA expression and copy number alterations of SOX2 in the subset of 172 oral squamous cell carcinoma patients from The Cancer Genome Atlas (TCGA) Head and Neck Squamous Cell Carcinoma cohort [23] using the platform cBioPortal. (**A**) Oncoprint and heatmap representations showing the percentage of cases with SOX2 gene amplification, mutation, and mRNA up-regulation. (**B**) SOX2 mRNA expression analysis in relation to the copy number alterations of SOX2 gene (RNA seq V2 RSEM) values were Log2 transformed (*y*-axis). Whiskers plot (min. to max.) with median values; ** $P < 0.01$ and *** $P < 0.001$, one-way ANOVA, Tukey's test. (**C**) Kaplan–Meier survival curves categorized by SOX2 mRNA expression (RNA seq V2 RSEM, z-score threshold ± 2) dichotomized as high mRNA levels (above the median) versus low mRNA levels (below the median), *P*-value estimated using the log-rank test.

4. Discussion

To the best of our knowledge this is the first study to investigate SOX2 protein expression along the different stages of oral carcinogenesis, from potentially malignant oral disorders, such as leukoplakia, to invasive carcinomas, to ascertain its contribution to tumor initiation and malignant transformation, and also late stages of disease progression.

Cancer stem cells (CSCs) are defined as a small subpopulation of cells in the tumors that possess the ability to initiate neoplasms and sustain tumor self-renewal [7]. *SOX2* gene mapping at 3q26 is frequently amplified in OSCC and other cancers. It has been established as an important CSC marker and a key molecule in the development of tumorigenesis in various cancers [13] and thus proposed as an oncogene [25,26]. Arnold et al. [27] reported that epithelial adult stem cells expressing SOX2 may be residual stem niches that originate from embryonic SOX2-positive tissue progenitors. Cai et al. [28] investigated the roles of OCT4 and SOX2 in the reprogramming of oral cancer stem cells. They immortalized oral epithelial cells by lentiviral transduction and found that double-transduced OCT4$^+$SOX2$^+$ cells were able to trigger tumor formation in immunodeficient mice; however, single-transduced OCT4$^+$ or SOX2$^+$ cells did not show tumorigenic capacity. They also stated that oral carcinogenesis may derive from OCT4$^+$SOX2$^+$ reprogrammed stem cells, in which SOX2 plays a major role in the regulation of the CSC niche [28]. Accordingly, it has been proposed that, in the absence of SOX2 expression, CSC self-renewal that sustains tumor growth could be abrogated,; therefore, supporting SOX2 inhibition as a potentially relevant therapeutic target for oral cancer [28].

OLK is the most frequent potentially malignant disorder in the oral cavity. Histologic grading of epithelial dysplasia in OLK is currently still the gold standard in the clinical practice to evaluate the risk of progression to invasive carcinoma [16]. However, the accuracy of the various grading systems so far developed have shown limited predictability, are largely subjective, and affected by a great inter- and intra-examiner variability [29]. According to the WHO 2017 classification there is not a unique criteria or grading system for oral epithelial dysplasia, and as a consequence diagnostic reproducibility is still limited. In addition to the currently accepted WHO three-tier classification for oral epithelial dysplasia, a binary grading system (low-grade vs. high-grade) has also been proposed [22]. However, this binary system has not yet been validated for use in the oral cavity. It is; therefore, of paramount importance to identify novel biomarkers that could provide complementary information to histology to more accurately predict the risk of malignant transformation of OLK. SOX2 has been demonstrated to play a central role in the maintenance of CSC pluripotency and self-renewal [27], thereby emerging as a promising marker for oral carcinogenesis. Luiz et al. [16] conducted a retrospective study to compare the expression of SOX2 in OLK with normal oral mucosa, and found that SOX2 expression was higher in OLK, although the relationship with oral cancer risk was not evaluated.

The presence of dysplastic features in the epithelium of the oral cavity are thought to be relevant to malignant transformation in OLK. In our study, patients harboring high-grade dysplasia indeed showed a significantly higher risk of malignant progression (HR = 19.08). SOX2 expression was also found to be a significant predictor of risk of cancer development (HRs of 6.13 and 5.75 depending on the cut-off used). Furthermore, dysplasia grading and SOX2 expression were both found to be significant independent predictors of oral cancer risk in multivariate analysis. Similarly, we recently demonstrated that NANOG expression was a novel cancer risk marker using the same subset of patients with oral epithelial dysplasia [21]. Importantly, this study further extends these data to show that patients harboring positive expression of both SOX2 and NANOG significantly exhibited a much higher risk of progression to oral cancer, thereby suggesting a cooperative oncogenic role of these two proteins in oral pathogenesis and malignant transformation.

SOX2 expression has also been analyzed during tumor progression. Freier et al. [30] reported SOX2 expression in 18% of OSCC, and other studies on SCC of larynx, pharynx, and oral cavity found frequencies up to 86% [22,31–33]. In our cohort of 125 OSCC cases, positive SOX2 expression was detected in 39% of cases, while the frequencies of SOX2 expression reported in a previous study, performed at our laboratory for other head and neck subsites, were 38% in hypopharynx, 42% in larynx, and 14% in sinonasal cancer. In consequence, SOX2 expression frequencies in the oral cavity were similar to those observed in neighboring tissues such as hypopharynx and larynx, but much higher than in the sinonasal tract. In silico analysis of RNAseq data in a subset of 172 OSCC patients from the TCGA HNSCC cohort [23] further contributed to demonstrate that SOX2 mRNA levels were up-regulated in 13% of OSCC patients. In addition, SOX2 gene amplification was also observed in 13% of cases; however, only 5% of cases harboring SOX2 amplification were concomitantly accompanied by increased gene expression, indicating that additional mechanisms must be contributing to SOX2 expression in OSCC. In this sense, there are various plausible transcriptional regulatory mechanisms, such as the transcription factors OCT4 and YAP1 or the hypoxic factor HIF1α, found to modulate SOX2 expression [34,35]. Moreover, SOX2 protein expression was detected at much higher percentages (38%) in our OSCC cohort than SOX2 mRNA expression (13%), thereby suggesting the possible involvement of post-transcriptional mechanisms. Similar observations have been reported for other CSC-related factors, such as NANOG or PDPN, detected in over 30% of OSCC patients at the protein level [21,36] compared to 3% at mRNA levels according to the TCGA data from 172 OSCC patients [21].

Studies published to date assessing the expression of SOX2 and its clinical and prognostic relevance in OSCC have led to contradictory results. It has been reported that a high expression of SOX2 was significantly associated with poorer survival in node-negative OSCC [18], while others [17,26,37] found that a high expression of SOX2 correlated with lymph node metastasis. The latter could be explained by the relevant role of SOX2 as an epithelial–mesenchymal transition (EMT) inducer, thereby acting through the Wnt/β-catenin and phosphoinositide 3-kinase/protein kinase B signaling pathways [38,39] to promote oncogenesis, invasion, and metastasis. Nonetheless, the opposite viewpoint is represented by patients with early-stage OSCC, where upregulation of SOX2 has been correlated with a lower incidence of lymph node metastasis [2,20]. These controversial results could be explained by the heterogeneity of different tumors, as well as by different molecular mechanisms underlying the complex process of metastasis. The majority of studies showed that SOX2 overexpression may promote cancer progression; however, it has also been reported that SOX2 overexpression inhibits cell proliferation [13]. Furthermore, SOX2 is thought to stabilize stem cell phenotype and prevent EMT [40], and even more, SOX2 could promote mesenchymal–epithelial transition (MET), attenuating the invasive phenotype [41]. In our study, the expression of SOX2 did not correlate to T classification, neck lymph node metastasis, disease stage, histological grade, tumor recurrence, and second primary carcinomas development.

Most studies showed that the survival rate of OSCC patients harboring low levels of SOX2 increased compared to those with high levels of SOX2 [13], although Züllig et al. [20] and Fu et al. [2] found that low expression of SOX2 was significantly associated with worse survival. Similarly, other studies on SCC from various locations described an association of elevated SOX2 expression with longer survival [19,42–44]. We herein observed that SOX2 expression was more frequently observed in early tumor stages as well as in N0 cases, and patients with positive SOX2 expression exhibited a better survival than those with negative expression, although differences did not reach statistical significance ($P = 0.07$).

It is possible that all these contradictory results are related to methodological differences in SOX2 IHC scoring, as well as in the heterogeneity of patient populations or even in the morphological and genetic heterogeneity observed in solid tumors. Another explanation may be that SOX2 copy number gain and expression are early tumor-initiating events, but this gene can lose its relevance in conveying an aggressive or metastasizing phenotype [22].

5. Conclusions

Together, our results reveal that SOX2 expression is a clinically relevant feature in early stages of oral tumorigenesis, and provide original evidence of its potential utility as biomarker for oral cancer risk assessment. According to these data, SOX2 expression emerges as an important determinant in the pathogenesis of a subset of OSCC, thereby contributing to tumor initiation and acquisition of an invasive phenotype rather than late stages of disease progression.

Supplementary Materials: The following are available online at http://www.mdpi.com/2077-0383/8/10/1744/s1, Table S1. Clinical and pathological characteristics of the 55 patients with oral epithelial dysplasia. Table S2. Clinical and pathological characteristics of the 125 patients with OSCC selected for study. Table S3. Relationship between the clinicopathological variables and SOX2 expression in OSCC patients. Table S4. Univariate Kaplan–Meier and Cox analysis to assess the association of clinicopathological variables on disease-specific survival in 125 OSCC patients.

Author Contributions: Conceptualization, J.C.d.V., J.P.R. and J.M.G.-P.; data curation, T.R.S.; formal analysis, J.C.d.V. and F.H.-P.; funding acquisition, J.P.R. and J.M.G.-P.; investigation, J.C.d.V., P.D.-P.d.M., J.P.R., E.A., R.G.-D., T.R.S. and J.M.G.-P.; methodology, J.C.d.V., P.D.-P.d.M., J.P.R. and E.A.; project administration, J.M.G.-P.; resources,

J.P.R.; supervision, J.C.d.V., and J.M.G.-P.; validation, T.R.S.; visualization, J.C.d.V., P.D.-P.d.M., J.P.R., F.H.-P. and J.M.G.-P.; writing—original draft, J.C.d.V. and J.M.G.-P.

Funding: This study was supported by grants from the Plan Nacional de I+D+I 2013-2016 ISCIII PI16/00280 and CIBERONC (CB16/12/00390), Fundación Merck Salud (17-CC-008), the Instituto de Investigación Sanitaria del Principado de Asturias (ISPA), PCTI-Asturias (GRUPIN14-003), Fundación Bancaria Caja de Ahorros de Asturias-IUOPA, and the FEDER Funding Program from the European Union.

Acknowledgments: We thank the samples and technical assistance kindly provided by the Principado de Asturias BioBank (PT13/0010/0046), financed jointly by Servicio de Salud del Principado de Asturias, Instituto de Salud Carlos III, and Fundación Bancaria Cajastur, and integrated in the Spanish National Biobanks Network. We also thank Juan Pérez Ortega for his excellent administrative support.

Conflicts of Interest: All the authors declare that they have no conflict of interest.

References

1. Torre, L.A.; Bray, F.; Siegel, R.L.; Ferlay, J.; Lortet-Tieulent, J.; Jemal, A. Global cancer statistics, 2012. *CA Cancer J. Clin.* **2015**, *65*, 87–108. [CrossRef]

2. Fu, T.Y.; Hsieh, I.C.; Cheng, J.T.; Tsai, M.H.; Hou, Y.Y.; Lee, J.H.; Liou, H.H.; Huang, S.F.; Chen, H.C.; Yen, L.M.; et al. Association of OCT4, SOX2, and NANOG expression with oral squamous cell carcinoma progression. *J. Oral Pathol. Med.* **2016**, *45*, 89–95. [CrossRef]

3. World Health Organization. Global Health Estimates (GHE) 2013: Global Deaths by Income Group, Cause and Year (2016–2060). Available online: https://www.who.int/healthinfo/global_burden_disease/projections/en/ (accessed on 23 June 2019).

4. Chhetri, D.K.; Rawnsley, J.D.; Calcaterra, T.C. Carcinoma of the buccal mucosa. *Otolaryngol. Head Neck Surg.* **2000**, *123*, 566–571. [CrossRef]

5. Yanik, E.L.; Katki, H.A.; Silverberg, M.J.; Manos, M.M.; Engels, E.A.; Chaturvedi, A.K. Leukoplakia, oral cavity cancer risk, and cancer survival in the U.S. elderly. *Cancer Prev. Res.* **2015**, *8*, 857–863. [CrossRef]

6. Califano, J.; Westra, W.H.; Meininger, G.; Corio, R.; Koch, W.M.; Sidransky, D. Genetic progression and clonal relationship of recurrent premalignant head and neck lesions. *Clin. Cancer Res.* **2000**, *6*, 347–352. [PubMed]

7. Costea, D.E.; Tsinkalovsky, O.; Vintermyr, O.K.; Johannessen, A.C.; Mackenzie, I.C. Cancer stem cells—New and potentially important targets for the therapy of oral squamous cell carcinoma. *Oral Dis.* **2006**, *12*, 443–454. [CrossRef] [PubMed]

8. Sinha, N.; Mukhopadhyay, S.; Das, D.N.; Panda, P.K.; Bhutia, S.K. Relevance of cancer initiating/stem cells in carcinogenesis and therapy resistance in oral cancer. *Oral Oncol.* **2013**, *49*, 854–862. [CrossRef] [PubMed]

9. Teodorczyk, M.; Kleber, S.; Wollny, D.; Sefrin, J.P.; Aykut, B.; Mateos, A.; Herhaus, P.; Sancho-Martinez, I.; Hill, O.; Gieffers, C.; et al. CD95 promotes metastatic spread via Sck in pancreatic ductal adenocarcinoma. *Cell Death Differ.* **2015**, *22*, 1192–1202. [CrossRef] [PubMed]

10. Hussenet, T.; Dali, S.; Exinger, J.; Monga, B.; Jost, B.; Dembelé, D.; Martinet, N.; Thibault, C.; Huelsken, J.; Brambilla, E.; et al. SOX2 is an oncogene activated by recurrent 3q26.3 amplifications in human lung squamous cell carcinomas. *PLoS ONE* **2010**, *5*, e8960. [CrossRef] [PubMed]

11. Takahashi, K.; Yamanaka, S. Induction of pluripotent stem cells from mouse embryonic and adult fibroblast cultures by defined factors. *Cell* **2006**, *126*, 663–676. [CrossRef]

12. Dalerba, P.; Cho, R.W.; Clarke, M.F. Cancer stem cells: Models and concepts. *Annu. Rev. Med.* **2007**, *58*, 267–284. [CrossRef] [PubMed]

13. Ren, Z.H.; Zhang, C.P.; Ji, T. Expression of SOX2 in oral squamous cell carcinoma and the association with lymph node metastasis. *Oncol. Lett.* **2016**, *11*, 1973–1979. [CrossRef] [PubMed]

14. Lim, Y.C.; Kang, H.J.; Kim, Y.S.; Choi, E.C. All-trans-retinoic acid inhibits growth of head and neck cancer stem cells by suppression of Wnt/β-catenin pathway. *Eur. J. Cancer* **2012**, *48*, 3310–3318. [CrossRef] [PubMed]

15. Mithani, S.K.; Mydlarz, W.K.; Grumbione, F.L.; Smith, I.M.; Califano, J.A. Molecular genetics of premalignant oral lesions. *Oral Dis.* **2007**, *13*, 126–133. [CrossRef]

16. Luiz, S.T.; Modolo, F.; Mozzer, I.; Dos Santos, E.C.; Nagashima, S.; Camargo Martins, A.P.; de Azevedo, M.L.V.; Azevedo Alanis, L.R.; Hardy, A.M.T.G.; de Moraes, R.S.; et al. Immunoexpression of SOX-2 in oral leukoplakia. *Oral Dis.* **2018**, *24*, 1449–1457. [CrossRef] [PubMed]

17. Michifuri, Y.; Hirohashi, Y.; Torigoe, T.; Miyazaki, A.; Kobayashi, J.; Sasaki, T.; Fujino, J.; Asanuma, H.; Tamura, Y.; Nakamori, K.; et al. High expression of ALDH1 and SOX2 diffuse staining pattern of oral

squamous cell carcinomas correlates to lymph node metastasis. *Pathol. Int.* **2012**, *62*, 684–689. [CrossRef] [PubMed]

18. Du, L.; Yang, Y.; Xiao, X.; Wang, C.; Zhang, X.; Wang, L.; Zhang, X.; Li, W.; Zheng, G.; Wang, S.; et al. Sox2 nuclear expression is closely associated with poor prognosis in patients with histologically node-negative oral tongue squamous cell carcinoma. *Oral Oncol.* **2011**, *47*, 709–713. [CrossRef]

19. Wilbertz, T.; Wagner, P.; Petersen, K.; Stiedl, A.C.; Scheble, V.J.; Maier, S.; Reischl, M.; Mikut, R.; Altorki, N.K.; Moch, H.; et al. SOX2 gene amplification and protein overexpression are associated with better outcome in squamous cell lung cancer. *Mod. Pathol.* **2011**, *24*, 944–953. [CrossRef]

20. Züllig, L.; Roessle, M.; Weber, C.; Graf, N.; Haerle, S.K.; Jochum, W.; Stoeckli, S.J.; Moch, H.; Huber, G.F. High sex determining region Y-box 2 expression is a negative predictor of occult lymph node metastasis in early squamous cell carcinomas of the oral cavity. *Eur. J. Cancer* **2013**, *49*, 1915–1922. [CrossRef]

21. de Vicente, J.C.; Rodríguez-Santamarta, T.; Rodrigo, J.P.; Allonca, E.; Vallina, A.; Singhania, A.; Donate-Pérez Del Molino, P.; García-Pedrero, J.M. The Emerging Role of NANOG as an Early Cancer Risk Biomarker in Patients with Oral Potentially Malignant Disorders. *J. Clin. Med.* **2019**, *8*, 1376. [CrossRef]

22. Takata, T.; Slootweg, P.J. Tumors of the oral cavity and mobile tongue. In *WHO Classification of Head and Neck Tumours*, 4th ed.; International Agency for Research on Cancer (IARC); IARC: Lyon, France, 2017; pp. 105–131.

23. Cancer Genome Atlas Network. Comprehensive genomic characterization of head and neck squamous cell carcinomas. *Nature* **2015**, *517*, 576–582. [CrossRef] [PubMed]

24. Cerami, E.; Gao, J.; Dogrusoz, U.; Gross, B.E.; Sumer, S.O.; Aksoy, B.A.; Jacobsen, A.; Byrne, C.J.; Heuer, M.L.; Larsson, E.; et al. The cBio cancer genomics portal: An open platform for exploring multidimensional cancer genomics data. *Cancer Discov.* **2012**, *2*, 401–404. [CrossRef] [PubMed]

25. Lu, Y.; Futtner, C.; Rock, J.R.; Xu, X.; Whitworth, W.; Hogan, B.L.; Onaitis, M.W. Evidence that SOX2 overexpression is oncogenic in the lung. *PLoS ONE* **2010**, *5*, e11022. [CrossRef] [PubMed]

26. Liu, X.; Qiao, B.; Zhao, T.; Hu, F.; Lam, A.K.; Tao, Q. Sox2 promotes tumor aggressiveness and epithelial-mesenchymal transition in tongue squamous cell carcinoma. *Int. J. Mol. Med.* **2018**, *42*, 1418–1426. [CrossRef] [PubMed]

27. Arnold, K.; Sarkar, A.; Yram, M.A.; Polo, J.M.; Bronson, R.; Sengupta, S.; Seandel, M.; Geijsen, N.; Hochedlinger, K. Sox2(+) adult stem and progenitor cells are important for tissue regeneration and survival of mice. *Cell Stem Cell* **2011**, *9*, 317–329. [CrossRef] [PubMed]

28. Cai, J.; He, B.; Li, X.; Sun, M.; Lam, A.K.; Qiao, B.; Qiu, W. Regulation of tumorigenesis in oral epithelial cells by defined reprogramming factors Oct4 and Sox2. *Oncol. Rep.* **2016**, *36*, 651–658. [CrossRef]

29. Abbey, L.M.; Kaugars, G.E.; Gunsolley, J.C.; Burns, J.C.; Page, D.G.; Svirsky, J.A.; Eisenberg, E.; Krutchkoff, D.J.; Cushing, M. Intraexaminer and interexaminer reliability in the diagnosis of oral epithelial dysplasia. *Oral Surg. Oral Med. Oral Pathol. Oral Radiol. Med.* **1995**, *80*, 188–191. [CrossRef]

30. Freier, K.; Knoepfle, K.; Flechtenmacher, C.; Pungs, S.; Devens, F.; Toedt, G.; Hofele, C.; Joos, S.; Lichter, P.; Radlwimmer, B. Recurrent copy number gain of transcription factor SOX2 and corresponding high protein expression in oral squamous cell carcinoma. *Genes Chromosomes Cancer* **2010**, *49*, 9–16. [CrossRef]

31. Hermsen, M.; Alonso Guervós, M.; Meijer, G.; van Diest, P.; Suárez Nieto, C.; Marcos, C.A.; Sampedro, A. Chromosomal changes in relation to clinical outcome in larynx and pharynx squamous cell carcinoma. *Cell Oncol.* **2005**, *27*, 191–198. [CrossRef]

32. Järvinen, A.K.; Autio, R.; Kilpinen, S.; Saarela, M.; Leivo, I.; Grénman, R.; Mäkitie, A.A.; Monni, O. High-resolution copy number and gene expression microarray analyses of head and neck squamous cell carcinoma cell lines of tongue and larynx. *Genes Chromosomes Cancer* **2008**, *47*, 500–509. [CrossRef]

33. Lin, S.C.; Liu, C.J.; Ko, S.Y.; Chang, H.C.; Liu, T.Y.; Chang, K.W. Copy number amplification of 3q26-27 oncogenes in microdissected oral squamous cell carcinoma and oral brushed samples from areca chewers. *J. Pathol.* **2005**, *206*, 417–422. [CrossRef] [PubMed]

34. Bora-Singhal, N.; Nguyen, J.; Schaal, C.; Perumal, D.; Singh, S.; Coppola, D.; Chellappan, S. YAP1 Regulates OCT4 Activity and SOX2 Expression to Facilitate Self-Renewal and Vascular Mimicry of Stem-Like Cells. *Stem Cells* **2015**, *33*, 1705–1718. [CrossRef] [PubMed]

35. Bae, K.M.; Dai, Y.; Vieweg, J.; Siemann, D.W. Hypoxia regulates SOX2 expression to promote prostate cancer cell invasion and sphere formation. *Am. J. Cancer Res.* **2016**, *6*, 1078–1088. [PubMed]

36. de Vicente, J.C.; Santamarta, T.R.; Rodrigo, J.P.; García-Pedrero, J.M.; Allonca, E.; Blanco-Lorenzo, V. Expression of podoplanin in the invasion front of oral squamous cell carcinoma is not prognostic for survival. *Virchows Arch.* **2015**, *466*, 549–558. [CrossRef]

37. Qiao, B.; He, B.; Cai, J.; Yang, W. The expression profile of Oct4 and Sox2 in the carcinogenesis of oral mucosa. *Int. J. Clin. Exp. Pathol.* **2013**, *7*, 28–37.

38. Ye, X.; Wu, F.; Wu, C.; Wang, P.; Jung, K.; Gopal, K.; Ma, Y.; Li, L.; Lai, R. β-Catenin, a Sox2 binding partner, regulates the DNA binding and transcriptional activity of Sox2 in breast cancer cells. *Cell Signal* **2014**, *26*, 492–501. [CrossRef]

39. Gen, Y.; Yasui, K.; Nishikawa, T.; Yoshikawa, T. SOX2 promotes tumor growth of esophageal squamous cell carcinoma through the AKT/mammalian target of rapamycin complex 1 signaling pathway. *Cancer Sci.* **2013**, *104*, 810–816. [CrossRef]

40. Thierauf, J.; Veit, J.A.; Hess, J. Epithelial-to-Mesenchymal Transition in the Pathogenesis and Therapy of Head and Neck Cancer. *Cancers* **2017**, *3*, 76. [CrossRef]

41. He, S.; Chen, J.; Zhang, Y.; Zhang, M.; Yang, X.; Li, Y.; Sun, H.; Lin, L.; Fan, K.; Liang, L.; et al. Sequential EMT-MET induces neuronal conversion through Sox2. *Cell Discov.* **2017**, *3*, 17017. [CrossRef]

42. Bass, A.J.; Watanabe, H.; Mermel, C.H.; Yu, S.; Perner, S.; Verhaak, R.G.; Kim, S.Y.; Wardwell, L.; Tamayo, P.; Gat-Viks, I.; et al. SOX2 is an amplified lineage-survival oncogene in lung and esophageal squamous cell carcinomas. *Nat Genet.* **2009**, *41*, 1238–1242. [CrossRef]

43. Ge, N.; Lin, H.X.; Xiao, X.S.; Guo, L.; Xu, H.M.; Wang, X.; Jin, T.; Cai, X.Y.; Liang, Y.; Hu, W.H.; et al. Prognostic significance of Oct4 and Sox2 expression in hypopharyngeal squamous cell carcinoma. *J. Transl. Med.* **2010**, *8*, 94. [CrossRef] [PubMed]

44. Zhang, X.; Yu, H.; Yang, Y.; Zhu, R.; Bai, J.; Peng, Z.; He, Y.; Chen, L.; Chen, W.; Fang, D.; et al. SOX2 in gastric carcinoma, but not Hath1, is related to patients' clinicopathological features and prognosis. *J. Gastrointest. Surg.* **2010**, *14*, 1220–1226. [CrossRef] [PubMed]

Journal of
Clinical Medicine

Article

The SRC Inhibitor Dasatinib Induces Stem Cell-Like Properties in Head and Neck Cancer Cells that are Effectively Counteracted by the Mithralog EC-8042

Francisco Hermida-Prado [1,2], M. Ángeles Villaronga [1,2], Rocío Granda-Díaz [1,2], Nagore del-Río-Ibisate [1,2], Laura Santos [1], Maria Ana Hermosilla [3], Patricia Oro [3], Eva Allonca [1,2], Jackeline Agorreta [2,4], Irati Garmendia [2,4], Juan Tornín [1], Jhudit Perez-Escuredo [3], Rocío Fuente [5], Luis M. Montuenga [2,4], Francisco Morís [3], Juan P. Rodrigo [1,2], René Rodríguez [1,2,*] and Juana M. García-Pedrero [1,2,*]

[1]   Department of Otolaryngology, Hospital Universitario Central de Asturias and Instituto de Investigación Sanitaria del Principado de Asturias; Instituto Universitario de Oncología del Principado de Asturias, University of Oviedo, 33011 Oviedo, Spain
[2]   Ciber de Cáncer, CIBERONC, 28029 Madrid, Spain
[3]   EntreChem SL, Vivero Ciencias de la Salud, 33011 Oviedo, Spain
[4]   Program in Solid Tumors, Center for Applied Medical Research (CIMA), Department of Pathology, Anatomy and Physiology, University of Navarra, and Navarra's Health Research Institute (IDISNA), 31008 Pamplona, Spain
[5]   Division of Pediatrics, Department of Medicine, Faculty of Medicine, University of Oviedo, 33006 Oviedo, Spain
*   Correspondence: renerg.finba@gmail.com (R.R.); juanagp.finba@gmail.com (J.M.G.-P.)

Received: 23 July 2019; Accepted: 31 July 2019; Published: 2 August 2019

Abstract: The frequent dysregulation of SRC family kinases (SFK) in multiple cancers prompted various inhibitors to be actively tested in preclinical and clinical trials. Disappointingly, dasatinib and saracatinib failed to demonstrate monotherapeutic efficacy in patients with head and neck squamous cell carcinomas (HNSCC). Deeper functional and mechanistic knowledge of the actions of these drugs is therefore needed to improve clinical outcome and to develop more efficient combinational strategies. Even though the SFK inhibitors dasatinib and saracatinib robustly blocked cell migration and invasion in HNSCC cell lines, this study unveils undesirable stem cell-promoting functions that could explain the lack of clinical efficacy in HNSCC patients. These deleterious effects were targeted by the mithramycin analog EC-8042 that efficiently eliminated cancer stem cells (CSC)-enriched tumorsphere cultures as well as tumor bulk cells and demonstrated potent antitumor activity in vivo. Furthermore, combination treatment of dasatinib with EC-8042 provided favorable complementary anti-proliferative, anti-invasive, and anti-CSC functions without any noticeable adverse interactions of both agents. These findings strongly support combinational strategies with EC-8042 for clinical testing in HNSCC patients. These data may have implications on ongoing dasatinib-based trials.

Keywords: head and neck squamous cell carcinoma; SRC; dasatinib; saracatinib; cancer stem cells; EC-8042

---

1. Introduction

Increased SRC expression and/or activity has been widely detected in a variety of human cancers, including head and neck squamous cell carcinomas (HNSCC) [1–3]. It has been demonstrated that aberrant SRC activity plays a central role in all stages of tumorigenesis from malignant transformation to tumor progression and ultimately development of metastatic disease [4,5].

SRC is a pleiotropic non-receptor tyrosine kinase that interacts with multiple receptor tyrosine kinases and modulates multiple oncogenic signaling pathways [6,7], thereby regulating a variety of cellular processes central to the malignant phenotype, including proliferation, survival, cell-cell adhesion, motility, invasion, and angiogenesis [8,9]. SRC induces disruption of focal adhesions by activation of focal adhesion kinase (FAK) [10]. FAK is associated with SRC in focal adhesions, which results in the phosphorylation of FAK followed by downstream activation of the Ras-mediated pathways [11]. In addition, SRC regulates the reorganization of the actin cytoskeleton through phosphorylation of p190 [12] and cortactin (CTTN), which results in increased motility [13].

The frequent dysregulation of SRC family kinases (SFK) in a variety of human cancers has led to a rapid development of multiple agents aimed at targeting SFK in cancer treatment [9]. Various inhibitors, in particular dasatinib (BMS-354825) and saracatinib (AZD0530), have been the subject of intense research in recent years [14–21]. Even though both drugs have proved powerful as antitumor agents in preclinical settings, their clinical efficacy in cancer patients has been limited if not disappointing [14–16,20,21]. In preclinical models of HNSCC, dasatinib and saracatinib showed potent effects on proliferation, cell migration, and invasion [22,23]; however, both inhibitors failed to demonstrate any significant activity as single agents in patients with recurrent and/or metastatic HNSCC [24,25].

In light of these data, it becomes highly desirable to establish relevant response biomarkers to improve patient stratification, treatment efficacy, and ultimately clinical outcome. In addition, a deeper functional and mechanistic knowledge of the actions of these drugs on tumor cells will contribute to deciphering not only anti-tumor functions but also unwanted pro-tumor actions, thus enabling the development of combination strategies that may prevent potential deleterious activities and provide anti-tumor complementary/synergistic strategies.

This study demonstrates that while dasatinib and saracatinib robustly blocked cell migration and invasion in HNSCC cell lines, both were found to strikingly enhance cancer stem cells (CSC) properties. These deleterious effects were effectively targeted by the low-toxicity mithramycin analog (mithralog) EC-8042 [26], which has previously demonstrated anti-stemness activity in other cancer cells [27]. Accordingly, EC-8042 efficiently inhibited the growth of both bulk tumor adherent cultures and CSC-enriched tumorsphere cultures in HNSCC-derived cells and demonstrated a robust antitumor activity in vivo. Moreover, combination treatment between dasatinib and EC-8042 benefits from complementary anti-tumor properties provided by each compound without any noticeable adverse interactions between them.

2. Materials and Methods

2.1. Drugs

EC-8042 (EntreChem, Oviedo, Spain), Dasatinib and Saracatinib (both from Selleck, Suffolk, UK) were prepared as 10 mM solutions in sterile DMSO or water, maintained at −20 °C and brought to the final concentration just before use.

2.2. Cell Culture

FaDu cells (male, hypopharyngeal squamous cell carcinoma) were purchased from the ATCC, and the HNSCC cell line UT-SCC38 derived from a laryngeal squamous carcinoma (T2N0M0) was kindly provided by R. Grenman (Department of Otolaryngology, University Central Hospital, Turku, Finland) [28]. Cells were grown in DMEM supplemented with 10% fetal bovine serum (FBS), 100 U/mL penicillin, 200 mg/mL streptomycin, 2 mmol/L L-glutamine, 20 mmol/L HEPES (pH 7.3), and 100 mmol/L non-essential amino acids. All the cells derived from HPV-negative primary HNSCC. All cell lines were periodically tested for mycoplasma contamination by PCR using the Biotools Detection kit (Madrid, Spain) specifically amplifying a conserved region of the mycoplasma 16S RNA gene. Cell line authentication was carried out by DNA (STR) profiling at the SCT Core Facilities (University of Oviedo, Spain).

2.3. Western Blotting Analysis

Cells were lysed in Laemmli sample buffer and sonicated before centrifugation. Protein lysates were separated by SDS–polyacrylamide gel electrophoresis (SDS–PAGE) and transferred to nitrocellulose membranes (Amersham Protran, GE Healthcare, Pittsburg, PA, USA). Subsequently, membranes were blocked for 1 h with Odyssey blocking buffer (LI-COR Biosciences, Lincoln, NE, USA) and incubated overnight with the indicated primary antibodies (Supplementary Methods) at 1:1000 dilution. The IRDye Infrared Fluorescent secondary antibodies anti-Rabbit and anti-Mouse IRDye 800CW and IRDye 680RD (LI-COR Biosciences) were used for detection. Membranes were scanned with the Odyssey Fc Dual-Mode Imaging System (LI-COR Biosciences) using the red (700 nm) and green (800 nm) channels, and signal analysis was performed using Image Studio Lite software (LI-COR, Nebraska). Results were normalized to GAPDH as loading control.

2.4. Cell Viability Assays

HNSCC cells were seeded into 96-well culture plates at a density of 2000 cells per well and incubated overnight. Drugs were serially diluted in medium over a range of concentrations and added to the cells. After 72 h treatment, cell viability was measured in quadruplicates using a MTS assay (CellTiter 96 Aqueous One Solution Cell Proliferation Assay from Promega, Madison, WI, USA) reading absorbance at 490 nm using a Synergy HT plate reader (BioTek, Winooski, VT, USA). The existence of synergy in drug combinations was determined by calculating the combination index (CI) as described in Supplementary Information.

2.5. Scratch-Induced Directional Migration Assay

Cells were plated in 24-well dishes with ibidi® culture inserts (ibidi LLC, Verona, WI, USA) at 80–90% confluence, and cell migration monitored as we previously described [29].

2.6. Three-Dimensional Spheroid Invasion Assays

Invasion assays using 3D spheroids were performed as previously described [30]. Cells were suspended in DMEM plus 5% Methyl cellulose (Sigma, St Louis, MO, USA) to form cell spheroids by serial pipetting into a non-adhesive Petri dish (2000 cells/spheroid), followed by overnight incubation in an inverted position. The next day, each cell spheroid was individually transferred to a 96-well plate, embedded into bovine collagen matrix (Advanced Biomatrix PureCol), and filled with 100 µL of complete media containing or not containing drugs. Cell invasion was monitored using a Zeiss Cell Observer Live Imaging microscope (Zeiss, Thornwood, NY, USA) and images acquired every 15 min for 24 h using a Zeiss AxioCam MRc camera. The invasive area was calculated as the difference between the final area (t = 24 h) and the initial area (t = 0 h) using image J analysis program, and data were normalized to control (vehicle-treated) cells. Three independent experiments were performed using quadruplicates for each condition.

2.7. Tumorsphere Formation Assay

HNSCC-derived cells lines were plated at a density of 500 cells/mL in 6-well tissue culture plates treated with a sterile solution of polyHEMA (10 g/L in 95% ethanol) (Sigma) to prevent cell attachment. Cells were grown in DMEM-F12 (GE Healthcare) supplemented with 1% Glutamax and 2% B27 Supplement from Life Technologies (Rockford, IL, USA), 10 ng/mL human bFGF and 20 ng/mL human EGF (PeproTech, London, UK) and 100 U/mL penicillin and 200 mg/mL streptomycin (Thermo, Waltham, MA, USA).

After 10–12 days, well-formed spheres were photographed in Leica Microsystems microscope DMIL T coupled with a Leica DC500 High-resolution Digital Camera (Leica Microsystems, Barcelona, Spain). Then, the tumorspheres were centrifuged at 300 rpm for 2 min, washed with PBS and either collected for RNA extraction or disaggregated with Gibco trypsin (0.25%)/EDTA for 15 min to measure cell viability by MTS.

2.8. RNA Extraction and Real-Time RT-PCR

Total RNA was extracted from HNSCC cells using Trizol reagent (Invitrogen Life Technologies, Carlsbad, CA, USA), and gene expression was analyzed by real-time RT-PCR as we previously reported [31] using SYBR Green Master Mix protocol (Applied Biosystems, Foster City, CA, USA) in a StepOnePlus Real-Time PCR System (Applied Biosystems, Foster City, CA, USA). Reactions were run in triplicates using the specific primers detailed in Supplementary Table S1, and the ribosomal coding gene RPL19 was used as endogenous control. The relative mRNA expression was calculated using the $2^{-\Delta\Delta CT}$ method, and the data were expressed as the fold-change normalized to RPL19 mRNA levels and relative to control (vehicle-treated) cells.

2.9. In Vivo Treatments of FaDu Xenografts

All experimental protocols were performed in accordance with the institutional guidelines of the University of Oviedo and approved by the Animal Research Ethical Committee of the University of Oviedo prior to the study. Female athymic NMRI-nude mice of 6–7 weeks old (Janvier Labs, St Berthevin, France) were subcutaneously inoculated (s.c.) with 1.5×10^6 FaDu cells mixed 1:1 with BD Matrigel Matrix High Concentration (BD Biosciences, Erembodegem, Belgium) previously diluted 1:1 in culture medium. Once tumors reached ~200 mm^3, mice were randomized into four treatment groups (n = 10 per group): (i) vehicle (saline solution intravenously (i.v.) as vehicle for EC-8042 and tartaric acid solution orally for dasatinib); (ii) dasatinib (10 mg/kg every day (16 doses) orally); (iii) EC-8042 (50 mg/kg every 7 days (4 doses) i.v.); and (iv) dasatinib plus EC-8042 combination.

Survival was represented using Kaplan–Meier analysis and the log-rank test to estimate significant differences among groups (PAST 3.01 software, University of Oslo, Norway). Tumor growth and drug efficacy (expressed as the percentage of tumor growth inhibition, %TGI) were calculated as indicated in Supplementary Information.

2.10. Tumorsphere Formation and Immunohistochemical Analyses of Tumors from FaDu Xenografts.

Upon removal, tumor samples were weighted and a portion of some tumors was disaggregated into single cell suspensions using MACS Tissue Dissociation Kit and the GentleMACS Dissociator system (Miltenyi Biotec, Bergisch Gladbach, Germany) as previously described [27], in order to perform tumorsphere formation assays after in vivo treatments. The remaining portion of the tumors were fixed in formol, embedded in paraffin, cut into 4-µm sections, and stained with hematoxylin and eosin (H&E). Immunohistochemical analyses were performed in an automatic workstation (Dako Autostainer Plus) with anti-Ki67 (Clone MIB-1 Dako # JR626, Prediluted), anti-active PARP (Abcam # 32064, at 1:500), anti-ALDH1 (BD Biosciences # 611195, at 1:500), anti-SOX2 (Merck Millipore # AB5603, at 1:1000), and phospho-FAK (Y861) (Invitrogen # 44-626G, at 1:100) using the Dako EnVision Flex + Visualization System (Dako Autostainer). The number of ALDH1-positive cells or SOX2-positive nuclei was counted at 40× in five independent microscopic fields per tissue section, and the mean of five fields was calculated for each treatment. p-FAK (Y861) staining intensity was evaluated, and the mean of five fields was calculated for each treatment. Quantification of staining for Ki67 proliferation index (number of positive cells per mm^2) and cleaved PARP (number of positive cells per mm^2) was automatically performed using the ImageJ software (National Institutes of Health, Bethesda, MD, USA) in six random images (×200) per sample.

2.11. Statistical Analyses

Statistical analysis was performed using GraphPad Prism version 6.0 (Graphpad Software Inc, La Jolla, CA, USA). Data are presented as the mean ± standard deviation (SD) of at least three independent experiments unless otherwise stated. Statistical significance will be determined either using a Student's unpaired t-test with two-tailed distribution for comparison across two groups or one/two-way ANOVA for comparing multiple samples/variables. In comparisons with control groups, the values of $p < 0.05$ were considered statistically significant (* $p < 0.05$; ** $p < 0.01$; *** $p < 0.001$; **** $p < 0.0001$).

3. Results

3.1. Dasatinib and Saracatinib Completely Blocked Migration and Invasion in HNSCC-Derived Cell Lines

We first evaluated the effect of dasatinib and saracatinib in the HNSCC-derived cell lines FaDu and UT-SCC38. As expected, both compounds decreased the phosphorylation levels of SRC at tyrosine 418 (Y418) and FAK at Y861 in FaDu and UT-SCC38 cells (Supplementary Figure S1A,B). Phospho-SRC Y418 levels rapidly decreased after 1 h treatment with saracatinib and dasatinib (Supplementary Figure S1C), and the phosphorylation levels of its downstream target FAK Y861 were efficiently targeted and durably reduced at 24 h. In addition, dasatinib (0.1 µM) and saracatinib (1 µM) completely blocked cell migration and invasion into 3D collagen matrices in both cell lines (Figure 1A,B, and Supplementary Materials: Videos S1 and S2). 24 h treatment with these concentrations of drugs had no significant effect on cell viability in UT-SCC38 and led to a 20% decrease in FaDu cells (Supplementary Figure S1D); however, this effect was very modest compared to the robust effects observed on cell migration (>90%) and invasion (>70%). Altogether, these data indicate that the potent anti-invasive effect observed upon dasatinib or saracatinib treatment cannot be attributed to the ability of these drugs to decrease cell viability. Nonetheless, longer treatments for 72 h with saracatinib and dasatinib led to a dose-dependent reduction of cell viability, with dasatinib having a more pronounced cytotoxic effect (Figure 1C).

3.2. Dasatinib and Saracatinib Promoted CSC-Like Phenotypes in HNSCC Cells

CSCs have been recognized to play critical roles driving tumor initiation, progression, recurrence, and treatment resistance. This prompted us to investigate whether dasatinib and saracatinib are able to target CSCs in HNSCC cells.

Clonal sphere-forming ability in non-adherent serum-free culture conditions (tumorsphere cultures) is a hallmark of self-renewal and CSC-related phenotype. Under these conditions, the expression of well-known CSC markers such as ALDH1A1 and SOX2 [32] was consistently and markedly enriched in FaDu and UT-SCC38 tumorspheres when compared to unselected adherent cultures (Supplementary Figure S2). Notably, dasatinib and saracatinib failed to eliminate CSC-enriched tumorsphere cultures of FaDu and UT-SCC38 cells (Figure 2A,B). Strikingly, we also found that both drugs significantly increased the expression of ALDH1A1 and SOX2 at both mRNA and protein levels (Figure 2C,D and Supplementary Figure S3A). Of note, specifically in UT-SCC38 cells, SOX2 expression was not affected by dasatinib, and consistent results were obtained at both mRNA and protein level. Nonetheless, NANOG1 and OCT4 levels also consistently increased upon dasatinib and saracatinib treatment in these two HNSCC-derived cell lines (Figure 2D), in good agreement with the previous paper by Koo et al. [33]. Furthermore, since the process of epithelial to mesenchymal transition (EMT) is known to play a critical role as a driver of tumor cell invasion and metastatic spreading in HNSCC [34–36], expression changes of EMT markers were also assessed. Our results consistently showed that the expression levels of the epithelial marker E-Cadherin increased upon treatment with dasatinib and saracatinib, while the expression of mesenchymal markers such as Vimentin and Snail decreased, thereby indicating that both compounds inhibit EMT in these two HNSCC cell lines (Supplementary Figure S4). In light of these data, the ability of dasatinib and saracatinib to revert EMT could be an additional mechanism (together with other invasion regulators targeted, e.g., SRC, FAK) to explain the potent inhibitory effect on cell invasiveness exhibited by these drugs.

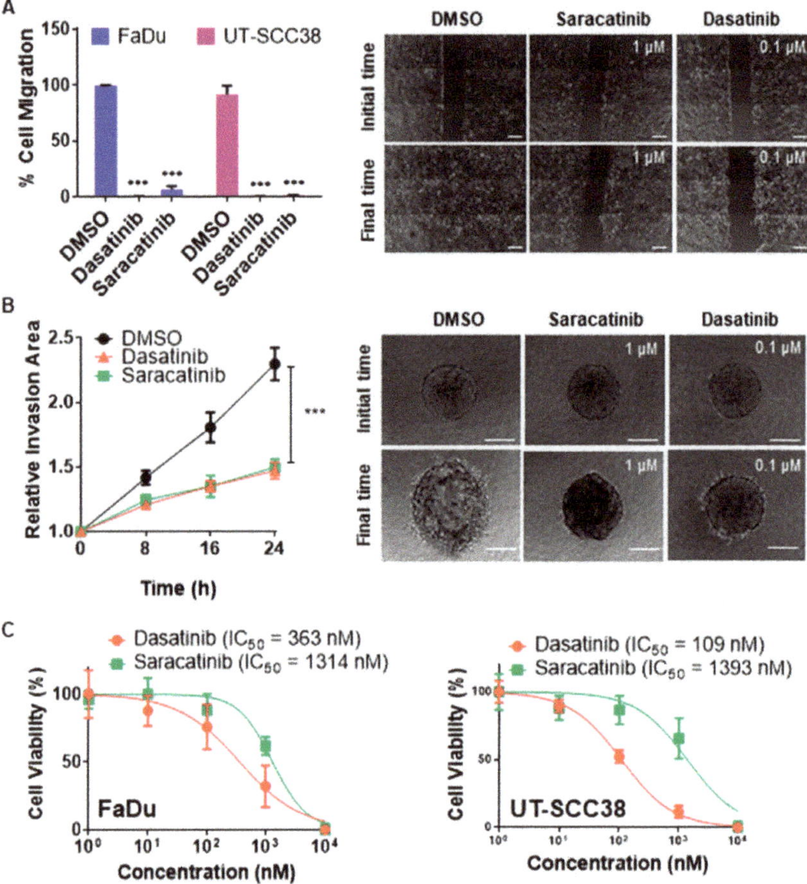

Figure 1. Effect of dasatinib and saracatinib on cell migration, invasion, and growth of head and neck squamous cell carcinomas (HNSCC)-derived cell lines. (**A**) Wound healing assays in FaDu and UT-SCC38 cells treated with either DMSO (vehicle), 0.1 μM dasatinib, or 1 μM saracatinib. The percentage of cell migration (left panel) and representative images showing the initial scratch (t = 0) area and the residual area at the final time (t = 15 h) in FaDu cells (right panel) are displayed. Scale bars = 200 μm. Data are expressed relative to vehicle-treated cells (mean ± SD, Student's *t*-test, *** $p < 0.001$). (**B**) 3D spheroid invasion assays in FaDu cells treated for 24 h with either DMSO (vehicle), 0.1 μM dasatinib, or 1 μM saracatinib. The quantification of the invasive area at the indicated times (left panel) and representative images of FaDu spheroids at initial (t = 0) and final time (t = 24 h) (right panel) for the different treatments are displayed. Scale bars = 200 μm. Data are expressed relative to DMSO-treated cells (mean ± SD, Student's *t*-test, *** $p < 0.001$). (**C**) Cell viability of FaDu (left panel) and UT-SCC38 (right panel) cells was measured by MTS assay after 72 h treatment with increasing doses of either dasatinib or saracatinib. IC_{50} values are shown.

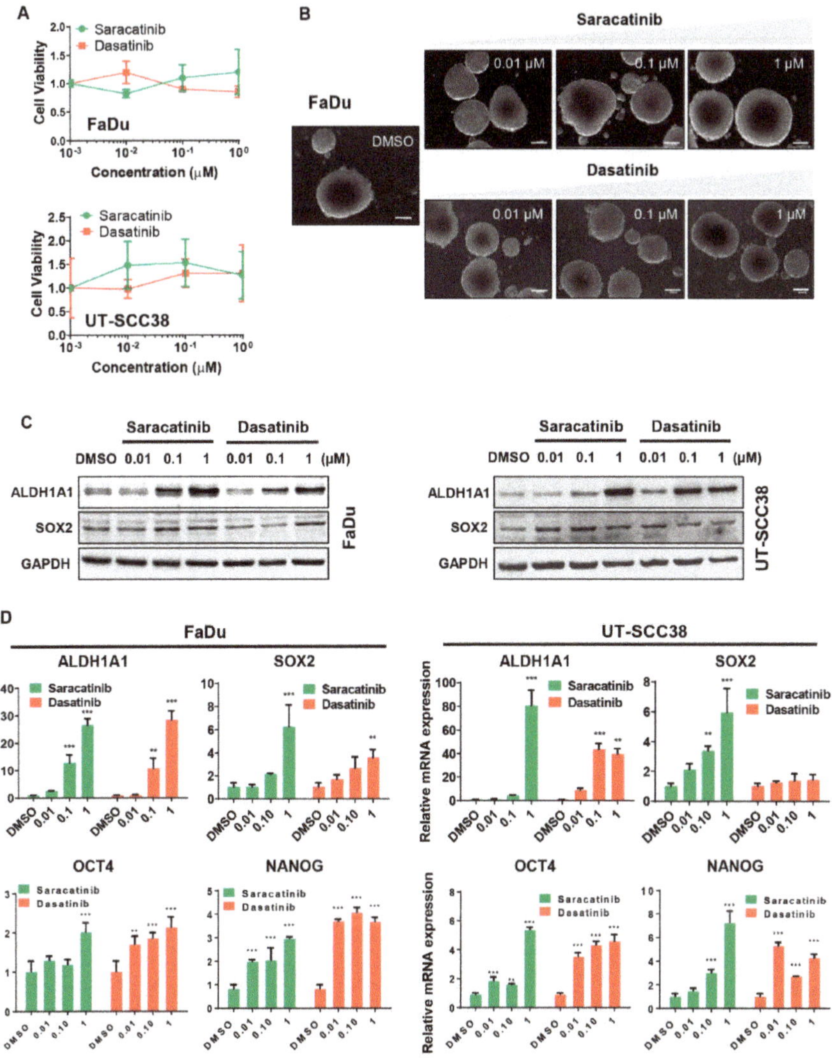

Figure 2. Effect of dasatinib and saracatinib on the cancer stem cells (CSC) properties of HNSCC-derived cell lines. (**A**,**B**) Cell viability (MTS assay) of UT-SCC38 and FaDu tumorspheres treated with increasing concentrations of dasatinib or saracatinib for 72 h. Quantification of cell viability (**A**) and representative images of FaDu tumorspheres upon the indicated treatments (**B**) are shown. (**C**,**D**) Analysis of the protein (Western blotting; C and quantification of IRdye fluorescent signals are plotted in Supplementary Figure S3A) and relative mRNA (RT-qPCR); (**D**) levels of the CSC-markers ALDH1A1, SOX2, OCT4, and NANOG in FaDu and UT-SCC38 cells treated with increasing concentrations of dasatinib and saracatinib for 72 h. Data were normalized to RPL19 levels and represented relative to vehicle-treated cells (mean ± SD, Student's *t*-test, ** $p < 0.01$, *** $p < 0.001$).

3.3. The Mithramycin Analog EC-8042 Effectively Abrogated the CSC Properties of HNSCC Cells

We next assessed the ability of the mithramycin analog EC-8042 to eradicate CSCs, since it has been described to potently reduce CSCs viability and expression of CSC-related markers in other cancers [27].

EC-8042 showed a much more robust cytotoxic effect on adherent cultures than the two SRC inhibitors (Figure 3A). Likewise, the treatment with EC-8042 was highly effective in reducing the number, size and the viability of CSC-enriched tumorsphere cultures (Figure 3B,C). Furthermore, pretreatment for 72 hours with EC-8042 (0.1 μM) but neither dasatinib (0.1 μM) nor saracatinib (1 μM) was also able to significantly prevent tumorsphere formation in FaDu cells (Figure 3D). In addition, the expression of the CSC-related markers ALDH1A1 and SOX2 was robustly and consistently reduced in both FaDu and UT-SCC38 cells after treatment with EC-8042 (Figure 3E).

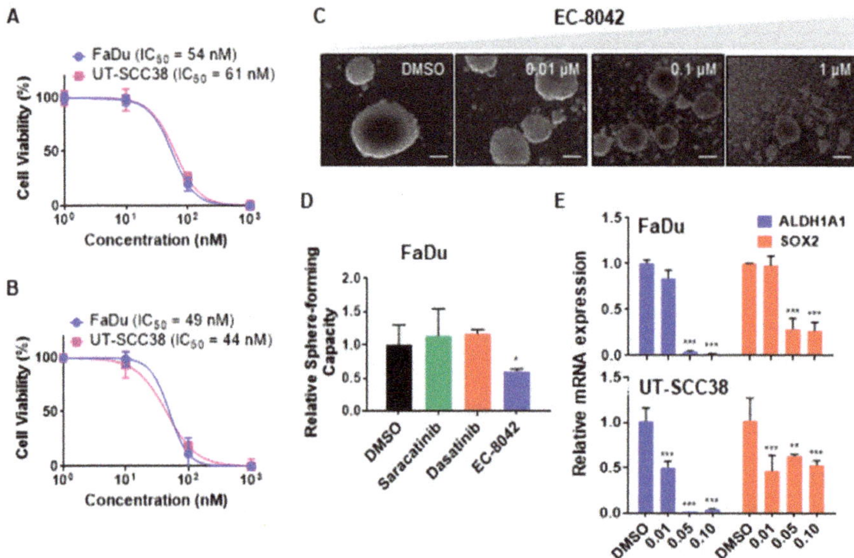

Figure 3. Effect of the mithralog EC-8042 on the viability and CSC properties of HNSCC-derived cell lines. (**A,B**) Dose-response curves of the cell viability by MTS in UT-SCC38 and FaDu adherent cells (**A**) or CSC-enriched tumorspheres (**B**) treated with increasing concentrations of EC-8042 for 72 h. IC$_{50}$ values are shown. (**C**) Representative images of FaDu tumorspheres upon treatment for 72 h with different doses of EC-8042. (**D**) Tumorsphere-forming ability of FaDu cells upon treatment for 72 h with either DMSO (vehicle), 0.1 μM dasatinib, 1 μM saracatinib, or 0.1 μM EC-8042. (**E**) RT-qPCR analysis of the CSC-markers ALDH1A1 and SOX2 in adherent cultures of FaDu and UTSCC38 cells treated with increasing concentrations of EC-8042 for 72 h. Data are expressed relative to DMSO-treated cells (mean ± SD, Student's *t*-test, * $p < 0.05$, ** $p < 0.01$, *** $p < 0.001$).

3.4. Combined Effects of Treatment with Dasatinib and EC-8042 in HNSCC Models

Next, we explored the effects of the combination treatment with dasatinib and EC-8042. First, we analyzed the effect of the combination on the activation of key signaling targets for both drugs (Figure 4A and Supplementary Figure S5). As shown previously in other tumor cell models [37], a 24-hour treatment with dasatinib efficiently inhibited SRC phosphorylation at Y418 while upregulating total SRC levels. SRC-dependent phosphorylation and activation of FAK at Y861, but not SRC-independent autophosphorylation at Y397, was completely and durably inhibited by dasatinib without affecting total FAK levels. The phosphorylation/activation of other downstream signaling targets, such as AKT at S473 or p44/42-MAPK, was also inhibited by dasatinib treatment (Figure 4A). EC-8042 treatment resulted in reduced levels of its target SP1 and also led to the down-regulation of both total FAK and p-FAK (Y397) levels. Importantly, the combination of both drugs produced an additive inhibitory effect on the phosphorylation/activation of most of the targets (Figure 4A).

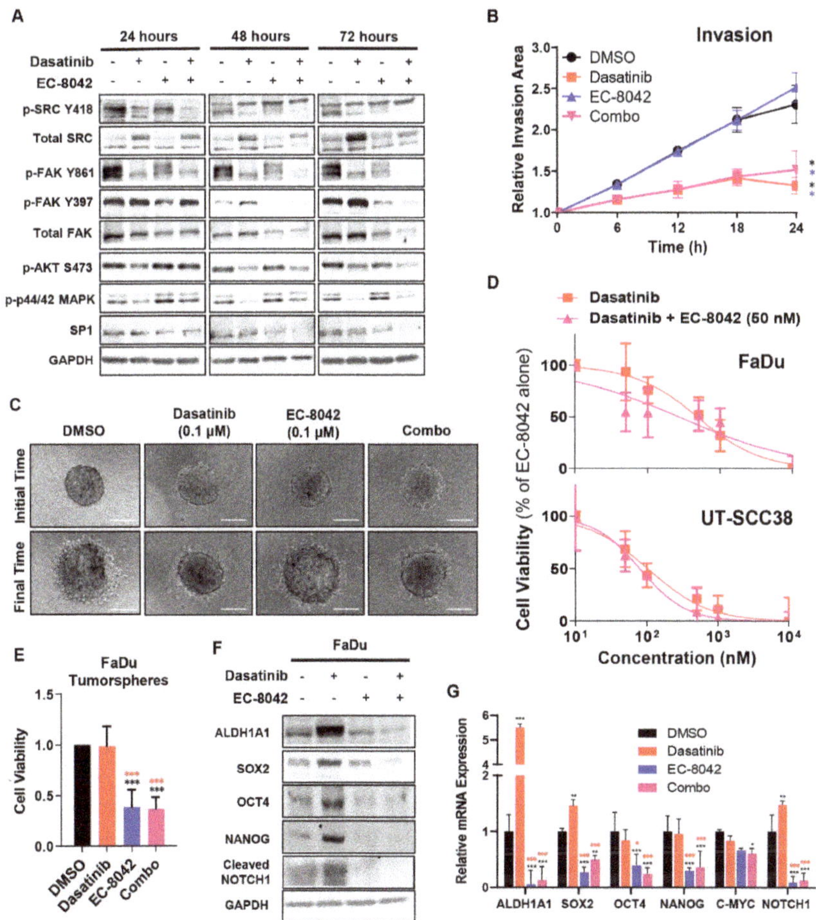

Figure 4. Functional effects of the combined treatment with dasatinib and EC-8042 in HNSCC cells. (**A**) Western blotting analysis of the expression/phosphorylation levels of the indicated proteins in FaDu cells treated for 24 h, 48 h, or 72 h with either 0.1 μM dasatinib or 0.1 μM EC-8042 alone, or in combination. Quantification of IRdye fluorescent signals are plotted in Supplementary Figure S5. (**B,C**) 3D spheroid invasion assay in FaDu cells treated for 24 h with 0.1 μM dasatinib, 0.1 μM EC-8042 alone, or combination. The quantification of the invasive area at the indicated times (**B**) and representative images of FaDu spheroids (**C**) at initial (t = 0) and final time (t = 24 h) for the different treatments are shown. Scale bars = 200 μm. Data are expressed relative to vehicle-treated cells (mean ± SD, Repeated Measures (RM)-one-way ANOVA, Tukey's test, * $p < 0.05$ vs. DMSO and EC-8042-treated cells). (**D**) Dose-response curves of cell viability by MTS in FaDu (upper panel) and UT-SCC38 (lower panel) cells treated with the indicated combinations of drugs for 72 h. In these graphs the dasatinib + EC-8042 series was normalized to the value observed after treatment with EC-8042 alone. The effect of EC-8042 alone is subtracted from the combination values, thus showing the shift in the IC$_{50}$ of dasatinib due to the combination. (**E**) Cell viability of CSC-enriched FaDu tumorspheres treated with 0.1 μM dasatinib, 0.1 μM EC-8042, or combination. Data are expressed relative to vehicle-treated

cells (mean ± SD, one-way ANOVA, Tukey's test, *** $p < 0.001$ vs. DMSO and dasatinib-treated cells). (**F**) Western blotting (left panel) and (**G**) RT-qPCR analysis (right panel) of the indicated oncogenic and CSC-related markers in adherent cultures of FaDu cells treated for 72 h with 0.1 μM dasatinib or 0.1 μM EC-8042 alone, or in combination. mRNA levels were normalized to RPL19 levels and the relative fold-change to vehicle-treated cells ± SD plotted, two-way ANOVA, Dunnett's test, * $p < 0.05$, ** $p < 0.01$, *** $p < 0.001$ vs. DMSO (black) and dasatinib-treated cells (red). Quantification of the IRdye fluorescent signals are plotted in Supplementary Figure S3B.

We next studied the ability of the combination to add jointly favorable anti-tumor effects with respect to single treatments. First, we found that the combined treatment maintained the ability of dasatinib to abolish invasion of HNSCC cells into 3D collagen matrices (Figure 4B,C, Supplementary Videos S1–S4). In addition, cell survival curves normalized to the effect of EC-8042 alone showed that the combination of dasatinib with EC-8042 did not produce a significant shift of the dasatinib IC_{50} values (Figure 4D). Furthermore, combination index values calculated according to the Chou and Talalay method [38] for EC-8042 and dasatinib combination were close to 1 (Supplementary Figure S6A), thus suggesting that this combination produce additive, rather than synergistic cytotoxic effects.

Moreover, the deleterious effects caused by dasatinib sustaining CSC-like phenotypes were effectively counteracted by EC-8042 co-treatment, thereby reducing the viability of CSC-enriched tumorspheres (Figure 4E) and the expression of various CSC-related factors such as ALDH1, SOX2, NANOG, OCT4, c-MYC, or NOTCH1 at mRNA and protein level (Figure 4F,G and Supplementary Figure S3B).

In addition, the effects of EC-8042 and combo treatments on the targeting of CSC subpopulations were also monitored by flow cytometry using the SORE6 reporter system, which allows dynamic monitoring of CSC subpopulation based on SOX2/OCT4 expression. This system has previously demonstrated its ability to detect CSC subpopulations in different tumor types [39]. Thus, we generated HNSCC-derived cell lines with stable expression of the SORE6 construct (UT-SCC38-SORE6-GFP and FaDu-SORE6-GFP) or their corresponding controls without the SORE6 response element (UT-SCC38-minCMV-GFP and FaDu-minCMV-GFP), which have been used as gating controls in flow cytometry analyses. FaDu-SORE6-GFP and UT-SCC38-SORE6-GFP displayed 28% and 22% of SORE6-positive cells, respectively. As expected for bona fide CSCs subpopulations, SORE-GFP+ subpopulations sorted from both cell lines by flow cytometry showed increased expression of SOX2 and OCT4 and also increased ability to grow as tumorspheres (Supplementary Figure S7A,B). To analyze the effect of the drugs on CSC subpopulations, cultures of SCC38-SORE6-GFP and FaDu-SORE6-GFP cells were treated with dasatinib, EC-8042 and the combination for 72 h. As shown by flow cytometry analysis, EC-8042 and the combined treatment, but not dasatinib alone, significantly decreased the SORE-GFP+ subpopulation in both cell lines (Supplementary Figure S7C,D), thus confirming the ability of EC-8042 alone or in combination with dasatinib to target CSC subpopulations in HNSCC cells.

Importantly, EC-8042 and the combined treatment demonstrated a profound antitumor activity in vivo on FaDu xenografts compared to vehicle- and dasatinib alone-treated groups (Figure 5). Thus, treatment with EC-8042 or EC-8042 plus dasatinib led to a significant reduction of tumor volume growth, presenting TGI percentages of 43% and 61%, respectively, and also a significant increase in survival (Figure 5A,B and Supplementary Figure S8A,B). Likewise, tumor weights in vehicle and dasatinib series doubled those of EC-8042 and combination series (Figure 5C) and even the combination regime was able to produce tumor regression in one case (Supplementary Figure S8A). Notably, none of the treatments caused loss of weight (Supplementary Figure S8C) or any other adverse effects in treated mice. In any case, a two-way ANOVA analysis of the relative tumor volumes at the experimental end point (Figure 5B) or the tumor weights (Figure 5C) show that the interaction between dasatinib and EC-8042 is considered not significant (Supplementary Figure S6B–D) and therefore, no synergistic effect was expected [40] in line with the CI values calculated in in vitro assays.

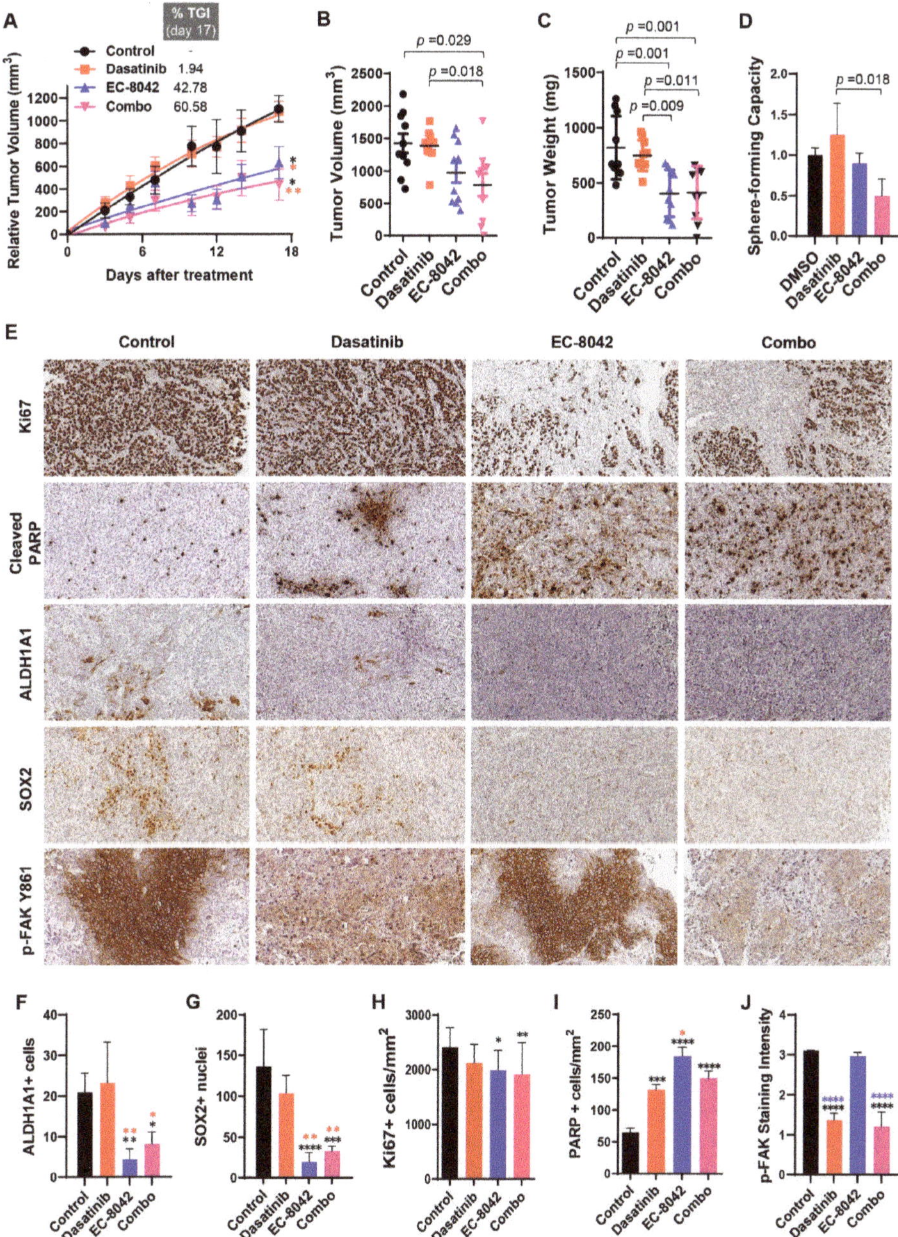

Figure 5. Effect of combination treatment with dasatinib and EC-8042 in FaDu xenografts. Mice with established FaDu xenografts were randomly assigned to four different treatment groups (*n* = 10 tumors per group) and treated with saline buffer intravenously (i.v.) and/or tartaric acid (orally) (control), dasatinib (orally) at a dose of 10 mg/Kg every day (16 doses), EC-8042 (i.v.) at a dose of 50 mg/Kg every 7 days (4 doses), or the dasatinib plus EC-8042 combination (combo). Animals were sacrificed 4 h after

the last treatment with dasatinib and 24 h after treatment with EC-8042. (**A**) Curves representing the mean tumor volume of FaDu xenografts during the treatments. Drug efficacy expressed as the percentage of tumor growth inhibition (%TGI) at the end of the experiment is indicated. Mean ± SD (n = 10 per treatment group), RM-one-way ANOVA, Tukey's test, * $p < 0.05$, ** $p < 0.01$ vs. control (black) and dasatinib-treated mice (red). (**B,C**) Distribution of tumor volumes (**B**) and tumor weights (**C**) at the end of the experiment. Mean ± SD (n = 10 per group), one-way ANOVA, Tukey's test, $p < 0.05$ between the indicated groups. (**D**) For the evaluation of CSC subpopulations after drug treatments, xenograft tumors were harvested and dissociated into single cells to assess the tumorsphere-forming ability after in vivo drug treatment. Mean ± SD (n = 3 per group), one-way ANOVA, Tukey's test. (**E**) Representative images and (**F–J**) graphs of the immunohistochemical analysis of Ki67, cleaved PARP, ALDH1, SOX2, and phospho-FAK (Y861) in paraffin-embedded tumors from FaDu xenografts after in vivo treatment with the indicated drugs. Scale bars = 100 µm. Mean ± SD (n = 5 per group), one-way ANOVA, Tukey's test, * $p < 0.05$, ** $p < 0.01$, *** $p < 0.001$, **** $p < 0.0001$ vs. control (black), dasatinib-(red) or EC-8042-treated (blue) mice.

The ability of drugs to target CSC subpopulations was also examined after in vivo treatments. Thus, a cohort of xenograft tumors was harvested at the end of the treatments, disaggregated into single cells and assayed for tumorsphere formation. Whilst tumorsphere formation slightly increased in dasatinib-treated tumors, we found an inhibition of tumorsphere-forming ability in tumors treated with EC-8042 alone, which reached statistical significance in combination with dasatinib (Figure 5D). Furthermore, immunohistochemical analysis also confirmed a significant decrease in the expression of both CSC markers ALDH1A1 and SOX2 in EC-8042-treated and combo-treated tumors (Figure 5E–G). We also found that EC-8042 and combination treatment led to a significant decrease in the percentage of Ki67-positive cells, as well as a significant increase in apoptotic cell death (cleaved-PARP staining) (Figure 5E,H,I). In addition, FAK (Y861) phosphorylation levels significantly diminished in dasatinib- and combo-treated tumors, as expected (Figure 5E,J). Taken together, these results demonstrate that EC-8042 and combination treatment efficiently targeted CSC properties and showed robust in vivo anti-tumor effects mainly due to reduced cell proliferation and increased apoptotic cell death.

Overall, we show evidence that HNSCC cells treated with a combination of dasatinib and EC-8042 benefit from favorable anti-tumor properties of both drugs while preventing the adverse pro-stemness effects induced by dasatinib. Even though our findings indicate the presence of additive cytotoxic effect in vitro but not in vivo synergistic effect, the combined treatment keeps the ability of dasatinib to abrogate relevant pro-tumor signaling (i.e., SRC/FAK signaling) and abolish invasive potential, while adding the potent capacity of EC-8042 to target CSC subpopulations and inhibit tumor growth, without any noticeable adverse interactions of both agents.

4. Discussion

Dasatinib and saracatinib have proven powerful antitumor activity in preclinical settings but rather limited clinical efficacy in cancer patients [14–16,21,22]. In the context of HNSCC, both drugs have failed to demonstrate any significant activity as single agents in patients with recurrent and/or metastatic disease [24,25]. We herein provide interesting new data uncovering important deleterious activities of dasatinib and saracatinib sustaining stem cell-like properties in HNSCC cell lines, which could be a plausible underlying reason to explain their lack of clinical efficacy as monotherapeutic agents in HNSCC patients.

Complete eradication of tumors requires therapies able to effectively eliminate the CSC subpopulations responsible for treatment resistance, relapse, and metastasis [41–43]. Indeed, we found that treatment of HNSCC cells with EC-8042, a mithramycin analog with reported anti-stemness activity in sarcomas [27], resulted in a potent elimination of both adherent cell cultures (bulk tumor population) and CSC-enriched tumorsphere cultures and demonstrated a robust antitumor activity in vivo in HNSCC xenograft models. EC-8042 was found to induce its anti-oncogenic effects through the inhibition of factors like SP1 [27]. In this regard, it has been demonstrated that SP1 is frequently upregulated in HNSCC and that a combined inhibition of SP1, using mithramycin, and TGFβ pathways induced cell death and prevented HNSCC recurrence [44].

In marked contrast to EC-8042, dasatinib completely abolished cell invasion but induced only a modest anti-proliferative effect in adherent HNSCC cultures. More importantly, dasatinib failed to eliminate CSC-enriched tumorspheres or to prove any significant antitumor activity in FaDu xenografts. In this line, it has been reported that dasatinib worsened the anti-tumor effects in combination with cetuximab or cetuximab and radiation therapy in FaDu-derived tumors [45]. In addition, saracatinib did not demonstrate a significant effect on HNSCC tumor growth in a mouse orthotopic model of tongue squamous cell carcinoma but impaired perineural invasion and cervical lymph node metastasis [46]. Similarly, results from a randomized trial aimed to test the treatment with erlotinib, dasatinib, or combination treatment in patients with operable HNSCC showed that erlotinib but not dasatinib significantly reduced tumor size [47].

The study of the effect of dasatinib and EC-8042 combination in HNSCC models shows non-synergistic but complementary effects, where both drugs exert complementary anti-tumor properties without antagonizing each other. Remarkably, combined treatment of dasatinib with EC-8042 efficiently counteracted the deleterious effects of dasatinib on CSC properties, thus showing potent anti-proliferative, anti-stemness, and anti-invasive effects in HNSCC cells and significant antitumor activity in xenografts. Clonal sphere-forming ability and also expression of well-known CSC markers were consistently and dramatically reduced both in vitro and in vivo by treatment with dasatinib and EC-8042, compared to dasatinib alone. Our results also evidenced additional beneficial effects of combination treatment, such as combined targeting of several factors and signaling pathways which are effectively inhibited by EC8042 (i.e., ALDH1, SOX2, NANOG, NOTCH1) and/or dasatinib (i.e., SRC/FAK/ERK and PI3K/AKT pathways), as well as complementary anti-proliferative, anti-stemness, and anti-invasive effects provided by each compound. It is therefore plausible that combined targeting of these signaling pathways may result in reducing more effectively the CSC subpopulations within the tumors.

The in vivo data clearly indicate that the reduction of tumor volume and weight observed appears to be caused mainly by EC-8042. According to our in vitro data, this is likely due to the potent anti-proliferative and cytotoxic effect of EC-8042 on both bulk tumor cells as well as CSC subpopulations. Contrasting this, dasatinib but not EC-8042 demonstrated robust anti-migratory and anti-invasive properties, and this beneficial effect was also observed by combined treatment in 3D spheroid invasion assays, probably mediated by inhibition of SRC/FAK signaling pathway and/or EMT. Notably, p-FAK Y861 was a major target of dasatinib but not EC-8042. Accordingly, p-FAK Y861 levels were found to robustly diminish by dasatinib or combo treatment both in vitro and in vivo. In marked contrast, p-FAK Y861 levels were not significantly affected by EC-8042 in vivo and only partially reduced in vitro, whereas p-FAK Y397 as well as total FAK levels were mainly targeted and robustly decreased by EC-8042. It seems quite reasonable to infer that this could exert a major influence on the ability of tumors to invade and metastasize. Altogether, these data suggest that patients treated with the combined treatment may benefit from the cytotoxic and anti-stemness effect of EC-8042 together with the prevention of cancer cell migration and invasion due to dasatinib action. These properties could be of particular importance for patients with advanced stages of disease who are at a higher risk of metastatic dissemination.

In contrast to these findings, it has been reported that dasatinib was able to inhibit Sox2 expression and tumorsphere formation in NSCLC cells [48,49]. These differences could be related to differences in cell and/or tissue context, since we consistently observed that dasatinib reduced the phosphorylation levels of SRC and also AKT whilst increasing endogenous Sox2 levels in HNSCC cells and also Sox2/Oct4 reporter activity. This highlights the need to establish accurate markers of drug response and adequate patient stratification to better select the patients who may benefit from the treatment with dasatinib.

5. Conclusions

This study unveils undesirable stem cell-promoting functions by dasatinib and saracatinib that could explain the lack of clinical efficacy of both drugs in HNSCC patients. Supporting this hypothesis, we show that HNSCC cells treated with a combination of dasatinib and EC-8042 benefit from complementary anti-invasive, anti-proliferative and anti-CSC functions. This combination counteracts the adverse pro-stemness effects induced by dasatinib and is therefore suggested as a novel therapeutic strategy for clinical testing in HNSCC patients. According to these data, novel combinational strategies with EC-8042 could contribute to improving treatment efficacy and long-term clinical outcomes in HNSCC patients.

Supplementary Materials: The following are available online at https://zenodo.org/record/3358610#.XUQDjRSYOpo. Supplementary Methods. Supplementary Table S1: Primers used for real-time RT-PCR. Figure S1: Western blot analyses and cell viability of FaDu and UT-SCC38 cells treated with either DMSO (vehicle), 0.1 μM dasatinib or 1 μM saracatinib. Figure S2: Analysis of the expression of the CSC markers ALDH1A1 and SOX2 in CSC-enriched tumorspheres and adherent cultures of FaDu and UT-SCC38 cells by Western-blot and RT-qPCR. Figure S3: Quantification of the infrared fluorescent signals from the Western blot analyses shown in Figures 2C and 4F. Figure S4: Effect of dasatinib and saracatinib on the expression of EMT markers in HNSCC-derived cell lines. Figure S5: Quantification of the infrared fluorescent signals from the Western blot analyses shown in Figure 4A. Figure S6: Analysis of synergism between Dasatinib and EC-8042. Figure S7: Effect of dasatinib, EC-8042 and combination treatment on the targeting of CSC subpopulations in HNSCC cell lines monitored by flow cytometry using the SORE6 reporter system. Figure S8: Pictures of tumors extracted from FaDu xenografts, in vivo survival analysis and mean body weight of mice during the in vivo treatments. Video S1: 3D spheroid invasion assay in FaDu cells treated with DMSO (vehicle). Video S2: 3D spheroid invasion assay in FaDu cells treated with 0.1 μM dasatinib. Video S3: 3D spheroid invasion assay in FaDu cells treated with 0.1 μM EC-8042. Video S4: 3D spheroid invasion assay in FaDu cells treated with 0.1 μM dasatinib plus 0.1 μM EC-8042.

Author Contributions: Conceptualization, R.R. and J.M.G.-P.; Data curation, F.H.-P., R.R. and J.M.G.-P.; Formal analysis, J.A., J.P.-E., F.M., R.R. and J.M.G.-P.; Funding acquisition, J.P.R., R.R. and J.M.G.-P.; Investigation, F.H.-P., M.A.V., R.G.-D., L.S., M.A.H., P.O., E.A., N.d.-R.-I., I.G., J.T. and R.F.; Methodology, F.H.-P., M.A.V., L.S., M.A.H., P.O., E.A., J.A. and I.G.; Resources, L.M.M., F.M. and J.P.R.; Software, J.A.; Supervision, J.M.G.-P.; Validation, R.G.-D., and N.d.-R.-I.; Visualization, F.H.-P., and R.R.; Writing—original draft, J.M.G.-P.; Writing—review & editing, F.H.-P., J.A., J.P.-E., L.M.M., F.M., J.P.R. and R.R.

Funding: This study was supported by grants from the Plan Nacional de I+D+I 2013-2016 [ISCIII (CP13/00013, PI13/00259 and PI16/00280 to JMGP and CPII16/00049 to RR), Red Temática de Investigación Cooperativa en Cáncer (RTICC) (RD12/0036/0015), CIBERONC (CB16/12/00390 to JPR and CB16/12/00443 to LM) and MINECO (SAF2013-42946-R and SAF2016-75286-R to RR)], the Instituto de Investigación Sanitaria del Principado de Asturias (ISPA), the Plan de Ciencia Tecnología e Innovación del Principado de Asturias (GRUPIN14-003 to JPR), Fundación Merck Salud (17-CC-008 to JPR), and the FEDER Funding Program from the European Union.

Acknowledgments: SORE6 lentiviral constructions were kindly donated by Lalage M. Wakefield (National Cancer Institute, Bethesda, MD, USA). We thank the technical assistance from the staff at the Bioterio and the Unidad de Histopatología Animal, IUOPA-Universidad de Oviedo. We also thank Juan Pérez Ortega for his excellent administrative support.

Conflicts of Interest: M.A.H, P.O. and J.P-E are employees of EntreChem SL. F.M. reports employment and ownership of stock in EntreChem SL. All other authors declare they have no competing interests.

References

1. Irby, R.B.; Yeatman, T.J. Role of Src expression and activation in human cancer. *Oncogene* **2000**, *19*, 5636–5642. [CrossRef] [PubMed]

2. Masaki, T.; Igarashi, K.; Tokuda, M.; Yukimasa, S.; Han, F.; Jin, Y.J.; Li, J.Q.; Yoneyama, H.; Uchida, N.; Fujita, J.; et al. pp60c-src activation in lung adenocarcinoma. *Eur. J. Cancer* **2003**, *39*, 1447–1455. [CrossRef]

3. Mandal, M.; Myers, J.N.; Lippman, S.M.; Johnson, F.M.; Williams, M.D.; Rayala, S.; Ohshiro, K.; Rosenthal, D.I.; Weber, R.S.; Gallick, G.E.; et al. Epithelial to mesenchymal transition in head and neck squamous carcinoma: association of Src activation with E-cadherin downregulation down-regulation, vimentin expression, and aggressive tumor features. *Cancer* **2008**, *112*, 2088–2100. [CrossRef] [PubMed]

4. Talamonti, M.S.; Roh, M.S.; Curley, S.A.; Gallick, G.E. Increase in activity and level of pp60c-src in progressive stages of human colorectal cancer. *J. Clin. Investig.* **1993**, *91*, 53–60. [CrossRef] [PubMed]

5. Summy, J.M.; Gallick, G.E. Src family kinases in tumor progression and metastasis. *Cancer Metastasis Rev.* **2003**, *22*, 337–358. [CrossRef] [PubMed]

6. Yeatman, T.J. A renaissance for SRC. *Nat. Rev. Cancer* **2004**, *4*, 470–480. [CrossRef] [PubMed]

7. Parsons, S.J.; Parsons, J.T. Src family kinases, key regulators of signal transduction. *Oncogene* **2004**, *23*, 7906–7909. [CrossRef] [PubMed]

8. Elsberger, B.; Stewart, B.; Tatarov, O.; Edwards, J. Is Src a viable target for treating solid tumours? *Curr. Cancer Drug Targets* **2010**, *10*, 683–694. [CrossRef]

9. Alvarez, R.H.; Kantarjian, H.M.; Cortes, J.E. The role of Src in solid and hematologic malignancies: Development of new-generation Src inhibitors. *Cancer* **2006**, *107*, 1918–1929. [CrossRef]

10. Avizienyte, E.; Frame, M.C. Src and FAK signaling controls adhesion fate and the epithelial-to-mesenchymal transition. *Curr. Opin. Cell Biol.* **2005**, *17*, 542–547. [CrossRef]

11. Schlaepfer, D.D.; Jones, K.C.; Hunter, T. Multiple Grb2-mediated integrin-stimulated signaling pathways to ERK2/mitogen-activated protein kinase: Summation of both c-Src and focal adhesion kinase initiated tyrosine phosphorylation events. *Mol. Cell. Biol.* **1998**, *18*, 2571–2585. [CrossRef] [PubMed]

12. Chang, J.H.; Gill, S.; Settleman, J.; Parsons, S.J. c-Src regulates the simultaneous rearrangement of actin cytoskeleton, p190RhoGAP, and p120RasGAP following epidermal growth factor stimulation. *J. Cell Biol.* **1995**, *130*, 355–368. [CrossRef] [PubMed]

13. Wang, W.; Liu, Y.; Liao, K. Tyrosine phosphorylation of cortactin by the FAK-Src complex at focal adhesions regulates cell motility. *BMC Cell Biol.* **2011**, *12*, 49. [CrossRef] [PubMed]

14. Creedon, H.; Brunton, V.G. Src kinase inhibitors: Promising cancer therapeutics? *Crit. Rev. Oncog.* **2012**, *17*, 145–159. [CrossRef] [PubMed]

15. Schuetze, S.M.; Bolejack, V.; Choy, E.; Ganjoo, K.N.; Staddon, A.P.; Chow, W.A.; Tawbi, H.A.; Samuels, B.L.; Patel, S.R.; von Mehren, M.; et al. Phase 2 study of dasatinib in patients with alveolar soft part sarcoma, chondrosarcoma, chordoma, epithelioid sarcoma, or solitary fibrous tumor. *Cancer* **2017**, *123*, 90–97. [CrossRef] [PubMed]

16. Scott, A.J.; Song, E.K.; Bagby, S.; Purkey, A.; McCarter, M.; Gajdos, C.; Quackenbush, K.S.; Cross, B.; Pitts, T.M.; Tan, A.C.; et al. Evaluation of the efficacy of dasatinib, a Src/Abl inhibitor, in colorectal cancer cell lines and explant mouse model. *PLoS ONE* **2017**, *12*, e0187173. [CrossRef] [PubMed]

17. Gore, L.; Kearns, P.R.; de Martino, M.L.; Lee.; De Souza, C.A.; Bertrand, Y.; Hijiya, N.; Stork, L.C.; Chung, N.G.; Cardos, R.C.; et al. Dasatinib in Pediatric Patients With Chronic Myeloid Leukemia in Chronic Phase: Results From a Phase II Trial. *J. Clin. Oncol.* **2018**, *36*, 1330–1338. [CrossRef] [PubMed]

18. Ocana, A.; Gil-Martin, M.; Martín, M.; Rojo, F.; Antolín, S.; Guerrero, Á.; Trigo, J.M.; Muñoz, M.; Pandiella, A.; Diego, N.G.; et al. A phase I study of the SRC kinase inhibitor dasatinib with trastuzumab and paclitaxel as first line therapy for patients with HER2-overexpressing advanced breast cancer. GEICAM/2010-04 study. *Oncotarget* **2017**, *8*, 73144–73153. [CrossRef] [PubMed]

19. Montemurro, M.; Cioffi, A.; Dômont, J.; Rutkowski, P.; Roth, A.D.; von Moos, R.; Inauen, R.; Toulmonde, M.; Burkhard, R.O.; Knuesli, C.; et al. Long-term outcome of dasatinib first-line treatment in gastrointestinal stromal tumor: A multicenter, 2-stage phase 2 trial (Swiss Group for Clinical Cancer Research 56/07). *Cancer* **2018**, *124*, 1449–1454. [CrossRef] [PubMed]

20. Kalinsky, K.; Lee, S.; Rubin, K.M.; Lawrence, D.P.; Iafrarte, A.J.; Borger, D.R.; Margolin, K.A.; Leitao, M.M., Jr.; Tarhini, A.A.; Koon, H.B.; et al. A phase 2 trial of dasatinib in patients with locally advanced or stage IV mucosal, acral, or vulvovaginal melanoma: A trial of the ECOG-ACRIN Cancer Research Group (E2607). *Cancer* **2017**, *123*, 2688–2697. [CrossRef] [PubMed]

21. Parseghian, C.M.; Parikh, N.U.; Wu, J.Y.; Jiang, Z.Q.; Henderson, L.; Tian, F.; Pastor, B.; Ychou, M.; Raghav, K.; Dasari, A.; et al. Dual Inhibition of EGFR and c-Src by Cetuximab and Dasatinib Combined with FOLFOX Chemotherapy in Patients with Metastatic Colorectal Cancer. *Clin. Cancer Res.* **2017**, *23*, 4146–4154. [CrossRef] [PubMed]

22. Johnson, F.M.; Saigal, B.; Talpaz, M.; Donato, N.J. Dasatinib (BMS-354825) tyrosine kinase inhibitor suppresses invasion and induces cell cycle arrest and apoptosis of head and neck squamous cell carcinoma and non-small cell lung cancer cells. *Clin. Cancer Res.* **2005**, *11*, 6924–6932. [CrossRef] [PubMed]

23. Green, T.P.; Fennell, M.; Whittaker, R.; Curwen, J.; Jacobs, V.; Allen, J.; Logie, A.; Hargreaves, J.; Hickinson, D.M.; Wilkinson, R.W.; et al. Preclinical anticancer activity of the potent, oral Src inhibitor AZD0530. *Mol. Oncol.* **2009**, *3*, 248–261. [CrossRef] [PubMed]

24. Brooks, H.D.; Glisson, B.S.; Bekele, B.N.; Johnson, F.M.; Ginsberg, L.E.; El-Naggar, A.; Culotta, K.S.; Takebe, N.; Wright, J.; Tran, H.T.; et al. Phase 2 study of dasatinib in the treatment of head and neck squamous cell carcinoma. *Cancer* **2011**, *117*, 2112–2119. [CrossRef] [PubMed]

25. Fury, M.G.; Baxi, S.; Shen, R.; Kelly, K.W.; Lipson, B.L.; Carlson, D.; Stambuk, H.; Haque, S.; Pfister, D.G. Phase II study of saracatinib (AZD0530) for patients with recurrent or metastatic head and neck squamous cell carcinoma (HNSCC). *Anticancer Res.* **2011**, *31*, 249–253.

26. Núñez, L.E.; Nybo, S.E.; González-Sabín, J.; Pérez, M.; Menéndez, N.; Braña, A.F.; Shaaban, K.A.; He, M.; Morís, F.; Salas, J.A.; et al. A novel mithramycin analogue with high antitumor activity and less toxicity generated by combinatorial biosynthesis. *J. Med. Chem.* **2012**, *55*, 5813–5825. [CrossRef]

27. Tornin, J.; Martinez-Cruzado, L.; Santos, L.; Rodriguez, A.; Núñez, L.E.; Oro, P.; Hermosilla, M.A.; Allonca, E.; Fernández-García, M.T.; Astudillo, A.; et al. Inhibition of SP1 by the mithramycin analog EC-8042 efficiently targets tumor initiating cells in sarcoma. *Oncotarget* **2016**, *7*, 30935–30950. [CrossRef]

28. Lansford, C.D.; Grenman, R.; Bier, H.; Somers, K.D.; Kim, S.Y.; Whiteside, T. Head and neck cancers. In *Human Cell Culture*; Masters, J.R.W., Plasson, B., Eds.; Kluwer Academic Press: Dordrecht, The Netherlands, 1999; pp. 185–255.

29. Alvarez-Teijeiro, S.; Menéndez, S.T.; Villaronga, M.A.; Rodrigo, J.P.; Manterola, L.; de Villalaín, L.; de Vicente, J.C.; Alonso-Durán, L.; Fernández, M.P.; Lawrie, C.H.; et al. Dysregulation of Mir-196b in Head and Neck Cancers Leads to Pleiotropic Effects in the Tumor Cells and Surrounding Stromal Fibroblasts. *Sci. Rep.* **2017**, *7*, 17785. [CrossRef]

30. Villaronga, M.A.; Hermida-Prado, F.; Granda-Diaz, R.; Menendez, S.T.; Alvarez-Teijeiro, S.; Quer, M.; Vilaseca, I.; Allonca, E.; Garzón-Arango, M.; Sanz-Moreno, V.; et al. Immunohistochemical Expression of Cortactin and Focal Adhesion Kinase Predicts Recurrence Risk and Laryngeal Cancer Risk Beyond Histologic Grading. *Cancer Epidemiol. Biomarkers Prev.* **2018**, *27*, 805–813. [CrossRef]

31. Menéndez, S.T.; Rodrigo, J.P.; Alvarez-Teijeiro, S.; Villaronga, M.Á.; Allonca, E.; Vallina, A.; Astudillo, A.; Barros, F.; Suárez, C.; García-Pedrero, J.M.; et al. Role of HERG1 potassium channel in both malignant transformation and disease progression in head and neck carcinomas. *Mod. Pathol.* **2012**, *25*, 1069–1078. [CrossRef]

32. Martinez-Cruzado, L.; Tornin, J.; Santos, L.; Rodriguez, A.; García-Castro, J.; Morís, F.; Rodriguez, R. Aldh1 Expression and Activity Increase During Tumor Evolution in Sarcoma Cancer Stem Cell Populations. *Sci. Rep.* **2016**, *6*, 27878. [CrossRef] [PubMed]

33. Koo, B.S.; Lee, S.H.; Kim, J.M.; Huang, S.; Kim, S.H.; Rho, Y.S.; Bae, W.J.; Kang, H.J.; Kim, Y.S.; Moon, J.H.; et al. Oct4 is a critical regulator of stemness in head and neck squamous carcinoma cells. *Oncogene* **2015**, *34*, 2317–2324. [CrossRef] [PubMed]

34. Thierauf, J.; Veit, J.A.; Hess, J. Epithelial-to-mesenchymal transition in the pathogenesis and therapy of head and neck cancer. *Cancers* **2017**, *9*, 76. [CrossRef] [PubMed]

35. Canel, M.; Serrels, A.; Frame, M.; Brunton, V.G. E-cadherin-integrin crosstalk in cancer invasion and metastases. *J. Cell Sci.* **2013**, *126*, 393–401. [CrossRef] [PubMed]

36. Lamouille, S.; Xu, J.; Derynck, R. Molecular mechanisms of epithelial-mesenchymal transition. *Nat. Rev. Mol. Cell Biol.* **2014**, *15*, 178–196. [CrossRef] [PubMed]

37. Tornin, J.; Hermida-Prado, F.; Padda, R.S.; Gonzalez, M.V.; Alvarez-Fernandez, C.; Rey, V.; Martinez-Cruzado, L.; Estupiñan, O.; Menendez, S.T.; Fernandez-Nevado, L.; et al. FUS-CHOP Promotes Invasion in Myxoid Liposarcoma through a SRC/FAK/RHO/ROCK-Dependent Pathway. *Neoplasia* **2018**, *20*, 44–56. [CrossRef] [PubMed]

38. Chou, T.C. Theoretical basis, experimental design, and computerized simulation of synergism and antagonism in drug combination studies. *Pharmacol. Rev.* **2006**, *58*, 621–681. [CrossRef]

39. Tang, B.; Raviv, A.; Esposito, D.; Flanders, K.C.; Daniel, C.; Nghiem, B.T.; Garfield, S.; Lim, L.; Mannan, P.; Robles, A.I.; et al. A Flexible Reporter System for Direct Observation and Isolation of Cancer Stem Cells. *Stem Cell Rep.* **2015**, *4*, 155–169. [CrossRef]

40. Slinker, B.K. The statistics of synergism. *J. Mol. Cell. Cardiol.* **1998**, *30*, 723–731. [CrossRef]

41. Li, Y.; Rogoff, H.A.; Keates, S.; Gao, Y.; Murikipudi, S.; Mikule, K.; Leggett, D.; Li, W.; Pardee, A.B.; Li, C.J. Suppression of cancer relapse and metastasis by inhibiting cancer stemness. *Proc. Natl. Acad. Sci. USA* **2015**, *112*, 1839–1844. [CrossRef]

42. He, K.; Xu, T.; Goldkorn, A. Cancer cells cyclically lose and regain drug-resistant highly tumorigenic features characteristic of a cancer stem-like phenotype. *Mol. Cancer Ther.* **2011**, *10*, 938–948. [CrossRef]

43. Reya, T.; Morrison, S.J.; Clarke, M.F.; Weissman, I.L. Stem cells, cancer, and cancer stem cells. *Nature* **2001**, *414*, 105–111. [CrossRef]

44. Citron, F.; Armenia, J.; Franchin, G.; Polesel, J.; Talamini, R.; D'Andrea, S.; Sulfaro, S.; Croce, C.M.; Klement, W.; Otasek, D.; et al. An Integrated Approach Identifies Mediators of Local Recurrence in Head and Neck Squamous Carcinoma. *Clin. Cancer Res.* **2017**, *23*, 3769–3780. [CrossRef]

45. Baro, M.; de Llobet, L.I.; Figueras, A.; Skvortsova, I.; Mesia, R.; Balart, J. Dasatinib worsens the effect of cetuximab in combination with fractionated radiotherapy in FaDu- and A431-derived xenografted tumours. *Br. J. Cancer* **2014**, *111*, 1310–1318. [CrossRef]

46. Ammer, A.G.; Kelley, L.C.; Hayes, K.E.; Evans, J.V.; Lopez-Skinner, L.A.; Martin, K.H.; Frederick, B.; Rothschild, B.L.; Raben, D.; Elvin, P.; et al. Saracatinib Impairs Head and Neck Squamous Cell Carcinoma Invasion by Disrupting Invadopodia Function. *J. Cancer Sci. Ther.* **2009**, *1*, 52–61. [CrossRef]

47. Bauman, J.E.; Duvvuri, U.; Gooding, W.E.; Rath, T.J.; Gross, N.D.; Song, J.; Jimeno, A.; Yarbrough, W.G.; Johnson, F.M.; Wang, L.; et al. Randomized, placebo-controlled window trial of EGFR, Src, or combined blockade in head and neck cancer. *JCI Insight* **2017**, *2*, e90449. [CrossRef]

48. Singh, S.; Trevino, J.; Bora-Singhal, N.; Coppola, D.; Haura, E.; Altiok, S.; Chellappan, S.P. EGFR/Src/Akt signaling modulates Sox2 expression and self-renewal of stem-like side-population cells in non-small cell lung cancer. *Mol. Cancer* **2012**, *11*, 73. [CrossRef]

49. Bhummaphan, N.; Pongrakhananon, V.; Sritularak, B.; Chanvorachote, P. Cancer Stem Cell Suppressing Activity of Chrysotoxine, a bibenzyl from Dendrobium pulchellum. *J. Pharmacol. Exp. Ther.* **2018**, *364*, 332–346. [CrossRef]

Journal of
Clinical Medicine

Article

Lipid Droplets Define a Sub-Population of Breast Cancer Stem Cells

Benjamin J. Hershey [†], Roberta Vazzana [†], Débora L. Joppi and Kristina M. Havas [*]

IFOM Foundation, FIRC Institute of Molecular Oncology, Via Adamello 16, 20139 Milan, Italy;
benjamin.hershey@ifom.eu (B.J.H.); Roberta.vazzana@ifom.eu (R.V.); deborajoppi@gmail.com (D.L.J.)
* Correspondence: kristina.havas@ifom.eu; Tel.: +39-0257-430-3278
† These authors contributed equally to this work.

Received: 29 October 2019; Accepted: 18 December 2019; Published: 29 December 2019

Abstract: Tumor recurrence is now the leading cause of breast cancer-related death. These recurrences are believed to arise from residual cancer stem cells that survive initial therapeutic intervention. Therefore, a comprehensive understanding of cancer stem cell biology is needed to generate more effective therapies. Here we investigate the association between dysregulation of lipid metabolism and breast cancer stem cells. Focusing specifically on lipid droplets, we found that the lipid droplet number correlates with stemness in a panel of breast cell lines. Using a flow cytometry-based method developed for this study, we establish a means to isolate cells with augmented lipid droplet loads from total populations and show that they are enriched in cancer stem cells. Furthermore, pharmacological targeting of fatty acid metabolism reveals a metabolic addiction in a subset of cell lines. Our results highlight a key role for the lipid metabolism in the maintenance of the breast cancer stem cell pool, and as such, suggest it as a potential therapeutic target.

Keywords: cancer stem cells; breast cancer; lipid; metabolism; therapeutic resistance

---

1. Introduction

Despite advancements in early detection, the World Health Organization reports that over 0.5 million women still succumb to breast cancer every year. The majority of these deaths are attributed to tumor recurrences, which are largely believed to arise from residual cancer stem cells (CSC) that survive the initial therapeutic intervention [1]. According to the CSC model of tumorigenesis, this population of cells is responsible for the origin, progression, and recurrence of the tumor, and therefore, for any therapy to have success, it must be able to effectively target this population [2]. Despite their central role in the development of malignant disease, CSCs remain poorly characterized. The characterization of this population has been hampered, in part, by the lack of robust markers for their identification. Thus, highlighting the necessity to identify more robust markers and therapeutic strategies for CSCs.

Given their central role in promoting tumor progression, the identification and characterization of breast CSCs remain an active area of research. To date, most rely upon the combinatorial expression of cell surface markers such as the cluster of differentiation (CD) 24 and CD44 for the identification of CSC populations [3]. Recently, metabolic stem cell markers, such as cytoplasmic aldehyde dehydrogenase (ALDH) A1 and A3 have been described as markers of adult stem cells and CSC in a number of tissue types [4,5]. Aldehydes are generated by the metabolism of a wide variety of xenobiotic and endobiotic compounds, including alcohols, amino acids, and anticancer drugs, as well as from lipid peroxidation. Aldehyde dehydrogenases function as detoxifying enzymes through their role in metabolizing aldehydes, thereby protecting cells from oxidative and electrophilic stress [6]. The finding that CSCs can, in part, be identified by a metabolic marker has stimulated an interest in the characterization of CSC metabolism.

In order to meet the increased biochemical demands that accompany increased proliferation, metabolic pathways are frequently dysregulated in cancer. The metabolic reprogramming that accompanies cancer onset is now understood to be essential for the pathogenesis of the disease and accordingly has been added to the list of cancer hallmarks [7,8]. Of the metabolic alterations reported in cancer, none is better studied than aerobic glycolysis, the production of lactate from glucose in the presence of oxygen [9,10]. Glycolytic metabolism has been shown to play an important role in supporting stemness in several cancer types. However, it is becoming increasingly clear that in addition to high glycolytic rates, tumorigenesis is supported by multiple aberrant metabolic processes, including dysregulation of lipid metabolism [11].

A coordinated dysregulation of lipid metabolism is observed in nearly all cancer types. In addition to fulfilling the basic requirements of structural lipids for membrane synthesis, lipids play important roles as signaling molecules and contribute significantly to energy homeostasis. As lipid metabolism affects multiple aspects of cellular biology, it is not surprising that alterations in lipid metabolism affect a diverse range of cellular processes including, growth, proliferation, differentiation, and motility.

In order to fulfill their heightened demand for lipids, cancer cells increase their uptake of exogenous fatty acids. This is achieved through increased surface expression of fatty acid translocase CD36. Interestingly, elevated CD36 expression has been reported in metastasis-initiating cell populations and is inversely correlated with survival prognosis [12]. Once taken up by the cells, the free fatty acids can be shuttled into the mitochondria to produce energy equivalents through fatty acid beta-oxidation(FAO). Maintenance of cellular energy stores by FAO has been shown to be fundamental to the survival of CSCs in scarce nutrient environments [13]. Multiple reports have highlighted the importance of increased exogenous fatty acid metabolism in CSCs [14].

In addition to the increased uptake of exogenous lipids, cancer cells also have the unique property of being able to synthesize lipids. Whilst fatty acid biosynthesis is normally restricted to hepatocytes, adipocytes, or mammary epithelium during lactation, several studies have now demonstrated the ability of cancer cells to preform de novo fatty acid biosynthesis [15,16]. Indeed, the expression of several enzymes involved in de novo fatty acid biosynthesis has been implicated in tumorigenesis and the maintenance of stemness in CSCs [17,18].

On a cellular level, excess fatty acids, either exogenous fatty acids taken up by the cells or products of de novo fatty acid biosynthesis, are processed into triacylglycerides and retained in specialized storage organelles called lipid droplets [19]. Lipid droplets are endoplasmic reticulum (ER)-derived organelles that are comprised of a phospholipid monolayer surrounding a core of neutral lipids, primarily triglycerides and sterol esters. They have long been considered to function as a primary store of energy but several recent reports have begun to highlight their role in a diverse range of cellular functions including ER stress, ROS detoxification, and protein dynamics [20].

Our goal in this study is to assess lipid metabolism as a therapeutic target in a panel of breast cancer cell lines. We report that stemness in breast cancer cell lines correlates with lipid droplet number. In line with this, we devised a FACS based strategy to isolate lipid droplet enriched populations and analyzed them for CSC markers. This analysis revealed that lipid droplet[hi] populations increase CSC markers, as well as increasing mammosphere-forming efficiency. Utilizing an inhibitor of fatty acid biosynthesis, we were able to effectively target the stem cell population in a subset of breast cancer cell lines, thus demonstrating the potential of lipid metabolism targeting compounds as adjuvants to traditional therapies.

2. Experimental Section

2.1. Cell Lines and Culture Conditions

In this study, the following human breast cancer cell lines were used: BT474, MCF7, T47D, and MDA-MB-231. BT474 cells were purchased from the Leibniz Institute DSMZ-German Collection of Microorganisms and Cell Cultures and maintained in Roswell Park Memorial Institute, RPMI 1640,

(Lonza, Basel, Switzerland) supplemented with 5% fetal bovine serum (Sigma-Aldrich, St. Louis, MO, USA), 2 mM L-Glutamine (Euroclone, Milano, Italy), and 10 µg/mL human insulin (Sigma-Aldrich, St. Louis, MO, USA). MCF7, T47D, and MDA-MB 231 were purchased from the National Cancer Institute, Bethesda, MD, USA (NCI) and maintained in RPMI 1640 supplemented with 10% fetal bovine serum and 2 mM L-Glutamine. All the cell lines were cultured at 37 °C in a humidified atmosphere, 5% CO_2 incubator.

Additionally, two cell lines, representative of normal breast epithelium, HMEC (Cambrex, East Rutherford, NJ, USA) and MCF10a (European Institute of Oncology, Milan, Italy) were used. HMEC cells were cultured in RPMI 1640 with 10% FBS and 2 mM L-glutamine. MCF10a cells were cultured in Dulbecco's modified Eagle medium/nutrient mixture F-12 (DMEM-Ham's F12), (Sigma Aldrich, St. Louis, MO, USA) containing: 5% horse serum (Life Technologies, Carlsbad, CA, USA), 10 µg/mL insulin, 20 ng/mL epidermal growth factor (Vinci Biochem, Florence, Italy), 500 ng/mL hydrocortisone (Sigma Aldrich, St. Louis, MO, USA), 100 ng/mL cholera toxin (Sigma Aldrich, St. Louis, MO, USA) and 2 mM L-glutamine.

2.2. 5-(Tetradecyloxy)-2-Furoic Acid (TOFA) Treatments

TOFA (Sigma-Aldrich, T6575, St. Louis, MO, USA) was resuspended in two milliliters of Dimethyl sulfoxide (DMSO) and stored at −20 °C. Experiments in which cells were treated with TOFA, unless otherwise specified, were cultured in the presence of 10 µM TOFA in their normal growth media for forty-eight hours prior to initiating measurements or assays.

2.3. Growth Curves for TOFA Sensitivity

To assess the effects of TOFA on cell proliferation each cell line was seeded across five, 96-well plates (Costar™, Corning, NY, USA) at the following cell densities, expressed as cells per well: MDA-MB-231 0.5×10^4, MCF7 1×10^4, BT474 and T47D 1.5×10^4. Additional wells were filled with growth media alone to act as plate blanks. All experimental conditions were set up in triplicate. Upon cell adhesion and spreading, a single plate was taken from each set of five plates, fixed and stained.

Fixation was done by removing the culture media and adding 100 µL of 4% paraformaldehyde solution, PFA, (HIMEDIA, Kennett Square, PA, USA) to each well for ten minutes. The PFA was then removed and 100 µL of Crystal Violet solution (Merck KGaA, Darmstadt, Germany) was added to each well. After twenty minutes, all non-cell associated Crystal Violet was removed by washing the wells with water.

Cells in the remaining sets of four plates were treated with either growth media containing 0.13% DMSO or growth media with 10 µM TOFA (Sigma-Aldrich, St. Louis, MO, USA) in DMSO. Every 24 h, for 96 h, a single plate was removed from each set and processed as described above.

Once in the entire time course, five plates in total per cell line were processed and completely dried. One hundred µL of acetic acid glacial was added to each well and incubated for twenty minutes at room temperature. The log optical density (OD) of each experimental well was then measured at a wavelength of 570 nm using a Wallace Victor 3™ multilabel reader (PerkinElmer Inc., Waltham, MA, USA). Reported OD values were generated by subtracting the average of the three experimental wells from the average of the three corresponding plate blank wells.

2.4. Fluorescence-Activated Cell Sorting

BODIPY™ 500/510 C_1, C_{12} (4,4-Difluoro-5-Methyl-4-Bora-3a,4a-Diaza-s-Indacene-3-Dodecanoic Acid) (Molecular Probes, Eugene, OR, USA) stock solution, 1 mg/mL, was prepared in ethanol and kept at −20 °C until used. For FACS sorting, the cells were incubated with 1 µg/mL BODIPY™, following an overnight incubation; they were washed twice, trypsinized, resuspended in Phosphate Buffered Saline(PBS) containing 2% FBS and 0.3% Gentamycin and sorted on a MoFlo Astrios cell sorter (Beckman Coulter, Pasadena, CA, USA).

2.5. FACS of Fatty Acid Loaded and TOFA Treated Cells

FACS analysis of fatty acid loaded cells was done by culturing cells to 30–40% confluence. The normal growth media was then spiked with 100 µM oleic acid (Sigma-Aldrich, O1383, St. Louis, MO, USA) and 50 µM palmitic acid (Sigma-Aldrich, P5585, St. Louis, MO, USA) for 48 h. Twelve hours prior to FACS, cells were treated with Bodipy at 1 µg/mL. Immediately before FACS, cells were then trypsinized and spun at 1200 rpm for five minutes. The pellet was gently washed with Dulbecco's Phosphate Buffered Saline (DPBS) and resuspended in DPBS containing 2%, 0.22 µm filtered, FBS. FACS was conducted using Attune NxT (Thermofisher Scientific Inc., Waltham, MA, USA). FACS data depicted represents analysis done on single, 4′,6-diamindino-2-phenylindole(DAPI)-negative cell populations. FlowJo version 10.4.2 (BD Life Sciences, Franklin Lakes, NJ, USA) was used for the analysis.

The same workflow, with the exception of the fatty acid loading step, was used to assess the impact of TOFA treatment on the lipid droplet content of the experimental cell lines.

2.6. Nile Red Staining and Lipid Droplets Quantification Analysis

The Nile Red staining was performed as previously reported [21]. Briefly, Nile red (Molecular Probes, Eugene, OR, USA) stock solution was made in DMSO at a concentration of 1 mg/mL. For cellular staining, cells were seeded on poly-lysine coated coverslips and then fixed in paraformaldehyde, 4%, for fifteen minutes. After washing, the cells were incubated for ten minutes in Nile Red, 1 µL of 1 mg/mL Nile Red stock in 10 mL of 150 mM NaCl, protected from light. Nuclei were stained using Dapi, and finally, coverslips were mounted onto glass slides.

Lipid droplet quantification was performed using a Fiji plug-in (developed by Martini E.). Briefly, after manually delineating an Region of Interest (ROI) around the cells, the plugin identified lipid droplets using the ImageJ's Find Maxima^2 algorithm on the maximum projection image after background removal (using the rolling ball algorithm^3) and noise reduction (with a median filter).

2.7. RNA Extraction and Quantitative RT-PCR Analysis

In order to perform the RT-PCR analysis, total RNA was first isolated using a trizol-chloroform extraction with TRIzol reagent (Life Technologies, Carlsbad, CA, USA) and chloroform. For the cells collected following cell sorting, RNA was extracted using the RNeasy mini kit (Qiagen, Hilden, Germany). After the extraction, the RNA was quantified by NanoDrop to assess both concentration and quality. Reverse transcription was performed using the SuperScript III reverse transcriptase kit (Invitrogen, Carlsbad, CA, USA). Gene expression was analyzed using the TaqMan gene expression analysis. The samples were amplified with primers for each gene; β-actin was used as the housekeeping gene. The primer assay IDs used in these experiments were: ACTB, Hs99999903_m1; SOX2, Hs01053049_s1; POU5F1, Hs00742896_s1; KLF4, Hs00358836_m1; ALDH1A1, Hs00946916_m1; ALDH1A3, Hs00167476_m1.

2.8. Mammosphere Formation Assay from Cell Lines

The primary mammospheres were produced as previously described [22]. The mammosphere media used in this assay was DMEM-F12 (Biowest, Nuaillé, France) supplemented with B27 supplement (Invitrogen, Carlsbad, CA, USA) and EGF 20 ng/mL (Vinci Biochem, Florence, Italy). To prepare non-adherent plates, standard 6-well plates were coated with 1 ml of 1% agarose solution in PBS. To produce the 1% agarose solution, agarose powder was dissolved in PBS and autoclaved.

Briefly, cells were centrifuged at 580× *g* for two minutes and resuspended in 2 mL PBS. To avoid cellular aggregates, a 22 G needle was used to syringe the cell suspension. We found out that 1×10^4 cells/well was a good seeding density for our cell lines. Cells were plated and incubated in a 5% CO_2 humidified incubator at 37 °C. After five days, all mammospheres larger than 50 µm were counted and the mammosphere formation efficiency (MFE) was calculated using the following formula:

mammosphere forming efficiency (%) = (number of mammospheres per well/number of cells seeded per well) × 100.

2.9. Assessment of Lipid Droplet Content Using CD44/CD24 Stem Cell Markers

MDA-MB-231 and BT474 cell lines were cultured in 6-well plates (Falcon®, Ref no. 353046, Corning, NY, USA). On the night prior to FACS analysis, cells were treated with BODIPY™ 500/510 C_1, C_{12}, as described in Section 2.4. Following incubation with BODIPY™ 500/510 C_1, C_{12}, the cells were harvested and incubated in 500 μL of a 1× DPBS, 5% BSA, blocking buffer for forty-five minutes at room temperature. The cells were then stained with Alexa Fluor® 647 mouse anti-human CD24 (BD Pharmingen, Material No. 561644, San Jose, CA, USA) and CD44-VioBlue® mouse anti-human CD44 (Miltenyi Biotec, Order No. 130-113-899, Bergisch Gladbach, Germany) for thirty minutes on ice. The antibody concentrations recommended on the accompanying data sheets were used for the stain. Following staining, the cells were pelleted and washed three times with a 1× DPBS, 1% BSA solution, prior to resuspension in a 1% FBS, 1× DPBS solution. The FACS was conducted using the Attune NxT (Thermofisher Scientific Inc., Waltham, MA, USA). FACS data depicted represents analysis done on single, propidium iodide negative, cell population. FlowJo version 10.4.2 (BD Life Sciences, Franklin Lakes, NJ, USA) was used for the analysis.

2.10. Fatty Acid Oxidation Assay

MDA-MB-231, MCF7, T47D, and BT474 cell lines were seeded into 96-well plates (Costar™, Corning, NY, USA) at 7×10^4 cells per well and treated with either the vehicle or 10 μM TOFA in DMSO. After approximately twenty hours, the cells were assessed using a fatty acid oxidation assay (Abcam, ab217602, Cambridge, United Kingdom) used in conjunction with an extracellular O_2 consumption assay (Abcam, ab197243, Cambridge, UK). The protocols accompanying the assays were followed to assess the cell lines after TOFA treatment. Experimental measurements were made using a Wallac Envision™ 2104 multilabel reader (Perkin-Elmer, Waltham, MA, USA), maintained at 37 °C throughout the course of the experiment. Excitation filter, UV (TRF) 340 and emission filter APC665 were used to assess the status of the oxygen-sensing probe used for the assay. Measurements of the oxygen-sensing probe were made every 90 s for one and a half hours.

2.11. Transmitted Light and Fluorescence Microscopy

Mammosphere images were acquired with an EVOS FL imaging system (Thermo Fisher Scientific, Inc., Waltham, MA, USA) transmitted light microscope. Fluorescent images were acquired with laser-scanning confocal microscopes: Leica TCS SP5 laser confocal scanner mounted on a Leica DMI 6000B inverted microscope equipped with motorized stage and HCX PL APO 63X/1.4NA oil immersion objective (Leica Mikrosysteme Vertrieb GmbH, Wetzlar, Germany) and Leica TCS SP2 AOBS laser confocal scanner mounted on a Leica DM IRE2 inverted microscope equipped with HCX PL APO 63X/1.4NA oil immersion objective (Leica Mikrosysteme Vertrieb GmbH, Wetzlar, Germany). For the excitation of fluorochromes dyes, 405, 488, 561, and 633 nm laser lines were used on Leica TCS SP5 and Leica TCS SP2 AOBS. The following settings were maintained for fluorescent images acquisition: digital zoom 2.5 and a 1024 × 1024 scan format.

2.12. Kaplan-Meier Plotter

Kaplan–Meier plots were generated using the Kaplan–Meier plotter found at http://kmplot.com/analysis/index.php?p=background [23]. This is an online platform that enables the user to assess the effect of 54,000 genes on survival in 21 cancer types. Prognostic values for PLIN2 mRNA (Affymetrix ID 209122_at) expression was evaluated for a cohort of 3951 breast cancer patients.

2.13. Statistical Analysis

All experiments were carried out at least three times unless otherwise indicated. Data were analyzed using GraphPad Prism version 8 statistical software (GraphPad Software, San Diego, CA, USA). Experimental results are reported as mean and standard deviation unless otherwise stated.

3. Results

3.1. Lipid Droplet Marker PLIN2 Expression Correlates with Disease Progression and Lipid Droplet Number

An increase in lipid droplet metabolism in therapeutic resistant cell populations has now been reported for several solid tumors, including breast [24,25]. This led us to ask whether there was a correlation between lipid droplets and disease progression in breast cancer. As a proxy for lipid droplets, we chose to use a member of the perilipin family, Perilipin 2 (PLIN2), a lipid-droplet associated structural protein. The over-expression of PLIN2 has previously been reported for several cancer types, although its prognostic value is not clear. Therefore, we asked whether PLIN2 over-expression has any prognostic value in breast cancer. Querying the Kaplan–Meier plotter [23], we have found a significant negative correlation between PLIN2 expression and relapse-free survival in breast cancer patients (Figure 1a). After stratifying for intrinsic breast cancer sub-type, we observed that the negative correlation was maintained in basal, luminal A and luminal B subtypes, although interestingly not in Her2+ breast cancers (Figure S1a).

Next, we set out to investigate whether PLIN2 expression correlates with an increase in lipid droplet number across a panel of breast cancer cell lines. Cell lines were selected to represent the major intrinsic subtypes as classified by their immuno-histological features are represented in Table 1.

We utilized 9-diethylamino-5H-benzo [alpha] phenoxazine-5-one (Nile Red) to assess the lipid droplet content across the panel (Figure 1b). Quantification of the baseline lipid droplet numbers in these cells revealed that although all breast cancer cell lines queried contain lipid droplets, BT474, and MB-MDA-231 had a significantly higher number of lipid droplets in comparison to the other cell lines tested (Figure 1c). Of note, we also quantified lipid droplet number in two control cell lines, MCF10a and HMECs. Despite being models of "healthy mammary epithelium," both cell lines had an elevated number of lipid droplets in comparison to the cancer cell lines (Figure S1b).

It has yet to be determined whether the increase in lipid droplets that is associated with therapeutic resistance is a consequence of a stress response or if it represents an enrichment of a therapy-resistant sub-clonal population. Considering the CSC theory of therapeutic resistance, we were prompted to ask whether there could be a correlation between lipid droplets and stemness. We began by evaluating the expression of stemness markers utilizing quantitative PCR. We found that the baseline, normalized values, of markers previously published to associated with CSCs including Octamer-binding transcription factor 4 (POUF51), SRY(Sex-determining region Y)-box2 (SOX2), ALDH1A3, and Kruppel -like factor 4 (KLF4) did not show a significant variation in between the cell lines tested (Figure S1c). Notably, the expression of ALDH1A1, which was also included in the panel, was limited to BT474. Breast CSCs can be enriched by culturing in non-adherent, non-differentiating culture conditions, where they grow as small clusters of CSC derived cells referred to as mammospheres. To ask whether there were discernible differences in CSC pools between the four cell lines, we performed mammosphere formation efficiency assays. Despite similar overall levels in the expression of stem cell markers, we observed clear variations in mammosphere formation efficiency between the four cell lines (Figures 1d and S1d).

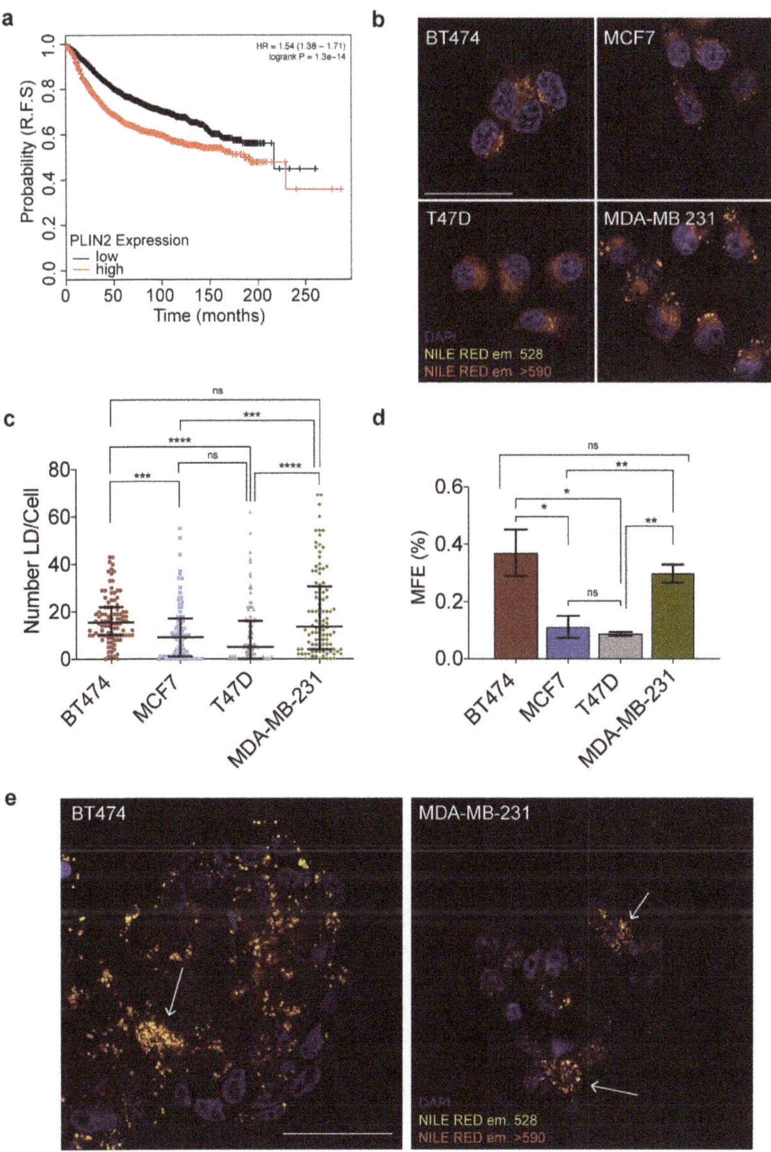

Figure 1. Lipid droplets correlate with breast cancer progression and stemness. (**a**) The effect of PLIN2 expression on relapse-free survival in breast cancer (**b**) Confocal imaging of Nile Red stained BT474, MCF7, T47D, and MDA-MB-231. The scale bar represents 50 μm. (**c**) Quantification of lipid droplets number/cell for 100 cells per cell line (right). (**d**) Mammosphere formation efficiency for the four cell lines used in this study. (**e**) Confocal imaging of mammospheres stained with Nile Red, arrows indicate lipid droplet enriched cells, scale bar represents 50 μm. Data are represented as mean ± SEM. Significance was calculated using two-tailed *t*-tests. * $p < 0.05$, ** $p < 0.01$, *** $p < 0.001$, **** $p < 0.0001$.

Table 1. Molecular classification of breast cancer cell lines.

Classification	ER	PR	HER2	Example Cell Lines [26,27]
Luminal A	+	+/-	-	**MCF7, T47D**, SUM185
Luminal B	+	+/-	+	**BT474**, MDA-MB-361
Claudin-low	-	-	-	**MDA-MB-231**, BT549
Basal	-	-	-	MDA-MB-468, SUM 190
HER2	-	-	+	SKBR3, MDA-MB-453

ER, estrogen receptor; PR, progesterone receptor; HER2, human epidermal growth factor receptor 2. The cell lines indicated in bold were used in this study.

The mammosphere formation assay suggested a correlation between lipid droplet number and stemness, but we now wanted to demonstrate that lipid droplet enriched populations are indeed found in mammospheres. Therefore we passaged mammospheres produced from MDA-MB-231 and BT474 for two generations and then visualized the lipid droplet containing cells using Nile Red. Interestingly, we found a sub-population of cells within the mammospheres to be highly enriched in lipid droplets (Figures 1e and S1e). Intriguingly, taken together, these observations suggested a link between lipid droplet load and stemness in breast cancer-derived cell lines.

3.2. Lipid Droplets Increase Cell Complexity and Provide a Means to Isolate Individual Populations

These observations led us to ask whether CSCs were enriched in the population of cells harboring high numbers of lipid droplets. To address this question, we needed a means to separate lipid droplet-enriched from lipid droplet-depleted populations. Having shown that the accumulation of lipid droplets within the cell is tractable through the use of lipid droplet specific dye Nile Red, we reasoned that we could utilize cell fluorescent fatty acid analogs to sort lipid droplet^{hi} populations from lipid droplet^{lo} populations.

In order to develop a florescenence-activated cell sorting (FACS) based approach for cell sorting, we needed a live cell marker of lipid droplets; we chose to use BODIPY™ 500/510(BODIPY™). BODIPY™ is a fluorescent fatty acid analog that can be used as a marker of lipid droplets and membranes [28]. To understand whether BODIPY™ would be suitable for a FACS based assay, we incubated cells in growth media containing BODIPY™ and evaluated the resulting fluorescence signal using confocal microscopy. Image analysis confirmed that the BODIPY™ was being taken up by the cells and incorporated into the neutral lipid species (triglycerides or sterol esters) that are stored in cytoplasmic lipid droplets. Additionally, membrane incorporation, whereas visible, was not the major contributor to the signal (Figure 2a).

We then proceeded to test whether we could use a FACS-based strategy to identify lipid droplet containing populations using BODIPY™. FACS-based analyses of the BODIPY™ loaded cells versus non-loaded controls revealed a strong shift in the fluorescence signal following excitation at 488 nm for all cell lines (Figure 2b). In order to demonstrate that the BODIPY™ signal being read by the FACS is attributable to the lipid droplet concentration, we tested the system by pre-incubating the cells in media containing additional fatty acids. The addition of exogenous fatty acids to the media resulted in their uptake and metabolism leading to a dramatic increase in lipid droplet number (Figure S2a).

Accordingly, we observed not only a significant increase in BODIPY™ signal in cells that had been pre-loaded with fatty acids but also a marked shift in the side scatter (SSC) measurements of the cells (Figure 2b). SSC at FACS is used as an indicator of the internal complexity or granularity of the cell. These parameters are often used to measure internal cellular components such as granules. Interestingly, increasing the number of lipid droplets also resulted in an increase in the SSC measurements for the treated population. Therefore, like granules, lipid droplets increase the complexity of the cytosol, in a way that was distinguishable at FACS. Taken together, the BODIPY™ driven FACS based strategy provided a robust method to stratify cells based on lipid droplet content.

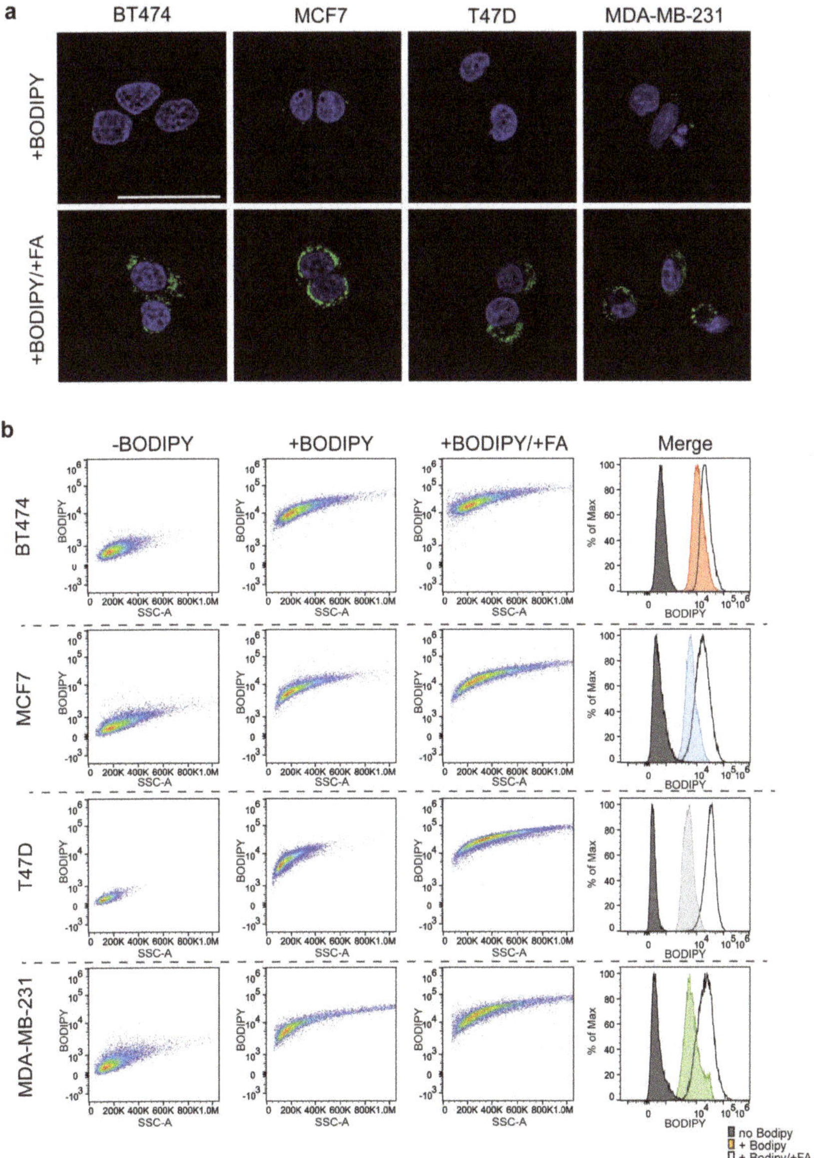

Figure 2. BODIPY^{TM} based FACS protocol distinguishes lipid droplet enriched cells. (**a**) Confocal imaging of BODIPY^{TM} 500,510 loaded cells. Upper panels are representative images from each of the four cell lines following a 12 h incubation with BODIPY^{TM}, lower panel representative image of cells incubated with palmitic acid, oleic acid and BODIPY^{TM}. The scale bar represents 50 μm. (**b**) Representative plots of BODIPY^{TM} signal in cells, − BODIPY^{TM}, + BODIPY^{TM,} and fatty acid loaded cells + BODIPY^{TM}. The corresponding histograms represent the FITC (BODIPY^{TM}) signal for each condition.

3.3. *Lipid Droplets Correlate with Stemness in a Subset of Breast Cancers*

The development of FACS based protocols for the identification and isolation of stem cells has been instrumental in shaping our understanding of their contribution to development and pathogenesis. In the breast, the cell population distinguished by CD24$^{lo/-}$/CD44^{hi} has been previously shown to be highly enriched in CSCs both in primary patient material as well as in established breast cancer cell lines [3,29]. The establishment of the BODIPY^{TM} based FACS protocol described above now allowed us to investigate whether CSCs populations, defined by CD24/CD44 levels, contain high levels of lipid droplets. In agreement with our previous data, we observed a 2.5–2.7 fold increase BODIPY^{TM} intensity in the stem enriched CD24$^{lo/-}$/CD44^{hi} pool (Figures 3a and S3). In addition to the observed increase in the BODIPY^{TM} signal, we also detected increased cytosolic complexity (SSC) in the CD24$^{lo/-}$/CD44^{hi} population. This is in accordance with the previous experiments that demonstrated that increasing lipid droplet load led to an increase in SSC. These results further strengthened the earlier observation that lipid droplet number correlates with stemness.

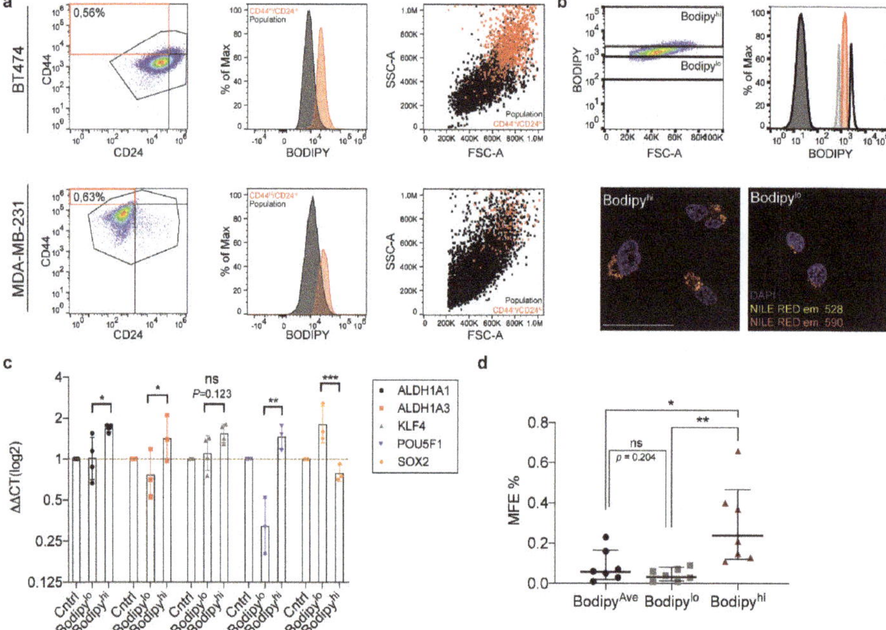

Figure 3. Lipid droplet^{hi} populations are enriched for stem cell markers in BT474 (**a**) FACS-based analysis of CD44^{hi}/CD24^{lo} populations in BT474 and MDA-MB-231. Scatter plot showing the distribution of the two surface markers (left), the histogram of MFI for BODIPY^{TM} in CD44^{hi}/CD24^{lo} (red) versus total population (black; middle), the dot plot to highlight SSC distribution of CD44^{hi}/CD24^{lo} (red) versus total population (black). (**b**) Scatter plot for FITC (BODIPY^{TM}) versus FSC illustrating the gating strategy for FACS sort, with histogram for FITC (BODIPY^{TM}) levels in sorted BT474 populations. Confocal microscopy of Nile Red stained sorted populations. The scale bar represents 50 μm. (**c**) Quantitative PCR to analyze the expression of a panel of genes associated with stemness for the three sorted populations from BT474. Data are represented as geometric mean ± SD (**d**) Mammosphere formation efficiency for the three sorted populations from BT474 ($n = 7$). Data are represented as geometric mean ± SD. Significance was calculated using the two-tailed *t*-test. * $p < 0.05$, ** $p < 0.01$, *** $p < 0.001$.

Having established multiple correlations between lipid droplet number and stemness features, we now wanted to demonstrate that populations of cells containing higher lipid droplet numbers were indeed enriched in CSCs. In order to isolate lipid droplet enriched populations from the total population, we applied the FACS based strategy described above. BT474 was chosen due to its propensity to show enrichment in lipid droplet number and mammosphere-forming efficiency. Three separate bins were created to enable sorting of the top (lipid droplet^{hi}) 5% and bottom (lipid droplet^{lo}) 5% of cells based upon the BODIPY™ fluorescence intensity. Cells were additionally collected from the average (lipid droplet^{ave}) BODIPY™ population to serve as a control group (Figure 3b). Following sorting, cells were assessed for lipid droplet number by fluorescence microscopy. Nile Red was again employed to visualize lipid droplets in the sorted cells. The staining confirmed enrichment of lipid droplets in the lipid droplet^{hi} population.

We proceeded to characterize the three populations isolated from the BT474 cell line for stemness traits. Quantitative PCR was performed for a panel of CSC markers. Interestingly, with the exception of SOX2, we saw a marked increase across all stem cell markers in the lipid droplet^{hi} population (Figure 3c), suggesting that the lipid droplet^{hi} population of BT474 was enriched for CSCs. Next, we tested the three isolated populations in mammosphere formation assays. The results demonstrated significant enrichment in mammosphere forming capacity in the lipid droplet^{hi} population (Figure 3d). Taken together, there is strong evidence that lipid droplet enrichment correlates with CSC traits in the BT474 breast cancer cell line.

3.4. Targeting Lipid Metabolism Directly Impact CSCs

Having observed an enhancement in the stem cell pool in lipid droplet^{hi} populations, we reasoned that if we were able to effectively target this population, we should affect the overall fitness of the cell lines. In order to target the metabolic pathways that lead to the accumulation of lipid droplets, we choose to use 5-tetradecyloxy-2-furoic acid (TOFA), an inhibitor of acetyl-CoA carboxylase-α(ACCA). Through inhibition of ACCA, the rate-limiting enzyme in long fatty acid biosynthesis, TOFA has been shown to effectively block fatty acid biosynthesis, therefore limiting FA and lipid droplet accumulation (Figure 4a and [30]). Therefore, we asked whether exposure to TOFA would impact lipid droplet level, cellular proliferation, and/or stemness in our cell line panel.

We began by evaluating the effect of TOFA on lipid metabolism and total lipid droplet number in our panel of cell lines. First, we evaluated the effects of TOFA treatment on FAO. With the exception of MDA-MB-231, we found a significant decrease in FAO following 24 h of TOFA treatment (Figure S4a). This could be attributed to the observed decrease in lipid droplets that were detected in the cells following treatment with 10 uM TOFA (Figure 4b). For a more quantitative evaluation of the effect of TOFA on the lipid droplet number across our cell lines, we used our previously established BODIPY FACS protocol. This analysis confirmed the efficacy of TOFA in reducing the lipid droplets in the cells post-treatment (Figure 4c). In line with the FAO and imaging data, MDA-MB-231 did not show a significant reduction in lipid droplets following TOFA treatment, suggesting that MDA-MB-231 is not reliant upon de novo fatty acid biosynthesis to meet its fatty acid requirements. Interestingly, BT474 did not have a significant shift in BODIPY^{TM} signal following sorting. This was unexpected given the clear differences seen both in the FAO levels and in the imaging. Despite the lack of change in the overall BODIPY^{TM} signal, we noticed a shift in cellular distribution in the SSC/FSC plots, which suggested an increase in cell size following treatment. This increase in cell size, with the accompanying increase in BODIPY^{TM} labeled cellular membranes, could account for the lack of difference observed in the histograms.

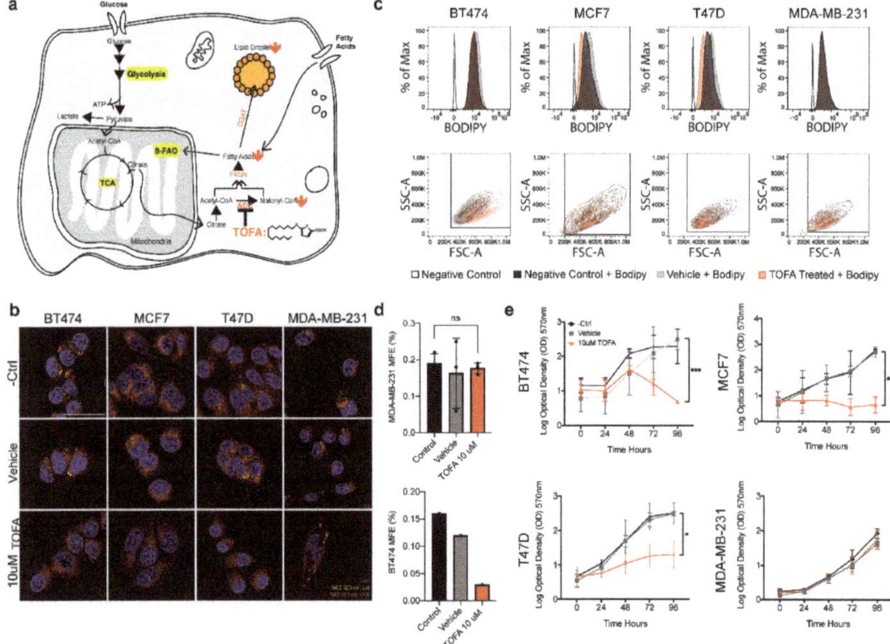

Figure 4. TOFA treatment reveals lipid metabolic addiction in a panel of breast cancer cell lines. (**a**) Schematic depicting the effects of TOFA treatment on lipid metabolism. (**b**) Confocal microscopy of lipid droplets in the cell lines used in this study. Lipid droplets were visualized for control, vehicle, and following 48 h of 10 µM TOFA treated with Nile Red. The scale bar represents 50 µm. (**c**) FACS profile of cells treated with media, vehicle or TOFA for 48 h. From top to bottom, BODIPYTM signal across the treatment groups, side scatter vs. forward scatter plots across the treatment groups. (**d**) Second generation mammosphere forming efficiency for MDA-MB-231 and BT474 following pre-treatment with TOFA, vehicle or regular growth media ($n = 2$). (**e**) Growth of each cell line over time in the presence of 10 µM TOFA, vehicle or growth media. Significance was calculated using the two-tailed *t*-test. * $p < 0.05$, ** $p < 0.01$, *** $p < 0.001$.

Having demonstrated that lipid droplet containing populations were enriched in CSCs, we asked whether treatment with TOFA was sufficient to deplete the CSC populations in the cell lines that were shown to respond to treatment. In order to address this, we performed a qPCR analysis on a panel of stem cell-associated genes in control and treated populations. Despite the global effects on lipid droplet number observed for BT474, T47D and MCF7, TOFA treatment only had marginal effects on the expression of stemness related genes in MCF7 and T47D, while affecting BT474 more significantly (Figure S4b). For a functional validation of these observations, we evaluated mammosphere forming efficiency for both BT474 and MDA-MB-231 for the following conditions: control, vehicle-treated and TOFA treated. Following 5 days of acute TOFA treatment, the surviving populations of BT474 and MDA-MB-231 were washed in TOFA-free growth media to remove any residual TOFA. They were then singularized and seeded in low adhesion, low differentiation conditions in the absence of TOFA and evaluated for mammosphere formation efficiency (Figures 4d and S4c). In line with the qPCR data and imaging data, we observed a significant decrease in second-generation mammosphere-forming capacity for BT474, while MDA-MB-231 had no marked effect on mammosphere generation following TOFA treatment.

The observation that TOFA acts as a potent inhibitor of lipid metabolism and stemness in a subset of cell lines prompted us to evaluate its impact upon proliferation. To do this, we monitored cell growth for 96 h across three treatment groups: growth media, vehicle, and 10 uM TOFA. With the exception of MDA-MB-231, TOFA treatment had a significant impact on the growth of all the cell lines in our panel, severely impeding proliferation (Figure 4d). Although the mammosphere formation assay suggested that TOFA negatively affected the stem cell pool in BT474, these results suggest a more global addiction to the ACCA-mediated fatty acid metabolism that may extend beyond the stem cell pool in these lines.

4. Discussion

The resurgence of interest in tumor metabolism has largely focused on the dysregulation of metabolic pathways in the primary tumor itself. However, there is increasing evidence that the metabolic hallmarks of singular populations could play a role in determining their response to therapeutic challenges. Indeed, over the past five years, numerous reports have emerged suggesting that alterations in lipid metabolism correlate with resistance [24,31]. The mechanisms by which dysregulated lipid metabolism contribute to resistance still remain largely unknown. In this study, we investigated the relationship between lipid metabolism and stemness traits by interrogating a panel of breast cancer cell lines.

One of the hallmarks of therapeutic resistant populations is an observed increase in cytoplasmic lipid droplets. Lipid droplets are the lipid storage organelle of the cell, in which neutral lipids such as triglycerides and sterol esters are sequestered. They are dynamic organelles that arise from pools of lipids within the ER in response to stress or alterations in metabolism. The first association of lipid droplets with tumor cells was first made over 50 years ago [32]. Recently, lipid droplets have been shown to correlate with stemness in colorectal cancer [33]. Despite that, surprisingly few have focused on the use of the lipid droplet as a biomarker in breast cancer.

The protein component of lipid droplets could be used as a proxy and facilitate their inclusion in histopathological analyses. In this study, we have focused on PLIN2. PLIN2 is a member of the perilipin family of proteins that includes PLIN2, -3, -4, and -5. This family of lipid-droplet associated proteins plays an instrumental role in lipid droplet formation, stability and trafficking [19,34]. Querying the Kaplan–Meier plotter, we observed a significant negative correlation between lipid droplet associated protein PLIN2 expression level and relapse-free survival in breast cancer patients [23]. This observation stimulated us to better understand the association. One of the outstanding questions in the field is whether the lipid droplets that have been observed in therapeutic resistant cell populations are the result of de novo synthesis in the face of a therapeutic challenge, or if they are an enrichment of a previously metabolically distinct population of cells. The finding that PLIN2 expression in the primary tumor correlates with a worse prognosis suggests that at least in a subset of patients, pre-existing alterations in lipid metabolism could correlate with disease progression.

Interestingly, evaluation of a panel of breast cancer cell lines revealed distinct differences in lipid droplet number between the cell lines. This is despite being maintained in similar media conditions, at comparable confluency. It is interesting to note, as beautifully cataloged in a recent study, that despite having been maintained for 40-plus years in culture, in non-physiological conditions, the metabolic profiles of these cell lines have retained some significant differences, suggesting perhaps a metabolic addiction [35]. In this regard, it is important to highlight that even within individual cell lines, we observed a high degree of heterogeneity in terms of lipid droplet loading within the population. This variation could reflect alterations in the cell cycle, or intriguingly, it could be a marker of metabolically distinct clonal populations, among which CSCs within the cell lines themselves.

One of the most elusive, yet highly sought-after contributors to heterogeneity is the population of CSCs. The CSC theory of tumor progression proposes that it is this population that resists therapeutic intervention, and acts as the seed for tumor recurrence. Therefore, efficient targeting of this population would lead to higher therapeutic efficacy and better long-term survival rates. However, for many tumor types, we still lack robust markers to enable the identification and characterization of the CSC population. Interestingly, we observed a strong correlation between lipid droplet number and stemness in the cell lines we used in this study.

Having devised a method to sort the population based upon lipid droplet number, we were able to demonstrate an increase in mammosphere forming capacity and stem cell markers in the lipid droplet[hi] segment of the population. An additional characteristic of lipid droplet containing cells which emerged from this study was the correlation between lipid droplets and increased cytosolic complexity, which was detected as increased SSC at FACS. The demonstration that lipid droplets contribute to increased side scatter provides an interesting stain-free mechanism to distinguish lipid droplet[hi] populations. Taken together, this data suggests that inclusion of lipid droplets into the current criteria for CSC identification could provide an additional, low-cost parameter for stem cell identification.

We also demonstrate that this feature of CSC metabolism can be exploited in therapeutic settings by targeting lipid metabolism. In line with previous reports for CRC, exposure to TOFA, an ACAA targeting drug, profoundly impacted upon lipid droplet persistence in our panel of cell lines [36]. Treatment with TOFA severely affected proliferation in three out of the four lines tested. In the case of BT474, for which we have shown a strong correlation between lipid droplet abundance and stemness, TOFA mediated lipid droplet depletion also resulted in a decrease in the expression of several markers of stemness. An interesting observation for further studies is the cell line MDA-MB-231 that appeared to be largely resistant to treatment by TOFA, as seen by the lack of impact upon lipid droplets number as quantified at FACS. In line with this, we observed no effect of TOFA treatment on viability or stemness in the MDA-MB-231, despite observing a drastic effect in all other cell lines. This data suggests that MDA-MB-231 is able to fulfill its fatty acid demands by other means. As TOFA specifically targets de novo fatty biosynthesis, it would be informative to perform a more inclusive study, using a panel of inhibitors targeting the various druggable components of the lipid droplet assembly cascade. However, despite the lack of efficacy on MDA-MB-231, the efficacy of TOFA in inhibiting proliferation of the other cell lines used in this study suggests that it would be interesting to explore the inclusion of lipid metabolic targeting compounds into current therapeutic regimes.

5. Conclusions

In this study we demonstrated a correlation between lipid droplet number and stemness features in a panel of breast cancer cell lines. Yet, our observations were limited to steady-state growth conditions. There is ample literature addressing the induction of lipid droplet accumulation in response to cellular stressors, including chemotherapeutic administration. Therefore, it would be critical to understand if and how lipid droplet biosynthesis is dysregulated following therapeutic exposure. For example, do lipid droplets accumulate in previously deplete populations? If so, what are the stress response pathways governing the expansion? Are lipid droplets dispensable for survival? It is our hope that the answers to these questions will reveal novel lipid metabolic targets in malignancy specific pathways.

Supplementary Materials: The following are available online at http://www.mdpi.com/2077-0383/9/1/87/s1, Figure S1: Lipid droplets correlate with stemness in breast cancer cell lines, Figure S2: BT474 and T47D cell lines internalize fatty acids and are able to form lipid droplets in a dose dependent manner, Figure S3: Gating Strategy for CD44hi/CD24low in BT474 and MDA-MB-231 cell lines, Figure S4: Inhibition of de novo fatty acid biosynthesis effects stemness traits in BT474.

Author Contributions: K.M.H. designed the research, and experiments were performed by B.J.H., R.V., and D.L.J.; all authors contributed to the preparation of the manuscript. All authors have read and agreed to the published version of the manuscript.

Funding: This research was supported by Fondazione Fiera (R.V.), Climbers Against Cancer (B.J.H.), Erasmus Plus (D.L.J.).

Acknowledgments: We thank I. Costa, F. Casagrande of the IFOM Imaging Technical Unit for assistance with imaging. We thanks Emanuele Grande of the IFOM Imaging Technical Unit for the development of the plug-in for assessing lipid droplet number. We thank Maria Grazia Totaro and Arianna Quintè of the IFOM Imaging Technical Unit for assistance with the FACS sort. In addition we would like to thank Laura Tizzoni and Valentina Dall'Olio for the technical support for the real time PCR analysis.

Conflicts of Interest: The authors declare no conflict of interest.

References

1. Ajani, J.A.; Song, S.; Hochster, H.S.; Steinberg, I.B. Cancer stem cells: The promise and the potential. *Semin. Oncol.* **2015**, *42*, S3–S17. [CrossRef] [PubMed]
2. Recasens, A.; Munoz, L. Targeting Cancer Cell Dormancy. *Trends Pharmacol. Sci.* **2019**, *40*, 128–141. [CrossRef] [PubMed]
3. Al-Hajj, M.; Wicha, M.S.; Benito-Hernandez, A.; Morrison, S.J.; Clarke, M.F. Prospective identification of tumorigenic breast cancer cells. *Proc. Natl. Acad. Sci. USA* **2003**, *100*, 3983–3988. [CrossRef] [PubMed]
4. Ricardo, S.; Vieira, A.F.; Gerhard, R.; Leitão, D.; Pinto, R.; Cameselle-Teijeiro, J.F.; Milanezi, F.; Schmitt, F.; Paredes, J. Breast cancer stem cell markers CD44, CD24 and ALDH1: Expression distribution within intrinsic molecular subtype. *J. Clin. Pathol.* **2011**, *64*, 937–946. [CrossRef] [PubMed]
5. Ginestier, C.; Hur, M.H.; Charafe-Jauffret, E.; Monville, F.; Dutcher, J.; Brown, M.; Jacquemier, J.; Viens, P.; Kleer, C.G.; Liu, S.; et al. ALDH1 is a marker of normal and malignant human mammary stem cells and a predictor of poor clinical outcome. *Cell Stem Cell* **2007**, *1*, 555–567. [CrossRef]
6. Singh, S.; Brocker, C.; Koppaka, V.; Chen, Y.; Jackson, B.C.; Matsumoto, A.; Thompson, D.C.; Vasiliou, V. Aldehyde dehydrogenases in cellular responses to oxidative/electrophilic stress. *Free Radic. Biol. Med.* **2013**, *56*, 89–101. [CrossRef]
7. Hanahan, D.; Weinberg, R.A. Hallmarks of cancer: The next generation. *Cell* **2011**, *144*, 646–674. [CrossRef]
8. Ward, P.S.; Thompson, C.B. Metabolic reprogramming: A cancer hallmark even warburg did not anticipate. In *Cancer Cell*; Elsevier Inc.: Amsterdam, The Netherlands, 2012; Volume 21, pp. 297–308.
9. Vander Heiden, M.G.; Cantley, L.C.; Thompson, C.B. Understanding the warburg effect: The Metabolic requirements of cell proliferation. *Science* **2009**, *324*, 1029–1033. [CrossRef]
10. Warburg, O. On respiratory impairment in cancer cells. *Science* **1956**, *124*, 269–270.
11. Currie, E.; Schulze, A.; Zechner, R.; Walther, T.C.; Farese Jr, R.V. Cellular fatty acid metabolism and cancer. *Cell Metab.* **2013**, *18*, 153–161. [CrossRef]
12. Pascual, G.; Avgustinova, A.; Mejetta, S.; Martín, M.; Castellanos, A.; Attolini, C.S.-O.; Berenguer, A.; Prats, N.; Toll, A.; Hueto, J.A.; et al. Targeting metastasis-initiating cells through the fatty acid receptor CD36. *Nature* **2016**, *541*, 41. [CrossRef] [PubMed]
13. Carracedo, A.; Cantley, L.C.; Pandolfi, P.P. Cancer metabolism: Fatty acid oxidation in the limelight. *Nat. Rev. Cancer* **2013**, *13*, 227–232. [CrossRef] [PubMed]
14. Wang, T.; Fahrmann, J.F.; Lee, H.; Li, Y.J.; Tripathi, S.C.; Yue, C.; Zhang, C.; Lifshitz, V.; Song, J.; Yuan, Y.; et al. JAK/STAT3-regulated fatty acid beta-oxidation is critical for breast cancer stem cell self-renewal and chemoresistance. *Cell Metab.* **2018**, *27*, 1357. [CrossRef] [PubMed]
15. Medes, G.; Thomas, A.; Weinhouse, S. Metabolism of neoplastic tissue. IV. A Study of lipid synthesis in neoplastic tissue slices In Vitro. *Cancer Res.* **1953**, *13*, 27–29. [PubMed]
16. Menendez, J.A.; Lupu, R. Fatty acid synthase and the lipogenic phenotype in cancer pathogenesis. *Nat. Rev. Cancer* **2007**, *7*, 763–777. [CrossRef] [PubMed]
17. Brandi, J.; Dando, I.; Pozza, E.D.; Biondani, G.; Jenkins, R.; Elliott, V.; Park, K.; Fanelli, G.; Zolla, L.; Costello, E.; et al. Proteomic analysis of pancreatic cancer stem cells: Functional role of fatty acid synthesis and mevalonate pathways. *J. Proteom.* **2017**, *150*, 310–322. [CrossRef] [PubMed]
18. Yasumoto, Y.; Miyazaki, H.; Vaidyan, L.K.; Kagawa, Y.; Ebrahimi, M.; Yamamoto, Y.; Ogata, M.; Katsuyama, Y.; Sadahiro, H.; Suzuki, M.; et al. Inhibition of fatty acid synthase decreases expression of stemness markers in glioma stem cells. *PLoS ONE* **2016**, *11*, e0147717. [CrossRef]
19. Olzmann, J.A.; Carvalho, P. Dynamics and functions of lipid droplets. *Nat. Rev. Mol. Cell Biol.* **2019**, *20*, 137–155. [CrossRef]
20. Welte, M.A.; Gould, A.P. Lipid droplet functions beyond energy storage. *Biochim. Biophys. Acta Mol. Cell Biol. Lipids* **2017**, *1862*, 1260–1272. [CrossRef]
21. Listenberger, L.L.; Brown, D.A. Fluorescent detection of lipid droplets and associated proteins. *Curr. Protoc. Cell Biol.* **2007**, *35*, 24.2.1–24.2.11. [CrossRef]

22. Shaw, F.L.; Harrison, H.; Spence, K.; Ablett, M.P.; Simoes, B.M.; Farnie, G.; Clarke, R.B. A detailed mammosphere assay protocol for the quantification of breast stem cell activity. *J. Mammary Gland Biol. Neoplasia* **2012**, *17*, 111–117. [CrossRef] [PubMed]
23. Gyorffy, B.; Lanczky, A.; Szallasi, Z. Implementing an online tool for genome-wide validation of survival-associated biomarkers in ovarian-cancer using microarray data from 1287 patients. *Endocr. Relat. Cancer* **2012**, *19*, 197–208. [CrossRef] [PubMed]
24. Havas, K.M.; Milchevskaya, V.; Radic, K.; Alladin, A.; Kafkia, E.; Garcia, M.; Stolte, J.; Klaus, B.; Rotmensz, N.; Gibson, T.J.; et al. Metabolic shifts in residual breast cancer drive tumor recurrence. *J. Clin. Investig.* **2017**, *127*, 2091–2105. [CrossRef] [PubMed]
25. Viale, A.; Pettazzoni, P.; Lyssiotis, C.A.; Ying, H.; Sanchez, N.; Marchesini, M.; Carugo, A.; Green, T.; Seth, S.; Giuliani, V.; et al. Oncogene ablation-resistant pancreatic cancer cells depend on mitochondrial function. *Nature* **2014**, *514*, 628. [CrossRef]
26. Neve, R.M.; Chin, K.; Fridlyand, J.; Yeh, J.; Baehner, F.L.; Fevr, T.; Clark, L.; Bayani, N.; Coppe, J.P.; Tong, F.; et al. A collection of breast cancer cell lines for the study of functionally distinct cancer subtypes. *Cancer Cell* **2006**, *10*, 515–527. [CrossRef]
27. Prat, A.; Parker, J.S.; Karginova, O.; Fan, C.; Livasy, C.; Herschkowitz, J.I.; He, X.; Perou, C.M. Phenotypic and molecular characterization of the claudin-low intrinsic subtype of breast cancer. *Breast Cancer Res. BCR* **2010**, *12*, R68. [CrossRef]
28. Kasurinen, J. A novel fluorescent fatty acid, 5-methyl-BDY-3-dodecanoic acid, is a potential probe in lipid transport studies by incorporating selectively to lipid classes of BHK cells. *Biochem. Biophys. Res. Commun.* **1992**, *187*, 1594–1601. [CrossRef]
29. Liu, S.; Cong, Y.; Wang, D.; Sun, Y.; Deng, L.; Liu, Y.; Martin-Trevino, R.; Shang, L.; McDermott, S.P.; Landis, M.D.; et al. Breast cancer stem cells transition between epithelial and mesenchymal states reflective of their normal counterparts. *Stem Cell Rep.* **2014**, *2*, 78–91. [CrossRef]
30. Halvorson, D.L.; McCune, S.A. Inhibition of fatty acid synthesis in isolated adipocytes by 5-(tetradecyloxy)-2-furoic acid. *Lipids* **1984**, *19*, 851–856. [CrossRef]
31. Shafi, A.A.; Putluri, V.; Arnold, J.M.; Tsouko, E.; Maity, S.; Roberts, J.M.; Coarfa, C.; Frigo, D.E.; Putluri, N.; Sreekumar, A.; et al. Differential regulation of metabolic pathways by androgen receptor (AR) and its constitutively active splice variant, AR-V7, in prostate cancer cells. *Oncotarget* **2015**, *6*, 31997–32012. [CrossRef]
32. Apffel, C.A.; Baker, J.R. Lipid droplets in the cytoplasm of malignant cells. *Cancer* **1964**, *17*, 176–184. [CrossRef]
33. Tirinato, L.; Liberale, C.; Di Franco, S.; Candeloro, P.; Benfante, A.; La Rocca, R.; Potze, L.; Marotta, R.; Ruffilli, R.; Rajamanickam, V.P.; et al. Lipid droplets: A new player in colorectal cancer stem cells unveiled by spectroscopic imaging. *Stem Cells* **2015**, *33*, 35–44. [CrossRef] [PubMed]
34. Gao, Q.; Goodman, J.M. The lipid droplet-a well-connected organelle. *Front. Cell Dev. Biol.* **2015**, *3*, 49. [CrossRef] [PubMed]
35. Li, H.; Ning, S.; Ghandi, M.; Kryukov, G.V.; Gopal, S.; Deik, A.; Souza, A.; Pierce, K.; Keskula, P.; Hernandez, D.; et al. The landscape of cancer cell line metabolism. *Nat. Med.* **2019**, *25*, 850–860. [CrossRef] [PubMed]
36. Wang, C.; Xu, C.; Sun, M.; Luo, D.; Liao, D.F.; Cao, D. Acetyl-CoA carboxylase-alpha inhibitor TOFA induces human cancer cell apoptosis. *Biochem. Biophys. Res. Commun.* **2009**, *385*, 302–306. [CrossRef] [PubMed]

Journal of
Clinical Medicine

Article

Stem Cells Inhibition by Bevacizumab in Combination with Neoadjuvant Chemotherapy for Breast Cancer

Renaud Sabatier [1,2,*], Emmanuelle Charafe-Jauffret [2,3], Jean-Yves Pierga [4], Hervé Curé [5], Eric Lambaudie [2,6], Dominique Genre [7], Gilles Houvenaeghel [2,6], Patrice Viens [1,2], Christophe Ginestier [2], François Bertucci [1,2], Patrick Sfumato [8], Jean-Marc Extra [1,†] and Anthony Gonçalves [1,2,†]

[1] Department of Medical Oncology, Institut Paoli-Calmettes, 13009 Marseille, France; viensp@ipc.unicancer.fr (P.V.); bertuccif@ipc.unicancer.fr (F.B.); extrajm@ipc.unicancer.fr (J.-M.E.); goncalvesa@ipc.unicancer.fr (A.G.)

[2] Aix Marseille Univ, CNRS U7258, INSERM U1068, Institut Paoli-Calmettes, CRCM, 13009 Marseille, France; jauffrete@ipc.unicancer.fr (E.C.-J.); lambaudiee@ipc.unicancer.fr (E.L.); houvenaeghelg@ipc.unicancer.fr (G.H.); CHristophe.ginestier@inserm.fr (C.G.)

[3] Department of Biopathology, Institut Paoli-Calmettes, 13009 Marseille, France

[4] Department of Medical Oncology, Institut Curie, Paris & St Cloud, Université Paris Descartes, 75005 Paris, France; jean-yves.pierga@curie.fr

[5] Department of Medical Oncology, Institut Jean Godinot, 51100 Reims, France; hcure@chu-grenoble.fr

[6] Department of Surgical Oncology, Institut Paoli-Calmettes, 13009 Marseille, France

[7] Department of Clinical Research and Innovation, Institut Paoli-Calmettes, 13009 Marseille, France; genred@ipc.unicancer.fr

[8] Department of Clinical Research and Innovation, Biostatistics unit, Institut Paoli-Calmettes, 13009 Marseille, France; sfumatop@ipc.unicancer.fr

* Correspondence: sabatierr@ipc.unicancer.fr; Tel.: +33-4-9122-3789; Fax: +33-4-9122-3670

† These authors contributed equally to this work.

Received: 8 April 2019; Accepted: 3 May 2019; Published: 6 May 2019

Abstract: Preclinical works have suggested cytotoxic chemotherapies may increase the number of cancer stem cells (CSC) whereas angiogenesis inhibition may decrease CSC proliferation. We developed a proof of concept clinical trial to explore bevacizumab activity on breast CSC. Breast cancer patients requiring preoperative chemotherapy were included in this open-label, randomized, prospective, multicenter phase II trial. All received FEC-docetaxel combination, and patients randomized in the experimental arm received concomitant bevacizumab. The primary endpoint was to describe ALDH1 (Aldehyde dehydrogenase 1) positive tumor cells rate before treatment and after the fourth cycle. Secondary objectives included safety, pathological complete response (pCR) rate, disease-free survival (DFS), relapse-free survival (RFS), and overall survival (OS). Seventy-five patients were included. ALDH1+ cells rate increase was below the predefined 5% threshold in both arms for the 32 patients with two time points available. Grade 3 or 4 adverse events rates were similar in both arms. A non-significant increase in pCR was observed in the bevacizumab arm (42.6% vs. 18.2%, $p = 0.06$), but survival was not improved (OS: $p = 0.89$; DFS: $p = 0.45$; and RFS: $p = 0.68$). The increase of ALDH1+ tumor cells rate after bevacizumab-based chemotherapy was less than 5%. However, as similar results were observed with chemotherapy alone, bevacizumab impact on breast CSC cells cannot be confirmed.

Keywords: early breast cancer; bevacizumab; neoadjuvant chemotherapy; cancer stem cells; ALDH1

1. Introduction

Breast cancer remains the first cancer in women in western countries [1]. Breast tumors can benefit from neoadjuvant chemotherapy to enhance the rate of conservative surgery [2–5]. Preoperative systemic treatments have also an impact on micro metastatic disease and have similar survival results than adjuvant systemic treatments [5,6]. It is also a way to assess the efficacy of new drugs or combinations and to develop translational research programs [7,8]. Major regimens used in this setting are sequential combinations of anthracyclines and taxanes [9–12], with the addition of trastuzumab and pertuzumab for HER2 (human epidermal growth factor receptor 2)-positive tumors [13,14].

Neoangiogenesis is a well-known hallmark of cancer [15], highly involved in tumorigenesis [16], and balanced by activating and inhibiting signals. Dysregulation of the equilibrium leads to an "angiogenic switch" allowing a tumor to grow over one to three mm [17]. The main pro-angiogenic and most specific factor is the vascular endothelial growth factor (VEGF). VEGF proteins are released by tumor cells, bind to their receptors (mainly VEGFR1 and VEGFR2), and are involved in cell division, endothelial cell migration, extracellular matrix modifications, vessels permeability, and vessels survival [18–20]. A combination of chemotherapy and an anti-VEGF monoclonal antibody (bevacizumab) has been shown to improve progression-free survival in first-line metastatic breast cancer [21–23] as well as pathological response rates after neoadjuvant treatment [24,25] but failed to improve overall survival in both early and advanced settings. Similar results have been obtained with other types of advanced cancer with improvements of progression-free survival with small or no clinical impact on overall survival [26–29].

The cancer stem cell theory has been largely developed in the last decade to explain resistance to cytotoxic agents [30]. Cancer stem cells display capacities of auto-renewal and differentiation and thus can drive resistance and disease recurrences [31–33].

Cancer stem cells have been described to display higher enzymatic activity of aldehyde dehydrogenase 1 (ALDH1) than differentiated cells, and can thus be identified using ALDH1 staining in various cancer localizations, including breast cancer [34–38]. Brain tumors models developed from cancer stem cells (CSC) synthetize higher levels of VEGF and can be inhibited with bevacizumab [39]. CSC secreting VEGF may also be localized in pro-angiogenic niches, with potentiation of antitumor effects when combining cytotoxic chemotherapy to antiangiogenic treatments [40,41]. Concerning breast cancer, in vitro models have suggested that CSCs express elevated levels of pro-angiogenic factors including VEGF, and that bevacizumab can decrease this pro-angiogenic effect [42].

As chemotherapy alone enriches breast CSC population after neoadjuvant treatment [30], we hypothesized that CSC inhibition by chemo-bevacizumab combinations can lead to a decrease or a stabilization of CSC rate within breast tumors. Differentiated cancer cells should indeed be killed by chemotherapy whereas CSC should be targeted by bevacizumab. We proposed to assess CSC inhibition by a chemotherapy–bevacizumab combination compared to chemotherapy alone for breast cancer patients treated in the neoadjuvant setting.

2. Experimental Section

2.1. Study Design

The AVASTEM trial (EudraCT Number: 2009-014773-40; ClinicalTrials.gov Identifier: NCT01190345) was conducted as an open-label, randomized (2:1), prospective, multicenter, phase II trial with chemotherapy with or without bevacizumab combination for the treatment of primary breast cancer candidate to receive preoperative chemotherapy. Its primary objective was to evaluate cancer stem cell (CSC) inhibition under treatment by assessing the percentage of residual ALDH1 positive cells after the fourth treatment cycle.

Secondary objectives were safety (as assessed according to CTCAE v3.0 criteria), pathological response (pCR) rate, disease-free survival (DFS), relapse-free survival (RFS), and overall survival (OS).

2.2. Patients

This trial was conducted in three French tertiary comprehensive cancer centers (Institut Paoli Calmettes, Marseille; Institut Curie, Paris; Institut Jean Godinot, Reims).

To be enrolled in this trial, patients had to have been diagnosed with a pathologically confirmed breast cancer, SBR grade 2 or 3 (except for inflammatory breast tumors), tumor size over 3 cm. Patients were 18 years or older; with a good performance status (ECOG $\leq$ 1); life expectancy of at least 3 months; adequate hematological, liver, and renal functions; without cardiac dysfunction (left ventricle ejection fraction $\geq$ 55%). Exclusion criteria were breast cancers of lobular subtype (except for HER2-positive, grade 3 or inflammatory breast tumors); prior treatment with bevacizumab; history of invasive cancer within the last 5 years (except skin carcinoma and in situ cervix cancers with appropriate treatment); uncontrolled high blood pressure (systolic > 150 mmHg and/or diastolic > 100 mmHg); history of congestive heart failure, unstable angina, myocardial infarction within the past 6 months; serious cardiac arrhythmia; thrombotic event within the last 6 months; history of coagulation disorder; history of digestive or non-digestive fistula or digestive perforation within the last 6 months; major surgery within the last 28 days; and history of anaphylaxis to monoclonal antibodies. All patients were affiliated to, or beneficiated to, a national insurance regimen. Patients were enrolled after signature of a written informed consent. All procedures were done in accordance with the 2008 Helsinki Declaration, after approval of the responsible ethics committees (institutional and national).

2.3. Treatments

Treatment scheme, dose reductions, and safety assessments were done according to protocol recommendations.

Patients randomized in arm A (experimental treatment) received four 21-day cycles of FEC100 IV infusions (5FU 500 mg/m^2, epirubicin 100 mg/m^2, and cyclophosphamide 500 mg/m^2) on day 1 plus bevacizumab (15 mg/kg on day 1). They then received four 21-day cycles of docetaxel (100 mg/m^2 on day 1) plus bevacizumab (15 mg/kg on day 1). Patients randomized in the control arm only received cytotoxic chemotherapy. Patients with HER2-positive tumors also received IV trastuzumab. No dose reduction was allowed for bevacizumab, and toxicities were managed according to EMA (European Medicines Agency) summary of product characteristics. Other drugs dose modifications were done in accordance to EMA guidelines.

2.4. Randomization Process

A stratified permuted block randomization method was used to allocate patients in a 2:1 ratio to receive FEC + bevacizumab or FEC alone. Block randomization lists were generated using a computer algorithm with randomly selected block sizes to ensure reproducibility and constant ratio of patients assigned between arms within blocks. The random allocation sequence and patients' assignment were performed by the biostatics unit from the Institut Paoli-Calmettes. Clinicians from the three participating centers enrolled the patients.

2.5. Primary Objective Analysis

Breast CSC and progenitors can be identified by exploring the expression of an enzyme involved in retinoic acid metabolism: ALDH1 [34,43]. CSC inhibition was assessed in the intention to treat population as the percentage of ALDH1-positive cells identified using immunohistochemistry (IHC) as previously described with the ALDH1 antibody (Becton Dickinson biosciences) used at a 1/50 dilution [34], both before treatment and after the fourth treatment cycle. All patients with ADLH1 assessment prior to and during treatment were included in primary objective analysis.

2.6. Analysis of Safety Data

Safety assessments consisted of recording all adverse events (AEs) (CTCAE v3.0 criteria and NYHA for cardiac AEs) and serious adverse events (SAEs) collected during the treatment period and within 28 days after treatment completion. Cardiac AEs were collected for 12 months after treatment discontinuation. The safety population was defined as the subjects who received at least a partial dose of treatment.

2.7. Analysis of Clinical Efficacy Data

The efficacy population was defined as the group of subjects who received at least one chemotherapy cycle. Pathological complete response after treatment was planned to be defined according to Sataloff classification [44]. However, as most of recent literature data are mainly based on the more consensual ypT0/is pN0 definition [45], we described both classifications. DFS was defined as the time from treatment initiation to disease recurrence wherever the localization, secondary cancer diagnosis, or death, whatever occurred first. RFS was defined as the time from treatment initiation to disease recurrence or death. OS was defined as the time from treatment initiation to death, whatever the cause.

2.8. Statistical Analysis

The number of subjects to be included in this trial was 75 with a 2:1 ratio (50 patients in the experimental arm) to show that the increase of the percentage of ADLH1-positive cells was not upper than 5% after the fourth treatment cycle in the experimental arm, with an 80% power and an alpha-risk of 5%. An exact Wilcoxon rank-signed test was performed to confirm the rejection of the hypothesis that ALDH1+ cells rate increase was upper than 5% after four chemotherapy cycles in the experimental arm. *p*-Values, estimation of location parameters (Hodges Lehmann pseudo-median) and their one sided 95% confidence intervals (CI) were determined using exactRankTests v0.8-29 R-package [46].

Descriptive statistics were used to summarize the frequency, severity, duration, and relationship to treatment for all AEs occurring after the initiation of treatment. Categorical variables were described using counts and frequencies, and quantitative variables were described using medians and ranges. Patients' characteristics were compared using exact Fisher's tests for qualitative variables and rank-Wilcoxon's tests for quantitative variables. Hazard ratios are provided with their bilateral confidence interval and Wald's test *p*-value for significance. Follow-ups were estimated using the inverse Kaplan–Meier method. Patients lost to follow-up or without event were censored at the date of last news. Survival curves were estimated using the Kaplan–Meier method, and the median DFS, RFS, and OS were calculated with their 95% confidence intervals. Both univariate and multivariate analyses were conducted using Cox's proportional hazard regression models including treatment arm, ADLH1 status at inclusion and pCR status as categorical explanatory variables. For Cox multivariate analysis, all the delays were considered from surgery date instead of inclusion due to an adjustment on pCR status (evaluated at surgery).

Except for the primary endpoint, all tests were two-sided. The level of statistical significance was set at $\alpha = 0.05$. Statistical analyses were carried out with the SAS® software version 9.3 (SAS Institute Inc., Cary, NC, USA) and the R software version 3.0.3 (https://www.R-project.org/). This study was performed according to the CONSORT statements [47], (see Supplementary checklist).

3. Results

3.1. Population Description

Seventy-five patients were included from March 2010 to July 2012 (Figure 1). Follow-up was stopped five years after inclusion of the last patients, i.e., in October 2017. Fifty patients received the experimental treatment (arm A) and twenty-five were randomized in the control arm (arm B). There was no imbalance between arms (Table 1). Median age at inclusion was 50.0 years for arm A versus 49.8 for arm B. Ten patients had de novo metastatic breast cancer (arm A = 14%, arm B = 12%). Out of non-metastatic cases, most of them were larger than 5cm (T3: 37.2% vs. 31.8%; T4a–b: 2.3% vs. 9.1%; T4d: 18.6% vs. 13.6%). There was a trend in higher rates of lymph node involvement for non-metastatic patients included in arm B (N1: 52.4% vs. 77.3%; N2: 16.7% vs. 4.5%; N3: 2.4% vs. 9.1%; $p = 0.06$). Pathologically-defined molecular subtypes were well balanced: fourteen cases (28%) were HER2-positive in arm A (including 6 HR-positive and 8 HR-negative tumors) versus eight (32%) in arm B (including 3 HR-positive and 5 HR-negative tumors). Triple negative tumors rates were also similar for both arms (42.9% vs. 40%).

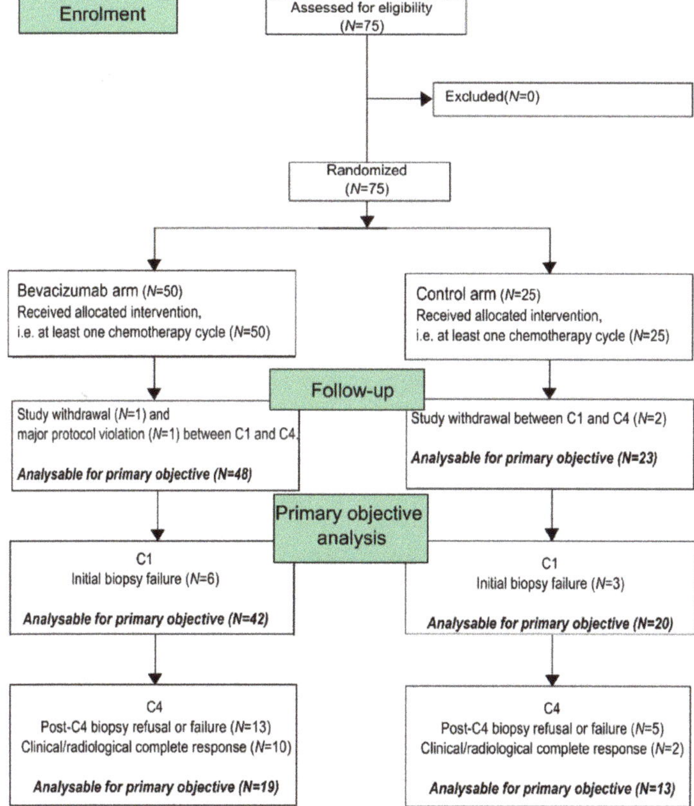

Figure 1. CONSORT flow diagram.

Table 1. Clinical and pathological features at inclusion. When required, results are notified as N (% of cases with data available).

		Bevacizumab Arm $(N = 50)$	Control Arm $(N = 25)$	p-Value
Clinical features				
Age at inclusion, years (median, min–max)		50.0 (24.3–66.9)	49.8 (28.4–68.5)	0.47
Menopausal		9 (18)	6 (24)	0.55
Tumor size	T2	21 (42)	10 (40)	1.00
	T3	17 (34)	9 (36)	
	T4	12 (24)	6 (24)	
Axillary lymph node positive		37 (75.5)	22 (88)	0.24
Metastatic disease at diagnosis		7 (14)	3 (12)	1
Pathological features at diagnosis				
HR status	Positive	21 (42)	10 (40)	1
	Negative	29 (58)	15 (60)	
HER2 status	Positive	14 (28)	8 (32)	0.86
	Negative	35 (70)	17 (68)	
	Equivocal	1 (2)		
Triple negative phenotype		21 (42.9)	10 (40)	1
SBR grade	1–2	12 (24)	9 (36)	0.41
	3	37 (76)	16 (64)	
Lymphovascular invasion positive		6 (13)	1 (4)	0.41

3.2. Treatment Administration and Safety

More than 80% of patients received eight chemotherapy cycles (A: 88%, B: 84%). Only five patients did not receive the four scheduled cycles of FEC. Seven patients (arm A) and 4 patients (arm B) did not receive all the planned docetaxel infusions. All the patients in the experimental arm received at least one cycle of bevacizumab. Out of them 43 (86%) received at least four cycles and 28 (56%) received the eight planned bevacizumab courses.

Concerning safety (Table 2), 94% of patients included in the experimental arm and 88% in the control arm experimented grade 3 to 4 adverse events. The most frequent were haematological disorders with 46 to 48% of grade 3 to 4 neutropenia, with 34% of febrile neutropenia in arm A and 8% in arm B. High grades of anaemia, asthenia, nausea, mucitis, and cutaneous disorders were also more frequent in the experimental arm (18% vs. 4%, 14% vs. 4%, 10% vs. 0%, 22% vs. 8%, and 20% vs. 12%, respectively). Regarding bevacizumab-related adverse events, severe high blood pressure increase was observed for two patients (4%) in arm A (vs. 0 in arm B); three cases of thrombotic event were observed for patients in arm A (none in the control arm).

No treatment-related death was reported during study duration.

Table 2. Most common adverse events (Common Terminology Criteria for Adverse Events version 3.0). Only adverse events (AEs) related to treatment and with an incidence ≥ 10% for all grades or ≥ 5% for grade ≥ 3 are presented.

	Bevacizumab Arm				Control Arm			
	All Grades		Grade ≥ 3		All Grades		Grade ≥ 3	
Adverse events	N	%	N	%	N	%	N	%
Haematological								
Anaemia	13	26	9	18	5	20	1	4
Lymphopenia	9	18	9	18	6	24	6	24
Neutropenia	24	48	23	46	12	48	12	48
Febrile neutropenia	17	34	17	34	2	8	2	8
Non-haematological								
Asthenia	37	74	7	14	18	72	1	4
Anorexia	9	18	2	4	4	16	1	4
Weight lost	13	26	0	0	1	4	0	0
Constipation	16	32	0	0	7	28	0	0
Diarrhoea	17	34	0	0	6	24	0	0
Nausea	43	86	5	10	20	80	0	0
Vomiting	18	36	3	6	2	8	0	0
Abdominal pain	7	14	1	2	2	8	0	0
Stomach pain	10	20	0	0	6	24	0	0
Dysphagia	8	16	0	0	1	4	0	0
Mucitis	39	78	11	22	17	68	2	8
Epistaxis	30	60	0	0	1	4	0	0
Arthralgia	10	20	1	2	11	44	1	4
Myalgia	18	36	0	0	12	48	0	0
Peripheral neuropathy	6	12	0	0	3	12	0	0
Cutaneous toxicities	43	86	10	20	21	84	3	12
Amenorrhea	10	20	5	10	9	36	6	24
High blood pressure	14	28	2	4	2	8	0	0
Headaches	15	30	0	0	1	4	0	0

3.3. Primary Endpoint: Percentage of ADLH1+ Cells

IHC analysis was performed to identify tumors with ALDH1 expression at inclusion. We evaluated the percentage of ALDH1+ cells for each sample at inclusion and after the fourth chemotherapy cycle. Absence of a large increase of ALDH1 expression was explored as a surrogate marker of CSC inhibition due to chemotherapy-bevacizumab combination. ALDH1 data before and after four chemotherapy cycles were available for 19 patients in arm A and 13 patients in arm B. The primary endpoint was thus not evaluable in 43 patients, of whom 12 patients were excluded from the analysis set due to complete remission at the end of the fourth cycle of chemotherapy (Figure 1). Most of the patients randomized in the experimental arm with both biopsy analyses available experimented a low increase (<5%) or a decrease of ALDH1+ cells rate (pseudo-median = −0.125, one-sided 95% CI ($-\infty$–0), p = 0.001, Wilcoxon test, N = 19), (Figure 2). ALDH1 expression was also significantly not higher after chemotherapy in the control arm (pseudo-median = −0.25, one-sided 95% CI ($-\infty$–0), p = 0.006, N = 13).

We then exploratory assessed ALDH1 expression levels on post-chemotherapy (after the 8th cycle) samples. ALDH1 pseudo-median of difference between post-chemo and pre-chemo samples was −0.125 (one-sided 95% CI ($-\infty$–0), p < 0.001, N = 19) in the experimental arm versus −0.25 (one-sided 95% CI ($-\infty$–10), p = 0.14, N = 7) for the chemotherapy only arm. When adding cases with complete pathological response after chemotherapy to patients without high ALDH1+ rate increase, we observed that these results were similar both for the experimental arm (pseudo-median = −0.125, one-sided 95% CI ($-\infty$–0), p < 0.001, N = 37) and the control arm (pseudo-median = −0.25, one-sided 95% CI ($-\infty$–0), p = 0.016, N = 11).

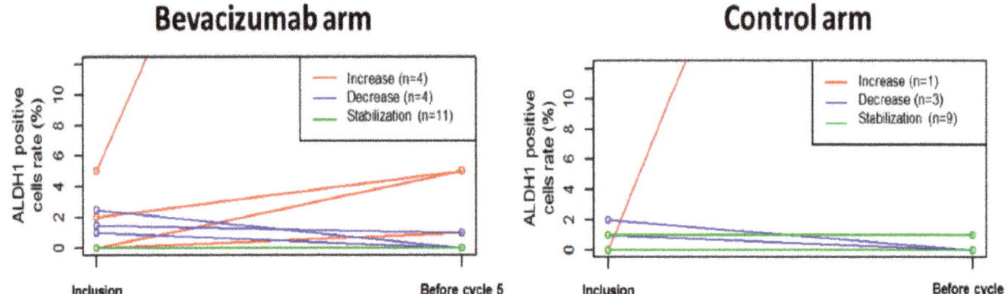

Figure 2. ADLH1 positive cells rates modifications after the first four cycles of treatment (each line represents one patient). Red lines represent patients presenting an increase of ALDH1+ cells rate. Blue lines represent patients presenting a decrease of ALDH1+ cells rate. Green lines represent patients without ALDH1+ cells rate modification.

3.4. Efficacy

3.4.1. Efficacy for the Whole Population

Regarding pathological response after chemotherapy (Table 3), a not statistically significant difference according to treatment arm was observed when using Sataloff classification (51.1% vs. 31.8%; OR = 2.24, 95% CI (0.77–6.54), p = 0.14). When using the usual definition of pathological complete response, there was a trend for pCR rate improvement for patients who received bevacizumab (42.6% vs. 18.2%; OR = 3.33, 95% CI (0.98–11.38), p = 0.06).

Table 3. Pathological response rates. Results are notified as N (% of cases with data available).

		Bevacizumab Arm (N = 50)	Control Arm (N = 25)	Odds Ratio (95% CI)	p-Value
Sataloff classification	pCR	23 (51.1)	7 (31.8)	2.24 (0.77–6.54)	0.14
	Non-pCR	22 (48.9)	15 (68.2)		
	Missing data	5	3		
ypT0/is pN0 definition	pCR	20 (42.6)	4 (18.2)	3.333 (0.98–11.38)	0.06
	RD	27 (57.5)	18 (81.8)		
	Missing data	3	3		

pCR: pathological complete response, RD: invasive residual disease in breast or lymph nodes.

Median follow-up was 60.9 months for the whole population (95% CI (60.4–61.8)) with no difference according to treatment arm (61.3 months, 95% CI (60.2–62.5), for arm A vs. 60.6 months, 95% CI (60.2–61.6), for arm B). The 5-y OS was 86% (95% CI (75–92)) for the whole cohort and 90% (95% CI (80–96)) for non-metastatic patients. There was no OS difference according to treatment arm (5-y OS = 85% for arm A vs. 87% for arm B; p = 0.89; HR = 1.10, 95% CI (0.28–4.25)), (Figure 3A). Considering non-metastatic patients, 5-y DFS was 73% (95% CI (60–82)) for the whole population, without improvement for patients who received bevacizumab (5-y DFS = 76% for arm A vs. 65% for arm B; p = 0.45; HR = 0.69, 95% CI (0.26–1.81)), (Figure 3B). Results were similar regarding RFS. Five-y RFS was 76% for arm A vs. 70% for arm B (p = 0.68; HR = 0.81, 95% CI (0.29–2.23)), (Figure 3C).

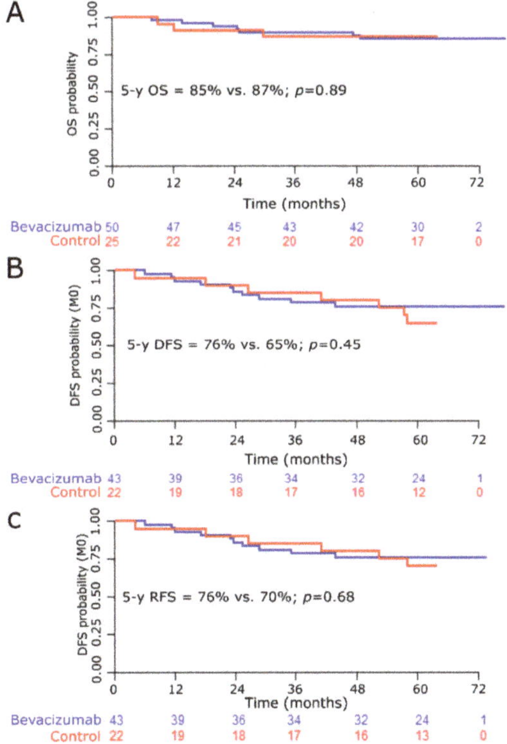

Figure 3. Kaplan–Meier curves. (**A**): Overall survival for the whole cohort. (**B**): Disease-free survival for non-metastatic patients. (**C**): Relapse-free survival for non-metastatic patients.

3.4.2. Efficacy according to ALDH1 Status

At inclusion, 21 of 63 patients with analyzable samples were ADLH1+ (33% in both treatment arms). ALDH1 status was also available for 32 patients after the fourth chemotherapy cycle. Of them, twenty-two percent had ADLH1+ tumors. Among the six ALDH1+ cases at inclusion in arm A, three were still ADLH1+ after four chemotherapy cycles (vs. three of four in arm B). Two of 13 ADLH1− cases became ADLH1+ in arm A versus one of nine in arm B. ADLH1 status at inclusion was not correlated to pCR (ypT0/is pN0 definition) after neoadjuvant treatment: OR (ALDH1− vs. ADLH1+) = 2.74, 95% CI (0.84–8.97), $p = 0.10$. Similar results were observed for OS (HR = 0.69, 95% CI = (0.15–3.06), $p = 0.62$), DFS (HR = 0.94, 95% CI = (0.37–2.40), $p = 0.90$), and RFS (HR = 1.10, 95% CI = (0.41–2.94), $p = 0.85$). Multivariate Cox regression analyses including ALDH1 status at inclusion, treatment arm, and pCR status failed to show survival benefits for ADLH1-positive tumors (Tables S1–S3).

4. Discussion

This randomized prospective phase II trial aimed to bring evidences of CSCs inhibition using a chemotherapy-bevacizumab combination. As hypothesized, we observed that ADLH1+ cells rate was not increased above the predefined threshold for patients receiving bevacizumab-based neoadjuvant chemotherapy. However, as similar results were obtained for the control group, an impact of angiogenesis inhibition on the CSC population cannot be confirmed.

Contrarily to our initial hypothesis, some preclinical results published after initiation of our trial intriguingly suggested that hypoxia generated by antiangiogenic agents, including bevacizumab, could stimulate the population of CSCs and contribute to the resistance of these drugs and the lack of survival benefit provided by these compounds in breast cancer [48]. An increase in CSCs was particularly noticeable when treatment was delivered in the context of large and established tumors, a setting comparable to that of neoadjuvant treatment, most likely in relation to a greater overall volume for hypoxia to occur. Nevertheless, these preclinical observations were not clearly validated in vivo, and we did not observe CSCs stimulation during bevacizumab treatment in the present trial.

The ADLH1 status at inclusion, which can be seen as a surrogate of the CSC population, was also neither prognostic nor predictive of pCR after treatment. These results are contradictory to other published data. In a Japanese cohort of more than 100 breast cancers receiving preoperative chemotherapy, ALDH1-positive cases were associated with a low pCR rate (9.5% vs. 32.2%; $p = 0.037$). Analyses done before and after neoadjuvant chemotherapy showed an increase of the proportion of ALDH1-positive tumor cells for patients who did not achieve pCR [49]. These results were confirmed in another set of 119 patients receiving anthracyclines and taxanes before breast surgery, for whom ALDH1 negativity at baseline was significantly associated with pCR, and ALDH1 expression was increased after chemotherapy [50]. In this study, ALDH1 expression in the residual tumor was also prognostic, with worse overall survival for ALDH1-positive cases. Similar results were observed in a retrospective work including more than 650 breast cancer patients [51]. We did not find such a correlation between ALDH1 expression at diagnosis and treatment efficacy in our study. Concerning pCR, the results were inconclusive ($p = 0.10$) probably because of a lack of power, and this trial was indeed not powered for pCR nor for survival analyses.

Pathological response rates were surprisingly high in our set with, according to the classification used, 42 to 51% of pCR in the bevacizumab arm and 18 to 32% in the control arm. These rates decreased our capacity to obtain post-treatment tumor samples for ALDH1 expression analysis. However, even though high, these rates are in accordance to the data published during the last years for the tumor phenotypes described in this set. Nearly 60% of patients included harbored HR-negative tumors, including more than 40% with triple-negative cancer. Triple negative breast cancer is now known to display higher rates of pCR, and our results are in accordance with what has been previously published with pCR rates ranging from 25% to 60% [51–53].

In the GeparQuinto trial, exploring the addition of VEGF-R inhibition to anthracycline and taxane-based neoadjuvant therapy, pCR rates in the TNBC subset were significantly higher with bevacizumab (39.3% vs. 27.9%, $p = 0.003$) [53], with results close to what we observed in our cohort. However, as we also notified, survival was not improved with the combination for TNBC (HR = 0.99 for DFS, $p = 0.94$; HR = 1.02 for OS, $p = 0.89$) nor for the whole HER2-negative cohort, (HR = 1.03 for DFS, $p = 0.78$; HR = 0.97 for OS, $p = 0.84$) [54]. To avoid this issue of TNBC excess leading to high pCR rates in both groups, we should have decreased the number of TNBC patients included. However, this would have led to insufficient enrolment within the predefined inclusion period.

This trial has some limits which deserve to be underlined. Firstly, the definition of the main endpoint was of importance at time of trial design but could be more criticized if this study takes place in 2019. Improvement of the surrogate markers of CSCs could have allowed us to reach our primary objective. Additional staining focused on the ALDH1A1 isoform, or other stem cells markers such as CD44+/CD24− may be of interest [55,56]. However recent studies suggest that CD44+ CD24− phenotype and ALDH1 phenotype may form part of two interconvert dynamic EMT states. Mesenchymal-like CSCs (CD44+/CD24− are more quiescent whereas epithelial-like BCSCs (ALDH1+) are more proliferative [57]. In the neoadjuvant setting, ALDH1 staining only displayed a prognostic value for patients in terms of pCR or metastasis-free survival [58]. We thus chose to monitor CSC during neoadjuvant chemotherapy using ALDH1 staining. The ALDEFLUOR assay, based on the assessment of ALDH enzymatic activity, could also have been evaluated, but is very unlikely to be developed in routine practice as it needs early processing of fresh tumor samples and flow cytometry experiments [55]. Secondly, even though the randomization was of importance to look for an association between bevacizumab treatment and CSC enrichment as suggested by

Li et al. [30], the lack of blind could have had an impact on efficacy data, including our primary endpoint. Thirdly, CSCs may not be a good candidate to predict neoadjuvant treatments efficacy. We have recently described that pCR rates after chemotherapy and bevacizumab-based neoadjuvant treatments in inflammatory breast cancer are correlated to circulating tumor cells (CTC) detection [59]. CTC-positive cases displayed worse three-year DFS (39% vs. 70%) and OS. CTC status at baseline was an independent prognostic feature in multivariate analysis. A team from Harvard has also recently reported that another type of tumor cells with (but not only with) CSC properties, called AKT1^{low} quiescent cancer cells, seem to be resistant to chemotherapy in the neoadjuvant setting for triple negative breast cancer [60]. Concerning the current study, as it was designed as a proof of concept trial with a biological primary endpoint, its sample-size was limited, response and survival endpoints were secondary, and definitive conclusions related to bevacizumab contribution to treatment efficacy can only be considered as preliminary. However, larger and more robust results have been published in randomized trials evaluating adjuvant bevacizumab efficacy for early breast cancers [61]. These data are in accordance with our results with no survival gain with bevacizumab-based combinations in the (neo)adjuvant setting, despite higher rates of pathological response [24,25,54,62]. Similar results have been obtained in the early setting for other tumor localizations, with no disease-free survival or overall survival improvements [63,64]. Even though pCR has been widely described to be prognostic in early breast cancer, survival gains secondary to treatment-induced pCR improvement is still debated. A large meta-analysis published a few years ago thus showed that trials describing pCR gains did not result in survival benefits, with the exception of the NOAH trial which introduced trastuzumab in the preoperative systemic regimen for HER2-positive cases [45].

5. Conclusions

Bevacizumab impact on CSC was not higher than that of chemotherapy alone for the cases enrolled in this trial. Moreover, despite that bevacizumab-chemotherapy combination enhanced pCR rates compared to chemotherapy alone for breast cancer receiving neoadjuvant chemotherapy, it did not improve survival. These observations warrant to be deeply explored in larger series and for other tumor localizations for which bevacizumab is more widely used such as ovarian, colorectal, and cervical cancers [27,28,65–67].

Supplementary Materials: The following are available online at http://www.mdpi.com/2077-0383/8/5/612/s1, Table S1: Overall survival for the whole population, Table S2: Disease-free survival for M0 patients, Table S3: Relapse-free survival for M0 patients.

Author Contributions: A.G. and J.-M.E. designed the study; E.C.-J. and C.G. performed stem cells analyses; R.S., J.-Y.P., H.C., E.L., G.H., P.V., F.B., J.-M.E., and A.G. enrolled patients and collected clinical data; D.G. was responsible of clinical trial management; P.S. performed all statistical analyses; all author contributed to data interpretation, manuscript writing, and approval of the submitted version.

Funding: This study was funded by INCa and DGOS (PHRC 2009, réf. n°24-005), Roche, and Institut Paoli-Calmettes.

Acknowledgments: The authors would like to thank the patients and their families for their participation in this phase II study.

Conflicts of Interest: A.G. received research grants and non-financial support from Roche. R.S. received personal fees and non-financial support from Roche. The other authors declare no conflict of interest.

References

1. Miller, K.D.; Siegel, R.L.; Lin, C.C.; Mariotto, A.B.; Kramer, J.L.; Rowland, J.H.; Stein, K.D.; Alteri, R.; Jemal, A. Cancer treatment and survivorship statistics, 2016. *CA Cancer J. Clin.* **2016**, *66*, 271–289. [CrossRef] [PubMed]
2. Hortobagyi, G.N.; Ames, F.C.; Buzdar, A.U.; Kau, S.W.; McNeese, M.D.; Paulus, D.; Hug, V.; Holmes, F.A.; Romsdahl, M.M.; Fraschini, G. Management of stage III primary breast cancer with primary chemotherapy, surgery, and radiation therapy. *Cancer* **1988**, *62*, 2507–2516. [CrossRef]

3. Bonadonna, G.; Valagussa, P.; Brambilla, C.; Ferrari, L.; Moliterni, A.; Terenziani, M.; Zambetti, M. Primary chemotherapy in operable breast cancer: Eight-year experience at the Milan Cancer Institute. *J. Clin. Oncol.* **1998**, *16*, 93–100. [CrossRef]

4. Mauri, D.; Pavlidis, N.; Ioannidis, J.P.A. Neoadjuvant versus adjuvant systemic treatment in breast cancer: A meta-analysis. *J. Natl. Cancer Inst.* **2005**, *97*, 188–194. [CrossRef] [PubMed]

5. Rastogi, P.; Anderson, S.J.; Bear, H.D.; Geyer, C.E.; Kahlenberg, M.S.; Robidoux, A.; Margolese, R.G.; Hoehn, J.L.; Vogel, V.G.; Dakhil, S.R.; et al. Preoperative chemotherapy: Updates of National Surgical Adjuvant Breast and Bowel Project Protocols B-18 and B-27. *J. Clin. Oncol.* **2008**, *26*, 778–785. [CrossRef]

6. Wolmark, N.; Wang, J.; Mamounas, E.; Bryant, J.; Fisher, B. Preoperative chemotherapy in patients with operable breast cancer: Nine-year results from National Surgical Adjuvant Breast and Bowel Project B-18. *J. Natl. Cancer Inst. Monographs* **2001**, *30*, 96–102. [CrossRef]

7. Forrest, A.P.; Levack, P.A.; Chetty, U.; Hawkins, R.A.; Miller, W.R.; Smyth, J.F.; Anderson, T.J. A human tumour model. *Lancet* **1986**, *2*, 840–842. [CrossRef]

8. Rapoport, B.L.; Demetriou, G.S.; Moodley, S.D.; Benn, C.A. When and how do I use neoadjuvant chemotherapy for breast cancer? *Curr. Treat. Options Oncol.* **2014**, *15*, 86–98. [CrossRef]

9. Miller, K.D.; McCaskill-Stevens, W.; Sisk, J.; Loesch, D.M.; Monaco, F.; Seshadri, R.; Sledge, G.W. Combination versus sequential doxorubicin and docetaxel as primary chemotherapy for breast cancer: A randomized pilot trial of the Hoosier Oncology Group. *J. Clin. Oncol.* **1999**, *17*, 3033–3037. [CrossRef] [PubMed]

10. Von Minckwitz, G.; Raab, G.; Caputo, A.; Schütte, M.; Hilfrich, J.; Blohmer, J.U.; Gerber, B.; Costa, S.D.; Merkle, E.; Eidtmann, H.; et al. Doxorubicin with cyclophosphamide followed by docetaxel every 21 days compared with doxorubicin and docetaxel every 14 days as preoperative treatment in operable breast cancer: The GEPARDUO study of the German Breast Group. *J. Clin. Oncol.* **2005**, *23*, 2676–2685. [CrossRef]

11. Bear, H.D.; Anderson, S.; Brown, A.; Smith, R.; Mamounas, E.P.; Fisher, B.; Margolese, R.; Theoret, H.; Soran, A.; Wickerham, D.L.; et al. The effect on tumor response of adding sequential preoperative docetaxel to preoperative doxorubicin and cyclophosphamide: Preliminary results from National Surgical Adjuvant Breast and Bowel Project Protocol B-27. *J. Clin. Oncol.* **2003**, *21*, 4165–4174. [CrossRef]

12. Diéras, V.; Fumoleau, P.; Romieu, G.; Tubiana-Hulin, M.; Namer, M.; Mauriac, L.; Guastalla, J.-P.; Pujade-Lauraine, E.; Kerbrat, P.; Maillart, P.; et al. Randomized parallel study of doxorubicin plus paclitaxel and doxorubicin plus cyclophosphamide as neoadjuvant treatment of patients with breast cancer. *J. Clin. Oncol.* **2004**, *22*, 4958–4965. [CrossRef]

13. Gianni, L.; Eiermann, W.; Semiglazov, V.; Manikhas, A.; Lluch, A.; Tjulandin, S.; Zambetti, M.; Vazquez, F.; Byakhow, M.; Lichinitser, M.; et al. Neoadjuvant chemotherapy with trastuzumab followed by adjuvant trastuzumab versus neoadjuvant chemotherapy alone, in patients with HER2-positive locally advanced breast cancer (the NOAH trial): A randomised controlled superiority trial with a parallel HER2-negative cohort. *Lancet* **2010**, *375*, 377–384.

14. Gianni, L.; Pienkowski, T.; Im, Y.-H.; Tseng, L.-M.; Liu, M.-C.; Lluch, A.; Starosławska, E.; de la Haba-Rodriguez, J.; Im, S.-A.; Pedrini, J.L.; et al. 5-year analysis of neoadjuvant pertuzumab and trastuzumab in patients with locally advanced, inflammatory, or early-stage HER2-positive breast cancer (NeoSphere): A multicentre, open-label, phase 2 randomised trial. *Lancet Oncol.* **2016**, *17*, 791–800. [CrossRef]

15. Hanahan, D.; Weinberg, R.A. Hallmarks of cancer: The next generation. *Cell* **2011**, *144*, 646–674. [CrossRef]

16. Weidner, N.; Folkman, J.; Pozza, F.; Bevilacqua, P.; Allred, E.N.; Moore, D.H.; Meli, S.; Gasparini, G. Tumor angiogenesis: A new significant and independent prognostic indicator in early-stage breast carcinoma. *J. Natl. Cancer Inst.* **1992**, *84*, 1875–1887. [CrossRef]

17. Folkman, J. Anti-angiogenesis: New concept for therapy of solid tumors. *Ann. Surg.* **1972**, *175*, 409–416. [CrossRef]

18. Ferrara, N.; Davis-Smyth, T. The biology of vascular endothelial growth factor. *Endocr. Rev.* **1997**, *18*, 4–25. [CrossRef]

19. Yoshiji, H.; Harris, S.R.; Thorgeirsson, U.P. Vascular endothelial growth factor is essential for initial but not continued in vivo growth of human breast carcinoma cells. *Cancer Res.* **1997**, *57*, 3924–3928.

20. Fontanini, G.; Vignati, S.; Lucchi, M.; Mussi, A.; Calcinai, A.; Boldrini, L.; Chiné, S.; Silvestri, V.; Angeletti, C.A.; Basolo, F.; et al. Neoangiogenesis and p53 protein in lung cancer: Their prognostic role and their relation with vascular endothelial growth factor (VEGF) expression. *Br. J. Cancer* **1997**, *75*, 1295–1301. [CrossRef]

21. Pivot, X.; Schneeweiss, A.; Verma, S.; Thomssen, C.; Passos-Coelho, J.L.; Benedetti, G.; Ciruelos, E.; von Moos, R.; Chang, H.-T.; Duenne, A.-A.; et al. Efficacy and safety of bevacizumab in combination with docetaxel for the first-line treatment of elderly patients with locally recurrent or metastatic breast cancer: Results from AVADO. *Eur. J. Cancer* **2011**, *47*, 2387–2395. [CrossRef]

22. Miller, K.; Wang, M.; Gralow, J.; Dickler, M.; Cobleigh, M.; Perez, E.A.; Shenkier, T.; Cella, D.; Davidson, N.E. Paclitaxel plus bevacizumab versus paclitaxel alone for metastatic breast cancer. *N. Engl. J. Med.* **2007**, *357*, 2666–2676. [CrossRef]

23. Miles, D.W.; Chan, A.; Dirix, L.Y.; Cortés, J.; Pivot, X.; Tomczak, P.; Delozier, T.; Sohn, J.H.; Provencher, L.; Puglisi, F.; et al. Phase III study of bevacizumab plus docetaxel compared with placebo plus docetaxel for the first-line treatment of human epidermal growth factor receptor 2-negative metastatic breast cancer. *J. Clin. Oncol.* **2010**, *28*, 3239–3247. [CrossRef]

24. Wedam, S.B.; Low, J.A.; Yang, S.X.; Chow, C.K.; Choyke, P.; Danforth, D.; Hewitt, S.M.; Berman, A.; Steinberg, S.M.; Liewehr, D.J.; et al. Antiangiogenic and antitumor effects of bevacizumab in patients with inflammatory and locally advanced breast cancer. *J. Clin. Oncol.* **2006**, *24*, 769–777. [CrossRef]

25. Bertucci, F.; Fekih, M.; Autret, A.; Petit, T.; Dalenc, F.; Levy, C.; Romieu, G.; Bonneterre, J.; Ferrero, J.-M.; Kerbrat, P.; et al. Bevacizumab plus neoadjuvant chemotherapy in patients with HER2-negative inflammatory breast cancer (BEVERLY-1): A multicentre, single-arm, phase 2 study. *Lancet Oncol.* **2016**, *17*, 600–611. [CrossRef]

26. Sandler, A.; Gray, R.; Perry, M.C.; Brahmer, J.; Schiller, J.H.; Dowlati, A.; Lilenbaum, R.; Johnson, D.H. Paclitaxel-carboplatin alone or with bevacizumab for non-small-cell lung cancer. *N. Engl. J. Med.* **2006**, *355*, 2542–2550. [CrossRef]

27. Oza, A.M.; Cook, A.D.; Pfisterer, J.; Embleton, A.; Ledermann, J.A.; Pujade-Lauraine, E.; Kristensen, G.; Carey, M.S.; Beale, P.; Cervantes, A.; et al. Standard chemotherapy with or without bevacizumab for women with newly diagnosed ovarian cancer (ICON7): Overall survival results of a phase 3 randomised trial. *Lancet Oncol.* **2015**, *16*, 928–936. [CrossRef]

28. Tewari, K.S.; Sill, M.W.; Long, H.J.; Penson, R.T.; Huang, H.; Ramondetta, L.M.; Landrum, L.M.; Oaknin, A.; Reid, T.J.; Leitao, M.M.; et al. Improved survival with bevacizumab in advanced cervical cancer. *N. Engl. J. Med.* **2014**, *370*, 734–743. [CrossRef]

29. Loupakis, F.; Cremolini, C.; Masi, G.; Lonardi, S.; Zagonel, V.; Salvatore, L.; Cortesi, E.; Tomasello, G.; Ronzoni, M.; Spadi, R.; et al. Initial therapy with FOLFOXIRI and bevacizumab for metastatic colorectal cancer. *N. Engl. J. Med.* **2014**, *371*, 1609–1618. [CrossRef]

30. Li, X.; Lewis, M.T.; Huang, J.; Gutierrez, C.; Osborne, C.K.; Wu, M.-F.; Hilsenbeck, S.G.; Pavlick, A.; Zhang, X.; Chamness, G.C.; et al. Intrinsic resistance of tumorigenic breast cancer cells to chemotherapy. *J. Natl. Cancer Inst.* **2008**, *100*, 672–679. [CrossRef]

31. Al-Hajj, M.; Becker, M.W.; Wicha, M.; Weissman, I.; Clarke, M.F. Therapeutic implications of cancer stem cells. *Curr. Opin. Genet. Dev.* **2004**, *14*, 43–47. [CrossRef] [PubMed]

32. Li, F.; Tiede, B.; Massagué, J.; Kang, Y. Beyond tumorigenesis: Cancer stem cells in metastasis. *Cell Res.* **2007**, *17*, 3–14. [CrossRef]

33. Korkaya, H.; Wicha, M.S. Selective targeting of cancer stem cells: A new concept in cancer therapeutics. *BioDrugs* **2007**, *21*, 299–310. [CrossRef]

34. Ginestier, C.; Hur, M.H.; Charafe-Jauffret, E.; Monville, F.; Dutcher, J.; Brown, M.; Jacquemier, J.; Viens, P.; Kleer, C.G.; Liu, S.; et al. ALDH1 is a marker of normal and malignant human mammary stem cells and a predictor of poor clinical outcome. *Cell Stem Cell* **2007**, *1*, 555–567. [CrossRef]

35. Lohberger, B.; Rinner, B.; Stuendl, N.; Absenger, M.; Liegl-Atzwanger, B.; Walzer, S.M.; Windhager, R.; Leithner, A. Aldehyde dehydrogenase 1, a potential marker for cancer stem cells in human sarcoma. *PLoS ONE* **2012**, *7*, e43664. [CrossRef]

36. Luo, Y.; Dallaglio, K.; Chen, Y.; Robinson, W.A.; Robinson, S.E.; McCarter, M.D.; Wang, J.; Gonzalez, R.; Thompson, D.C.; Norris, D.A.; et al. ALDH1A isozymes are markers of human melanoma stem cells and potential therapeutic targets. *Stem Cells* **2012**, *30*, 2100–2113. [CrossRef]

37. Jiang, F.; Qiu, Q.; Khanna, A.; Todd, N.W.; Deepak, J.; Xing, L.; Wang, H.; Liu, Z.; Su, Y.; Stass, S.A.; et al. Aldehyde dehydrogenase 1 is a tumor stem cell-associated marker in lung cancer. *Mol. Cancer Res.* **2009**, *7*, 330–338. [CrossRef]

38. Li, T.; Su, Y.; Mei, Y.; Leng, Q.; Leng, B.; Liu, Z.; Stass, S.A.; Jiang, F. ALDH1A1 is a marker for malignant prostate stem cells and predictor of prostate cancer patients' outcome. *Lab. Invest.* **2010**, *90*, 234–244. [CrossRef]

39. Bao, S.; Wu, Q.; Sathornsumetee, S.; Hao, Y.; Li, Z.; Hjelmeland, A.B.; Shi, Q.; McLendon, R.E.; Bigner, D.D.; Rich, J.N. Stem cell-like glioma cells promote tumor angiogenesis through vascular endothelial growth factor. *Cancer Res.* **2006**, *66*, 7843–7848. [CrossRef]

40. Calabrese, C.; Poppleton, H.; Kocak, M.; Hogg, T.L.; Fuller, C.; Hamner, B.; Oh, E.Y.; Gaber, M.W.; Finklestein, D.; Allen, M.; et al. A perivascular niche for brain tumor stem cells. *Cancer Cell* **2007**, *11*, 69–82. [CrossRef] [PubMed]

41. Folkins, C.; Man, S.; Xu, P.; Shaked, Y.; Hicklin, D.J.; Kerbel, R.S. Anticancer therapies combining antiangiogenic and tumor cell cytotoxic effects reduce the tumor stem-like cell fraction in glioma xenograft tumors. *Cancer Res.* **2007**, *67*, 3560–3564. [CrossRef] [PubMed]

42. Sun, H.; Jia, J.; Wang, X.; Ma, B.; Di, L.; Song, G.; Ren, J. CD44+/CD24- breast cancer cells isolated from MCF-7 cultures exhibit enhanced angiogenic properties. *Clin. Transl. Oncol.* **2013**, *15*, 46–54. [CrossRef]

43. Charafe-Jauffret, E.; Ginestier, C.; Iovino, F.; Tarpin, C.; Diebel, M.; Esterni, B.; Houvenaeghel, G.; Extra, J.-M.; Bertucci, F.; Jacquemier, J.; et al. Aldehyde dehydrogenase 1-positive cancer stem cells mediate metastasis and poor clinical outcome in inflammatory breast cancer. *Clin. Cancer Res.* **2010**, *16*, 45–55. [CrossRef]

44. Sataloff, D.M.; Mason, B.A.; Prestipino, A.J.; Seinige, U.L.; Lieber, C.P.; Baloch, Z. Pathologic response to induction chemotherapy in locally advanced carcinoma of the breast: A determinant of outcome. *J. Am. Coll. Surg.* **1995**, *180*, 297–306. [PubMed]

45. Cortazar, P.; Zhang, L.; Untch, M.; Mehta, K.; Costantino, J.P.; Wolmark, N.; Bonnefoi, H.; Cameron, D.; Gianni, L.; Valagussa, P.; et al. Pathological complete response and long-term clinical benefit in breast cancer: The CTNeoBC pooled analysis. *Lancet* **2014**, *384*, 164–172. [CrossRef]

46. Hothorn, T.; Hornik, K. ExactRankTests: Exact Distributions for Rank and Permutation Tests.; R package version 0.8-29. Available online: https://rdrr.io/cran/exactRankTests/ (accessed on 1 April 2019).

47. Schulz, K.F.; Altman, D.G.; Moher, D.; CONSORT Group. CONSORT 2010 statement: Updated guidelines for reporting parallel group randomised trials. *PLoS Med.* **2010**, *7*, e1000251. [CrossRef] [PubMed]

48. Conley, S.J.; Gheordunescu, E.; Kakarala, P.; Newman, B.; Korkaya, H.; Heath, A.N.; Clouthier, S.G.; Wicha, M.S. Antiangiogenic agents increase breast cancer stem cells via the generation of tumor hypoxia. *Proc. Natl. Acad. Sci. USA* **2012**, *109*, 2784–2789. [CrossRef]

49. Tanei, T.; Morimoto, K.; Shimazu, K.; Kim, S.J.; Tanji, Y.; Taguchi, T.; Tamaki, Y.; Noguchi, S. Association of breast cancer stem cells identified by aldehyde dehydrogenase 1 expression with resistance to sequential Paclitaxel and epirubicin-based chemotherapy for breast cancers. *Clin. Cancer Res.* **2009**, *15*, 4234–4241. [CrossRef] [PubMed]

50. Alamgeer, M.; Ganju, V.; Kumar, B.; Fox, J.; Hart, S.; White, M.; Harris, M.; Stuckey, J.; Prodanovic, Z.; Schneider-Kolsky, M.E.; et al. Changes in aldehyde dehydrogenase-1 expression during neoadjuvant chemotherapy predict outcome in locally advanced breast cancer. *Breast Cancer Res.* **2014**, *16*, R44. [CrossRef]

51. Kida, K.; Ishikawa, T.; Yamada, A.; Shimada, K.; Narui, K.; Sugae, S.; Shimizu, D.; Tanabe, M.; Sasaki, T.; Ichikawa, Y.; et al. Effect of ALDH1 on prognosis and chemoresistance by breast cancer subtype. *Breast Cancer Res. Treat.* **2016**, *156*, 261–269. [CrossRef] [PubMed]

52. Von Minckwitz, G.; Untch, M.; Blohmer, J.-U.; Costa, S.D.; Eidtmann, H.; Fasching, P.A.; Gerber, B.; Eiermann, W.; Hilfrich, J.; Huober, J.; et al. Definition and impact of pathologic complete response on prognosis after neoadjuvant chemotherapy in various intrinsic breast cancer subtypes. *J. Clin. Oncol.* **2012**, *30*, 1796–1804. [CrossRef] [PubMed]

53. Gerber, B.; Loibl, S.; Eidtmann, H.; Rezai, M.; Fasching, P.A.; Tesch, H.; Eggemann, H.; Schrader, I.; Kittel, K.; Hanusch, C.; et al. Neoadjuvant bevacizumab and anthracycline-taxane-based chemotherapy in 678 triple-negative primary breast cancers; results from the geparquinto study (GBG 44). *Ann. Oncol.* **2013**, *24*, 2978–2984. [CrossRef] [PubMed]

54. Von Minckwitz, G.; Loibl, S.; Untch, M.; Eidtmann, H.; Rezai, M.; Fasching, P.A.; Tesch, H.; Eggemann, H.; Schrader, I.; Kittel, K.; et al. Survival after neoadjuvant chemotherapy with or without bevacizumab or everolimus for HER2-negative primary breast cancer (GBG 44—GeparQuinto). *Ann. Oncol.* **2014**. [CrossRef]

55. Charafe-Jauffret, E.; Ginestier, C.; Birnbaum, D. Breast cancer stem cells: Tools and models to rely on. *BMC Cancer* **2009**, *9*, 202. [CrossRef]

56. Yamauchi, T.; Fernandez, J.R.E.; Imamura, C.K.; Yamauchi, H.; Jinno, H.; Takahashi, M.; Kitagawa, Y.; Nakamura, S.; Lim, B.; Krishnamurthy, S.; et al. Dynamic changes in CD44v-positive cells after preoperative anti-HER2 therapy and its correlation with pathologic complete response in HER2-positive breast cancer. *Oncotarget* **2018**, *9*, 6872–6882. [CrossRef]

57. Liu, S.; Cong, Y.; Wang, D.; Sun, Y.; Deng, L.; Liu, Y.; Martin-Trevino, R.; Shang, L.; McDermott, S.P.; Landis, M.D.; et al. Breast cancer stem cells transition between epithelial and mesenchymal states reflective of their normal counterparts. *Stem Cell Reports* **2014**, *2*, 78–91. [CrossRef]

58. Lee, H.E.; Kim, J.H.; Kim, Y.J.; Choi, S.Y.; Kim, S.-W.; Kang, E.; Chung, I.Y.; Kim, I.A.; Kim, E.J.; Choi, Y.; et al. An increase in cancer stem cell population after primary systemic therapy is a poor prognostic factor in breast cancer. *Br. J. Cancer* **2011**, *104*, 1730–1738. [CrossRef] [PubMed]

59. Pierga, J.-Y.; Bidard, F.-C.; Autret, A.; Petit, T.; Andre, F.; Dalenc, F.; Levy, C.; Ferrero, J.-M.; Romieu, G.; Bonneterre, J.; et al. Circulating tumour cells and pathological complete response: Independent prognostic factors in inflammatory breast cancer in a pooled analysis of two multicentre phase II trials (BEVERLY-1 and -2) of neoadjuvant chemotherapy combined with bevacizumab. *Ann. Oncol.* **2017**, *28*, 103–109. [CrossRef]

60. Kabraji, S.; Solé, X.; Huang, Y.; Bango, C.; Bowden, M.; Bardia, A.; Sgroi, D.; Loda, M.; Ramaswamy, S. AKT1low quiescent cancer cells persist after neoadjuvant chemotherapy in triple negative breast cancer. *Breast Cancer Res.* **2017**, *19*. [CrossRef]

61. Bell, R.; Brown, J.; Parmar, M.; Toi, M.; Suter, T.; Steger, G.G.; Pivot, X.; Mackey, J.; Jackisch, C.; Dent, R.; et al. Final efficacy and updated safety results of the randomized phase III BEATRICE trial evaluating adjuvant bevacizumab-containing therapy in triple-negative early breast cancer. *Ann. Oncol.* **2016**. [CrossRef]

62. Von Minckwitz, G.; Eidtmann, H.; Rezai, M.; Fasching, P.A.; Tesch, H.; Eggemann, H.; Schrader, I.; Kittel, K.; Hanusch, C.; Kreienberg, R.; et al. Neoadjuvant chemotherapy and bevacizumab for HER2-negative breast cancer. *N. Engl. J. Med.* **2012**, *366*, 299–309. [CrossRef]

63. Kerr, R.S.; Love, S.; Segelov, E.; Johnstone, E.; Falcon, B.; Hewett, P.; Weaver, A.; Church, D.; Scudder, C.; Pearson, S.; et al. Adjuvant capecitabine plus bevacizumab versus capecitabine alone in patients with colorectal cancer (QUASAR 2): An open-label, randomised phase 3 trial. *Lancet Oncol.* **2016**, *17*, 1543–1557. [CrossRef]

64. Wakelee, H.A.; Dahlberg, S.E.; Keller, S.M.; Tester, W.J.; Gandara, D.R.; Graziano, S.L.; Adjei, A.A.; Leighl, N.B.; Aisner, S.C.; Rothman, J.M.; et al. Adjuvant chemotherapy with or without bevacizumab in patients with resected non-small-cell lung cancer (E1505): An open-label, multicentre, randomised, phase 3 trial. *Lancet Oncol.* **2017**, *18*, 1610–1623. [CrossRef]

65. Pujade-Lauraine, E.; Hilpert, F.; Weber, B.; Reuss, A.; Poveda, A.; Kristensen, G.; Sorio, R.; Vergote, I.; Witteveen, P.; Bamias, A.; et al. Bevacizumab combined with chemotherapy for platinum-resistant recurrent ovarian cancer: The AURELIA open-label randomized phase III trial. *J. Clin. Oncol.* **2014**, *32*, 1302–1308. [CrossRef]

66. Aghajanian, C.; Blank, S.V.; Goff, B.A.; Judson, P.L.; Teneriello, M.G.; Husain, A.; Sovak, M.A.; Yi, J.; Nycum, L.R. OCEANS: A randomized, double-blind, placebo-controlled phase III trial of chemotherapy with or without bevacizumab in patients with platinum-sensitive recurrent epithelial ovarian, primary peritoneal, or fallopian tube cancer. *J. Clin. Oncol.* **2012**, *30*, 2039–2045. [CrossRef]

67. Yamazaki, K.; Nagase, M.; Tamagawa, H.; Ueda, S.; Tamura, T.; Murata, K.; Eguchi Nakajima, T.; Baba, E.; Tsuda, M.; Moriwaki, T.; et al. Randomized phase III study of bevacizumab plus FOLFIRI and bevacizumab plus mFOLFOX6 as first-line treatment for patients with metastatic colorectal cancer (WJOG4407G). *Ann. Oncol.* **2016**, *27*, 1539–1546. [CrossRef] [PubMed]

Journal of
Clinical Medicine

Article

Cadherin 11 Inhibition Downregulates β-catenin, Deactivates the Canonical WNT Signalling Pathway and Suppresses the Cancer Stem Cell-Like Phenotype of Triple Negative Breast Cancer

Pamungkas Bagus Satriyo [1,2], Oluwaseun Adebayo Bamodu [3,4], Jia-Hong Chen [5,6], Teguh Aryandono [7], Sofia Mubarika Haryana [8], Chi-Tai Yeh [1,3,4,5,*] and Tsu-Yi Chao [1,3,4,5,6,*]

[1] International Ph.D. Program in Medicine, College of Medicine, Taipei Medical University, Taipei City 11031, Taiwan; pbagus57@yahoo.com
[2] Doctorate Program of Medical and Health Science, Faculty of Medicine Public Health and Nursing, Universitas Gadjah Mada, Yogyakarta 55281, Indonesia
[3] Department of Hematology & Oncology, Taipei Medical University-Shuang Ho Hospital, New Taipei City 23561, Taiwan; dr_bamodu@yahoo.com
[4] Department of Medical Research & Education, Taipei Medical University-Shuang Ho Hospital, New Taipei City 23561, Taiwan
[5] Graduate Institute of Clinical Medicine, College of Medicine, Taipei Medical University, Taipei City 11031, Taiwan; ndmc_tw.tw@yahoo.com.tw
[6] Division of Medical Oncology and Hematology, Tri-Service General Hospital, National Defense Medical Centre, Taipei 11409, Taiwan
[7] Department of Surgery, Faculty of Medicine Public Health and Nursing, Universitas Gadjah Mada, Yogyakarta 55281, Indonesia; teguharyandono@yahoo.com
[8] Department of Histology and Cellular Biology, Faculty of Medicine Public Health and Nursing, Universitas Gadjah Mada, Yogyakarta 55281, Indonesia; sofia.mubarika@gmail.com
* Correspondence: ctyeh@s.tmu.edu.tw (C.-T.Y.); 10575@s.tmu.edu.tw (T.-Y.C.);
Tel.: +886-2-2490088 (ext. 8881) (C.-T.Y.); +886-2-2490088 (ext. 8885) (T.-Y.C.);
Fax: +886-2-2248-0900 (C.-T.Y. & T.-Y.C.)

Received: 6 December 2018; Accepted: 22 January 2019; Published: 27 January 2019

Abstract: Background: Cancer stem cells (CSCs) promote tumor progression and distant metastasis in breast cancer. Cadherin 11 (CDH11) is overexpressed in invasive breast cancer cells and implicated in distant bone metastases in several cancers. The WNT signalling pathway regulates CSC activity. Growing evidence suggest that cadherins play critical roles in WNT signalling pathway. However, CDH11 role in canonical WNT signalling and CSCs in breast cancer is poorly understood. Methods: We investigated the functional association between CDH11 and WNT signalling pathway in triple negative breast cancer (TNBC), by analyzing their expression profile in the TCGA Breast Cancer (BRCA) cohort and immunohistochemical (IHC) staining of TNBC samples. Results: We observed a significant correlation between high CDH11 expression and poor prognosis in the basal and TNBC subtypes. Also, CDH11 expression positively correlated with β-catenin, wingless type MMTV integration site (WNT)2, and transcription factor (TCF)12 expression. IHC results showed CDH11 and β-catenin expression significantly correlated in TNBC patients ($p < 0.05$). We also showed that siRNA-mediated loss-of-CDH11 (siCDH11) function decreases β-catenin, Met, c-Myc, and matrix metalloproteinase (MMP)7 expression level in MDA-MB-231 and Hs578t. Interestingly, immunofluorescence staining showed that siCDH11 reduced β-catenin nuclear localization and attenuated TNBC cell migration, invasion and tumorsphere-formation. Of translational relevance, siCDH11 exhibited significant anticancer efficacy in murine tumor xenograft models, as demonstrated by reduced tumor-size, inhibited tumor growth and longer survival time. Conclusions: Our findings indicate that by modulating β-catenin, CDH11 regulates the canonical WNT signalling pathway. CDH11 inhibition suppresses the CSC-like phenotypes and tumor growth of TNBC cells and represents a novel therapeutic approach in TNBC treatment.

Keywords: Cadherin 11; WNT signaling; β-catenin; cancer stem cells; TNBC

---

1. Introduction

Triple negative breast cancer (TNBC) constitutes 10−20% of all breast cancer cases and exhibits more aggressive traits and worse patient prognosis than hormone receptor- and HER2- positive breast cancer [1]. To date, TNBC lacks US-FDA-approved target therapy [1,2]. Chemotherapy remains the primary treatment of choice. Despite a documented record of TNBC sensitive to chemotherapy, only 30% of such cases show pathologic complete responses, while the less- or non- sensitive patients exhibit a high probability of developing distant metastases and relapses [2]. Metastatic breast cancer is currently an incurable disease, with palliative treatment being the therapy of last resort to alleviate pain, enhance quality of life and help patients live longer [3]. The discovery and development of novel efficacious targeted therapy to treat TNBC and prevent metastatic breast disease are important issues.

Since the discovery of CSCs in solid tumors, namely, breast cancer in 2003 [4], there is growing evidence of CSCs existence or presence in other solid tumors [5]. The CSCs possess self-renewal capacity and differentiate into heterogeneous tumor cells. They have been implicated in metastatic dissemination of cancerous cells in many cancer models including breast cancer and are shown to be more resistant to chemotherapy and lead to cancer relapse [6]. Targeting these cells in cancer treatment may be more beneficial for TNBC patients.

The WNT signalling pathway is a principal regulator of self-renewal and helps maintain the undifferentiated state of normal mesenchymal stem cells and CSCs [7,8]. The presence of the WNT ligand has been shown to promote β-catenin translocation to the nucleus and activate target gene of WNT signalling pathway [9]. Knocking down β-catenin reduced stem cell-like cell population and their chemoresistance to doxorubicin; the β-catenin-silenced TNBC cell lines generated smaller tumors than control cells when inoculated into the mammary fat pad of tumor xenograft mice models [10]. In the membrane, β-catenin binds to different cadherins in the adhesion junction complex and become stabilized; however, β-catenin in the cytoplasm is degraded by the APC-Axin-GSK3β complex in the absence of WNT ligands [11]. In fact, cadherins inhibit activation of the canonical WNT signalling pathway by keeping β-catenin in the membrane. β-catenin translocation to the nucleus is prevented in the presence of WNT ligands [12]. During epithelial-to-mesenchymal transition (EMT), cadherins act as a pool of calcium-dependent adhesion-competent β-catenin, which is a principal mediator of canonical WNT signalling activation, thus, depleting cadherins reduces β-catenin in the membrane and suppresses WNT signalling activation despite the presence of β-catenin in the cytoplasm [13]. Cadherin 11, a type II cadherin and mesenchymal protein marker, is overexpressed in the invasive breast cancers [14]. In prostate and renal cancers, CDH11 plays an important role in distant bone metastasis. Expression of CDH11 in metastatic bone tumors is higher than in primary tumor sites [15,16]. The brain, lungs, and bones are the most common sites of distant metastasis in TNBC patients [17,18]. Interestingly, these organs have higher expression of CDH11 than the breast itself in normal condition [19]. Inhibition of CDH11 in breast cancer cell lines reduced migration and invasion ability [20], suggesting that it may play an important role in TNBC metastasis.

In this present study, we hypothesized and validated our hypothesis that CDH11 regulates the canonical WNT signalling pathway and the CSCs-like phenotypes. The inhibition of the CSCs-like phenotypes of TNBC cells through CDH11 repression represents a novel therapeutic approach in TNBC treatment. Thus, we highlight a novel aspect of TNBC biology and provide a molecular basis for further exploration of CDH11-mediated anti-TNBC targeted therapy.

2. Materials and Methods

2.1. Public Dataset Analyses

The TCGA Breast Cancer (BRCA) cohort datasets were downloaded from UCSC Xena Browser platform. This cohort contains several datasets from 1247 samples of breast cancer patients. We also made use of the TNBC cohort data of the Molecular Taxonomy of *Breast Cancer* International Consortium (METABRIC) cohort dataset ($n = 1904$) downloaded from the European Genome-Phenome archive (EGAS00000000098). The METABRIC study classifies breast tumors into subcategories, based on genetic fingerprints and molecular signatures which are intended to help predict therapeutic response and determine the optimal course of treatment. The gene expression RNAseq-IlluminaHiSeq and Phenotypes datasets were downloaded and used for further analysis. The PAM50 mRNA nature2012 clinical parameter was used for classifying breast cancer patients into luminal A, luminal B, Her2-enriched and basal-like (BL) subgroups. The status of ER, PR and Her2 were used to determine the triple negative breast cancer subgroup. To establish correlation between CDH11 and prognosis of breast cancer patient for each subgroup, we performed Kaplan Meier (KM) overall survival analysis using the "R2: Genomics Analysis and Visualization Platform". For the low/high expression group dichotomization, we did not use the traditional median or mean cutoff values, rather we employed a bioinformatics approach using the automated 'Kaplan scan' cutoff function of the R2 genomic interface platform. The 'Kaplan scan' generates a KM plot based on the most optimal mRNA cut-off expression level to discriminate between a good (low expression) and bad (high expression) prognosis cohort. This was followed by the Bonferroni test for statistical significance (*p*-value) of the KM plot.

2.2. Immunohistochemistry

Triple negative breast cancer tissue samples ($n = 38$) were obtained from the Shuang Ho Hospital (SHH) breast cancer cohort. Ethical approval for the study was obtained from Joint Institutional Review Board (JIRB) of the Taipei Medical University (approval number: N201603028). Tissue sections (4 μm) were deparaffinized and rehydrated in gradually decreased concentration of methanol (100%, 95%, and 70%). Antigen retrieval was carried out by boiling slides in pressure cooker containing TrilogyTM buffer (Sigma-920P-06, Cell Marque, Sigma-Aldrich, Inc. St. Louis, MO, USA) for 5 min, and followed by incubation in hydrogen peroxide blocking solution (TA-125-H2O2Q, Thermo Fisher Scientific, Waltham, MA, USA) for 10 min. Nonspecific binding was blocked with Ultra V Block (TA-125-PBQ, Thermo Fisher Scientific, Waltham, MA, USA) for 10 min. The slides were incubated in primary antibodies against cadherin 11 (polyclonal antibody, 71-7600, Thermo Fisher Scientific, Waltham, MA, USA) and β-catenin (H-102: sc-7199, Santa Cruz Biotechnology, Santa Cruz, CA, USA) with working dilution 1:100 and 1:50, respectively for overnight at 4 °C. Later, tissue slides were incubated in Primary Antibody Amplifier Quanto (TL-125-QPB, Thermo Fisher Scientific, Waltham, MA, USA) for 10 min, in Horseradish peroxidase (HRP) Polymer Quanto (TL-125-QPH, Thermo Fisher Scientific, Waltham, MA, USA) for 10 min and then in DAB Quanto Chromogen (TA-004-QHCX, Thermo Fisher Scientific, Waltham, MA, USA) diluted 3:100 in DAB Quanto Substrate (TA-125-QHSX, Thermo Fisher Scientific, Waltham, MA, USA) for 3 min. Slides were counterstained with hematoxylin. The immunoreactive score system (IRS) was used to measure the expression level of protein of interest as previously described [21]. For final analyses, negative-mild staining was categorized as "Low" while moderate-strong positive staining was categorized as "High".

2.3. Cell Culture

TNBC cell lines, MDA-MB-231 and Hs578T were purchased from American Type Culture Collection (ATCC, Manassas, VA, USA). MDA-MB-231, derived from the pleural effusion and metastatic site of a female patient with breast adenocarcinoma, constitutively express WNT7B, EGF and TGFα, and forms poorly differentiated adenocarcinoma (grade III) in experimental mice models. Hs578T, however, is from a female patient with primary breast carcinoma and is non-tumorigenic in immunosuppressed mice.

The selection of the 2 cell lines provided a basis for phenotypic and functional comparison between two variants of TNBC cells. The cell lines used in this study were periodically tested and confirmed to be free from mycoplasma and/or cross-contamination with cells derived for a different origin during laboratory manipulation or processing. Both cell lines were cultured in DMEM) (DML10-1000ML, Caisson Labs, Smithfield, UT, USA) supplemented with 10% heat-inactivated fetal bovine serum (FBS), 100 IU/mL Penicillin and 100 µg/mL Streptomycin, in 5% CO_2 humidified atmosphere at 37 °C. For maintenance, both cell lines were sub-cultured every 48–72 h.

2.4. CDH11 Knockdown

For CDH11 loss-of-function studies, we used On-Target plus Human CDH11 siRNA-Smart pool (L-013493-00-0005, Dharmacon, Lafayette, CO, US) that contains 4 siRNA mix. CDH11 #1: 5′-GUGAGAACAUCAUUACUUA-3′. CDH11 #2: 5′-GGACAUGGGUGGACACAUAUG-3′. CDH11 #3: 5′-GGAAAUAGCGCCAAGUUAG-3. CDH11 #4: 5′-CCUUAUGACUCCAUUCAAA-3′. A day before transfection, MDA-MB-231 cell line was seeded into 6-well cell plate at a density of 3×10^5 cells/well in DMEM containing 10% FBS without antibiotics. The Hs578t cell line was seeded in the same condition except the density (2×10^5 cells/well). After 24 hours incubation, 30–40% confluent cells were transfected with siRNA using Lipofectamine 3000 (L3000008, Thermo Fisher Scientific, Waltham, MA, USA) in serum-free DMEM. 80 nM of siRNA was used as final concentration. Six hours after initial transfection, the medium was replaced with fresh DMEM containing 10% FBS. Next day, cells were re-transfected with same siRNA, following same transfection steps. The siRNA-transfected cells were incubated for 48 h since the first transfection and used for western blot, and functional assays.

2.5. Western Blotting

Triple negative breast cancer cells were harvested by scraping method then lyzed with RIPA lysis buffer that supplemented with protease inhibitor (1X, Cat# 78430, Halt Protease Inhibitor Single-Use Cocktail, Thermo Fisher Scientific, Waltham, MA, USA) and phosphatase inhibitor (0.5X, Pierce™ Phosphatase Inhibitor Mini Tablets, Thermo Fisher Scientific, Waltham, MA, USA). After homogenization with pellet pestle, the cells were incubated on ice for 30 min. The re-suspended cells were centrifuged at 12.000 rpm for 10 min; then, supernatants were collected as total protein samples. Pierce™ BCA Protein Assay Kit (Thermo Fisher Scientific, Waltham, MA, USA) was used to determine final protein concentration. After boiling with sample buffer for 10 min and centrifuged for 10 min, the total protein samples were loaded onto and ran by sodium dodecyl sulfate-polyacrylamide gel electrophoresis (SDS-PAGE), then blots transferred onto polyvinylidene fluoride (PVDF) membranes, followed by blocking with 5% skimmed milk or 5% BSA in TBS-Tween-20 (TBST) for 1 h. The membranes were then incubated in primary antibodies against CDH11 (polyclonal antibody, 1:1000, Cat# 71-7600, Thermo Fisher Scientific, Waltham, MA, USA), β-catenin (monoclonal antibody, 1:1000, Cat# 9582P, Cell Signalling Technology, Danvers, MA, USA), c-Myc (monoclonal antibody, 1:500, Cat# sc-40, Santa Cruz Biotechnology, Santa Cruz, CA, USA), Met (monoclonal antibody, 1:1000, Cat# 8198, Cell Signalling Technology, Danvers, MA, USA), SOX2 (monoclonal antibody, 1:1000, Cat# 3579P, Cell Signalling Technology, Danvers, MA, USA), CD44 (monoclonal antibody, 1:1000, Cat# 3570, Cell Signalling Technology, Danvers, MA, USA), KLF4 (polyclonal antibody, 1:1000, Cat# 4038P, Cell Signalling Technology, Danvers, MA, USA), c-Jun (monoclonal antibody, 1:1000, Cat# 9165, Cell Signalling Technology, Danvers, MA, USA), MMP7 (monoclonal antibody, 1:1000, Cat# 3801, Cell Signalling Technology, Danvers, MA, USA),and β-actin (monoclonal antibody, 1:10.000, Cat# ab6276, Abcam Biotechnology Company, Cambridge, MA, UK) at 4 °C overnight. After washing with TBST, membranes were probed with an HRP-labeled anti-rabbit or anti-mouse IgG secondary antibody as appropriate. Protein bands were visualized by UVP Biospectrum Imaging System (Vision Works LS 6.8, Level Biotechnology Inc. New Taipei City, Taiwan) using ECL reagents (Thermo Fisher Scientific, Waltham, MA, USA).

2.6. Transswell Matrigel Invasion and Migration Assay

We used the 24-well plate transwell system to evaluate migration ability. The upper chambers of the transwell were pre-coated with solubilized Matrigel (BD Bioscience). After the matrigel polymerized, MDA-MB-231 and Hs578t cell lines (2 x 10^4 cells/well) were seeded into the upper chamber containing 200 µL serum-free DMEM, while 500 µL DMEM with 10% FBS in the lower chamber served as chemoattractant. The medium was discarded after 24 h incubation and cells on the membrane were fixed using formaldehyde (10%) for 20 min at room temperature before crystal violet staining for 20 min. The cells on the upper side of the membrane were carefully removed with a cotton swab. The migrated cells were observed under microscope and the total number of cells on the lower surface counted. Migration assay was performed following same steps as the invasion assay, but without Matrigel on the upper chamber.

2.7. Immunofluorescence Assay

The wild-type and siCDH11 MDA-MB-231 and Hs578t cells were seeded in 96-well plates and incubated for 24 h, washed with 200 µL ice-cold PBS, then fixed using 200 µL pre-chilled 4% formaldehyde for 15 min at room temperature before being blocked with 200 µL blocking buffer (1x PBS/ 5% BSA/ 0.3% Triton^{TM} X-100) for 1 h at room temperature. After washing with PBS, the cells were incubated with 100 µL primary antibodies against CDH11 (monoclonal antibody, Cat#, 5B2H5, 1:50, Thermo Fisher Scientific, Waltham, MA, USA), or β-catenin (monoclonal antibody, Cat# 9582P, 1:50, Cell Signalling Technology, Danvers, MA, USA) at 4 °C overnight in a dark room. Next day, cells were washed with 0.1% PBS, incubated with 100 µL Alexa-Flour-conjugated secondary antibodies against rabbit and mouse for 1 h at room temperature in the dark. In this experiment, the primary and secondary antibodies were diluted in antibody dilution buffer (1x PBS/ 1% BSA/ 0.3% Triton^{TM} X-100). Next, cells were washed with PBS and were counterstained with DAPI (0.5 µg/mL, Vector Laboratories, Burlingame, CA, USA) and evaluated using fluorescence microscope.

2.8. Mammosphere Formation Assay

The wild-type and siCDH11 MDA-MB-231 and Hs578t cells were harvested and seeded in 24-well plates (Cat# CLS3473, Corning Costar Ultra-Low Attachment Multiple Well Plates, Sigma-Aldrich, St. Louis, MO, USA) that contained 500 µL serum-free DMEM medium, supplemented with B27 (Invitrogen, Carlsbad, CA, USA), and 10 ng/mL epidermal growth factor (BD Biosciences, Palo Alto, CA, USA). The cells were seeded in 3 different density (1000 cells/cm^2, 2000 cells/cm^2, and 4000 cells/cm^2) for each group. After 5 days incubation, formed mammospheres were observed and evaluated under microscope.

2.9. In Vivo Study

The MDA-MB-231 and Hs578t cells were treated with siRNA specific for CDH11 for 48h. NOD/SCID mice were randomized into wild type (n = 5) and siCDH11 (n = 5) group for each cell line. The mice were inoculated subcutaneously with 2 × 10^6 wild type (WT) or siCDH11 MDA-MB-231 or Hs578t cells in their hind flank. Mice tumor sizes were measured on days 6, 9, 12, 15, 18, and 27 after TNBC cell inoculation using callipers and tumor volumes calculated with a standard formula: length × width^2 × 0.5. The tumor-bearing mice were sacrificed, and the tumor mass were observed and measured on day 27 post-inoculation. In parallel in vivo studies following same steps except sacrificing on day 27, the mice were observed until day 45 post-inoculation to assess the effect of altered CDH11 expression on the survival rates. For maintenance of siRNA effect, mice in the siCDH11 group were injected intra-tumorally with 10 µmol/L of 'siCDH11-atelocollagen' complex on days 9, 18, 27, and 36 after TNBC cell inoculation. The siCDH-atelocollagen complex, usually prepared the preceding evening and stored at 4 °C before use, consisted of equal volumes of well mixed siCDH11 and atelocollagen (Koken Co. Ltd., Tokyo, Japan). All tumor xenograft animal studies were approved by the TMU-SHH Joint Institutional Review Board and performed in accordance with protocol approved by the TMU Institutional Animal Care and Use Committee (approval number: LAC-2015-0383).

2.10. Statistical Analysis

Kaplan Meier overall survival analysis were performed using "R2: Genomics Analysis and Visualization Platform" and GraphPad Prism for Windows version 7.00 software (GraphPad Software Inc., San Diego, CA, USA). Statistical significance was assumed when $p \leq 0.05$. For correlation studies between CDH11 and several components of the WNT signalling pathway SigmaPlot for Windows version 10.0 software (Systat Software Inc., San Jose, CA, USA) was used. All cell-based assays were performed at least three times in triplicates and results expressed as Mean $\pm$ SD. Quantitative and qualitative analyses of western blot, mammosphere formation, migration and invasion assays data were performed using ImageJ version 1.51j8 (Wayne Rasband National Institutes of Health, Bethesda, MD, USA) and the graph results were created using SigmaPlot.

3. Results

3.1. High CDH11 Expression has Significant Positive Correlation with Poor Overall Survival in Patients with Basal-Like and Triple Negative Breast Cancer

To determine existing correlation or association between CDH11 expression and prognosis of breast cancer patients, we assessed TCGA Breast Cancer (BRCA) cohort datasets from UCSC Xena Browser (https://xenabrowser.net/datapages/). We divided the breast cancer patients into luminal A, luminal B, Her2-enriched, BL and TNBC subgroup. We observed that high CDH11 expression correlated with significantly better survival rates in the luminal A ($p = 0.0381$) and Her-2 enriched ($p = 0.0319$) subgroups, while no statistically significant correlation was observed between CDH11 expression and luminal B breast cancer patient survival ($p = 0.2831$) (Figure 1A–C). However, in contrast, high CDH11 expression in the BL ($p = 0.0049$) and TNBC ($p = 0.045$) subgroups significantly correlated with worse overall survival (Figure 1D,E). Our data is consistent with the aggressive phenotype of TNBC cells and their poor prognosis despite initial good response to therapy and corroborate significant overlap between the TNBC and BL breast cancer molecular subtypes [22,23]. These results indicate the prognostic nature of CDH11 expression profile and suggest its role as a potential candidate for anti-TNBC targeted therapy.

3.2. CDH11 Positively Modulates β-Catenin Expression and Is Associated with Activation of the Canonical WNT Signalling Pathway

β-catenin is a key component of the canonical WNT signalling pathway and has been shown to directly bind to the cytoplasmic domain of CDH11 (13). We hypothesized that CDH11 plays an important role in enhancing the stemness, migration, and invasion potential of TNBC cells through activation of the canonical WNT signalling pathway. Using the METABRIC-Breast Cancer (IlluminaHiSeq) cohort dataset ($n = 1904$), we investigated the correlation between CDH11 and several components of the WNT signalling pathway in the TNBC subgroup. We found that CDH11 expression positively correlated with β-catenin, WNT2, and TCF2 expression with pearson's r = 0.29 ($p = 8.3 \times 10^{-30}$), 0.60 ($p = 7.8 \times 10^{-183}$) and 0.48 ($p = 1.3 \times 10^{-101}$), respectively (Figure 2A–C). Corroborating the results above, our IHC data demonstrated that tissues from TNBC patients concurrently overexpressed CDH11 and β-catenin proteins, compared to the weak/non-expression in non-tumor tissues (Figure 2D–F). In fact, correlative analysis of our local TNBC cohort ($n = 38$) indicate a strong positive correlation between CDH11 and β-catenin (r = 0.62; $p < 0.0001$; Figure 2F, *lower panel*). To reconfirm these findings and establish the functional significance of CDH11 expression and/or activity in TNBC cells, we knocked-down CDH11 in MDA-MB-231 and Hs578t TNBC cell lines and examined its effect on WNT signalling using western blot and immunofluorescence staining assays. Reduced CDH11 expression in the TNBC cells altered their morphology from spindle-like to polygonal, reduced cell viability, and increased cell doubling time (Figure 3A,B). Additionally, our western blot data showed that MDA-MB-231 and Hs578t cells transfected with siCDH11 exhibited significantly decreased expression levels of CDH11, β-catenin, Met, c-Myc, c-Jun, and MMP7 (Figure 3C,D). Furthermore, we observed that siCDH11 in MDA-MD-231 and Hs578t cells reduced β-catenin nuclear localization and expression (Figure 4). These data corroborate the modulatory role of CDH11 on the WNT/β-catenin signalling pathway and suggest that CDH11 plays an important role in the maintenance of β-catenin protein in the MDA-MB-231 and Hs578t TNBC cell lines.

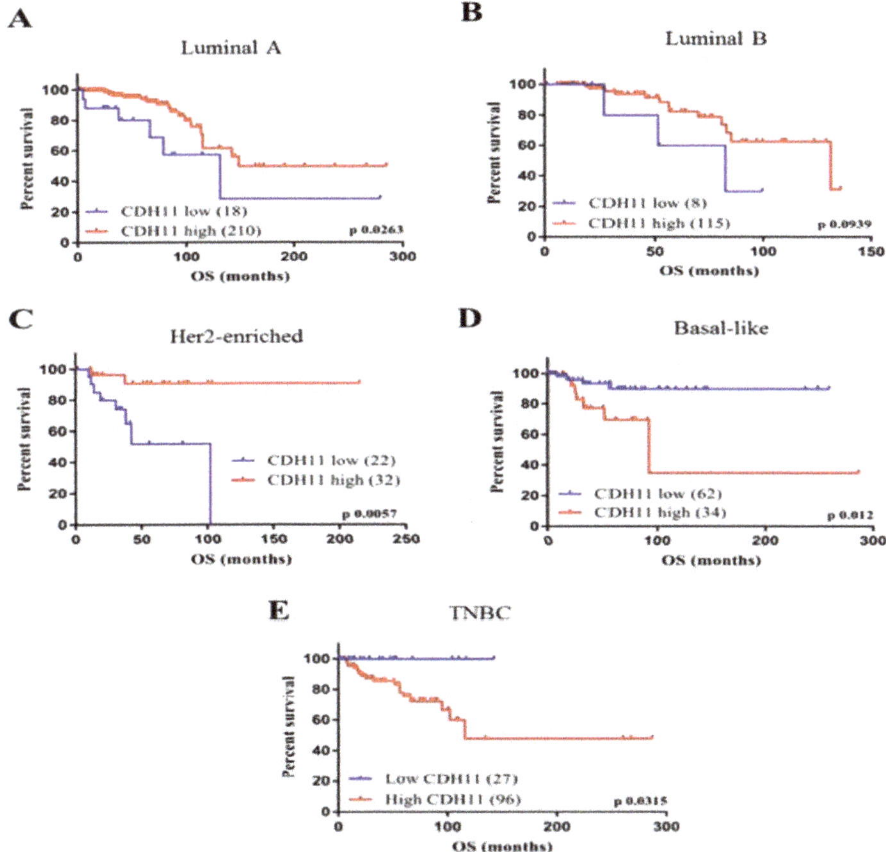

Figure 1. Patients with TNBC and basal-like breast cancer exhibiting high expression of CDH11 are characterized by worse prognosis. Kaplan-Meier plots based on analyses of breast cancer cohort (*n* = 1247) using the TCGA Breast Cancer (BRCA) cohort datasets show the effect of CDH11 expression on the overall survival rates in patients regardless of molecular subtype, (**A**) Luminal A, (**B**) Luminal B, (**C**) Her2-enriched, (**D**) Basal-like, or (**E**) Triple negative breast cancer.

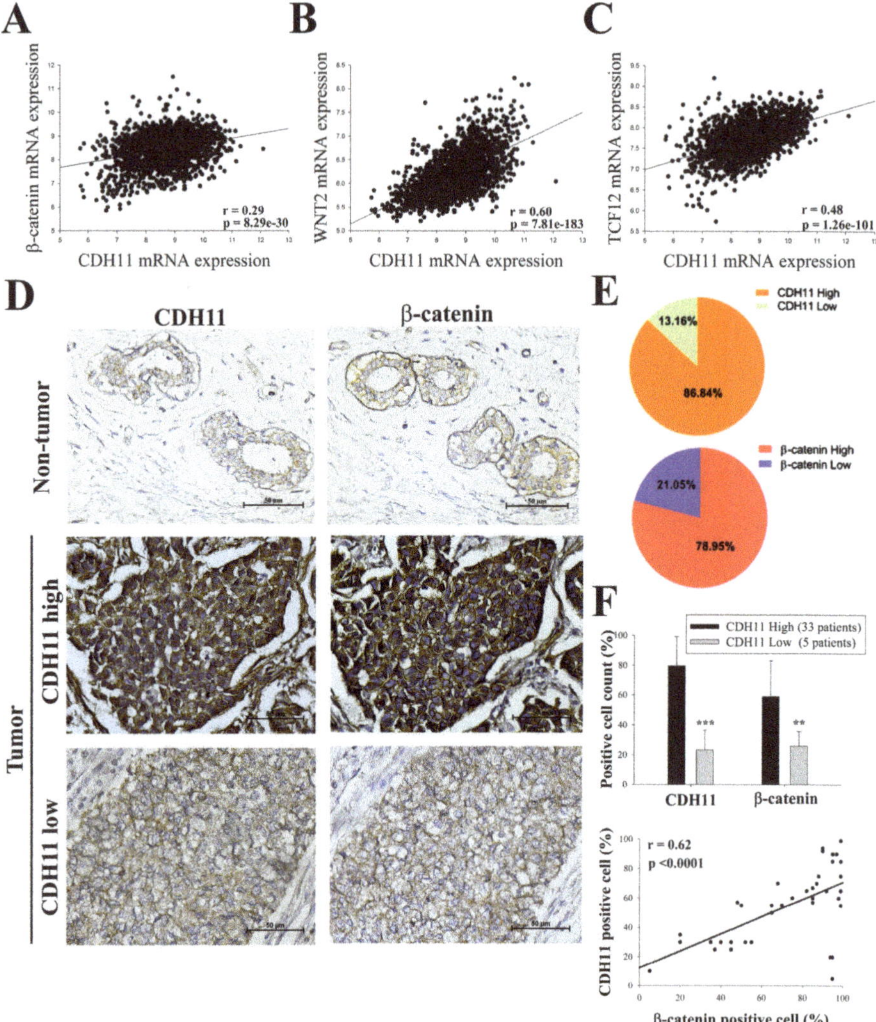

Figure 2. CDH11 positively modulates β-catenin expression and activates the canonical WNT signalling pathway. Statistical analyses of the TNBC sub-group of the METABRIC Breast Cancer cohort (*n* = 1904) show positive correlation between expression of CDH11 and (**A**) β-catenin, (**B**) WNT2, and (**C**) TCF12 mRNA. (**D** and **E**) Differential expression of CDH11 and β-catenin in non-tumor and TNBC tissue samples demonstrated by representative IHC images and percentages of patients. Scale bar: 50 μm. (**F**) Histogram showing positive correlation between CDH11 and β-catenin protein expression level in our TNBC cohort (*n* = 38). * *p* < 0.05, ** *p* < 0.01, *** *p* < 0.001

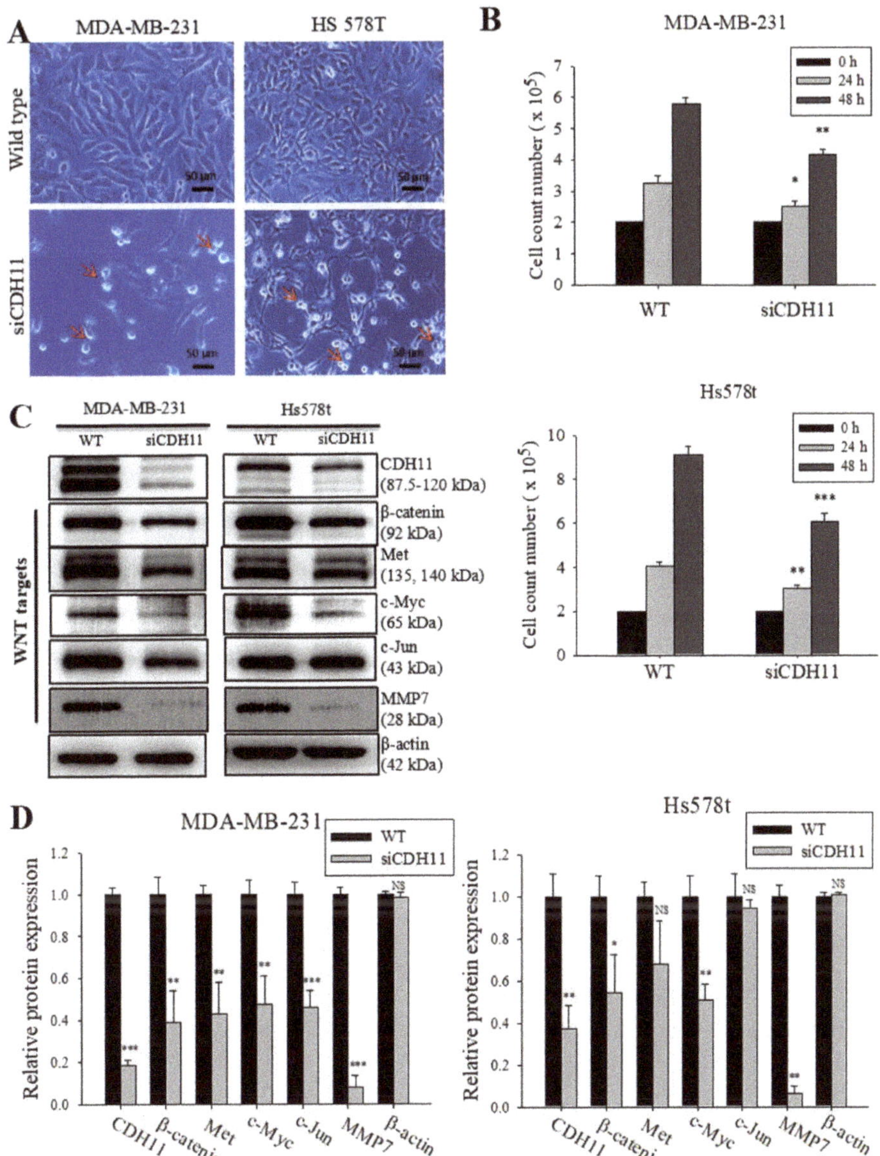

Figure 3. Loss of CDH11 function using siRNA changes cell morphology and inhibits activation of the canonical WNT signalling pathway in TNBC cells. (**A**) The cell morphology of MDA-MB-231 and Hs578t changed from spindle-like to polygonal after loss of CDH11 expression. (**B**) Graphical representation of the effect of siCDH11 on the doubling time of MDA-MB-231 and Hs578t cells as determined by trypan-blue exclusion. (**C**) Western blot results show CDH11 knockdown decreased β-catenin, Met, c-Myc, c-Jun, and MMP7 protein expression level in MDA-MB-231 and Hs578t cell lines. (**D**) Graphical representation of C. Results expressed as mean ± SD of assays done 3 times in triplicate. * $p < 0.05$, ** $p < 0.01$, *** $p < 0.001$; NS, not significant; Red arrow, floating non-viable cells.

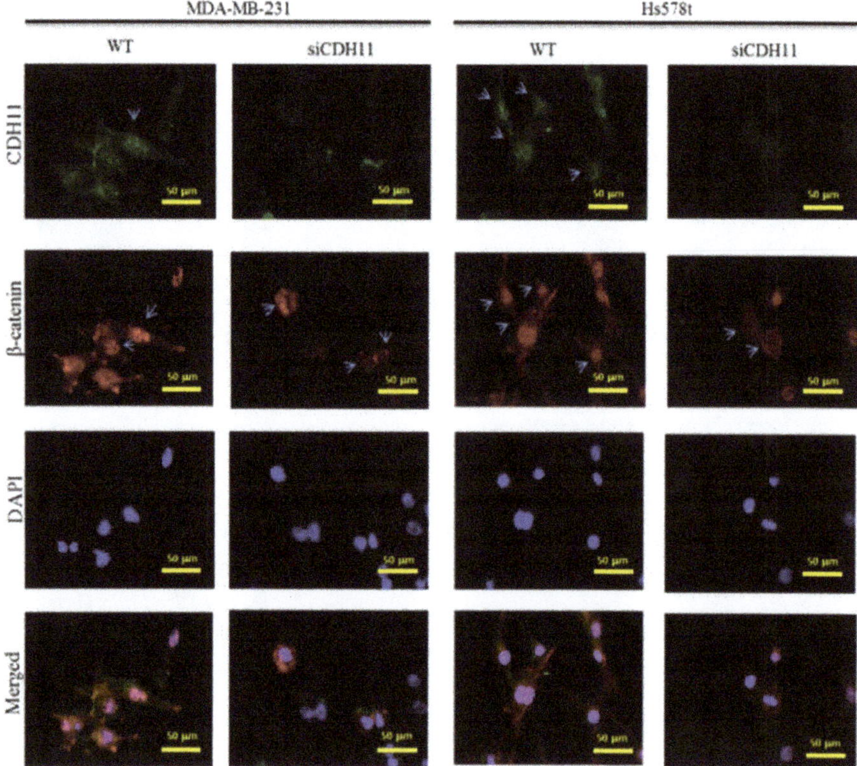

Figure 4. Knockdown of CDH11 using siRNA reduces β-catenin nuclear co-localization in TNBC cells. Immunofluorescence staining of MDA-MB-231 and Hs578t cells with CDH11 (green), β-catenin (red), and DAPI (blue). The results show that silencing CDH11 concurrently reduced CDH11 and β-catenin protein expression level and their nuclear localization. The blue arrows indicate presence and absent of CDH11 and β-catenin in the nuclear region. Images are representative of four separate assays in triplicate. Scale bar: 50 μm.

3.3. CDH11 Inhibition Suppresses the Stem Cell-Like Phenotype of TNBC Cells

CSCs constitute a subset of cancer cells with self-renewal ability and enhanced resistance to some chemotherapy agents [6]. We investigated whether CDH11 plays an important role in TNBC stemness. Tumorsphere formation assays using wild-type (WT) and siCDH11 MDA-MB-231 and Hs578t cells show that the siCDH11 cells displayed lesser tumorsphere formation efficiency compared to the WT cells (Figure 5A–E). Measurement of tumorsphere sizes showed that all the siCDH11 tumorspheres were significantly smaller than those formed by their WT counterparts of both cell lines except in the lowest seeding cells of Hs578t (Figure 5B,C). The siCDH11 cells produced fewer tumorspheres than WT cells at all seeding density, except 4000 cells/cm^2 (Figure 5D,E). To further confirm the important role of CDH11 in TNBC stemness, we assessed several CSCs markers. We observed significantly downregulated Sox2, KLF4, CD44, and c-Myc protein expression levels in the siCDH11 group compared to the WT group (Figure 5F and Figure S1A). These data demonstrate that tumorsphere formation efficiency is suppressed in the CDH11-deficient MDA-MB-231 and Hs578t TNBC cell lines and suggest a role for CDH11 expression in the modulation of the stem cell-like phenotype of TNBC cells.

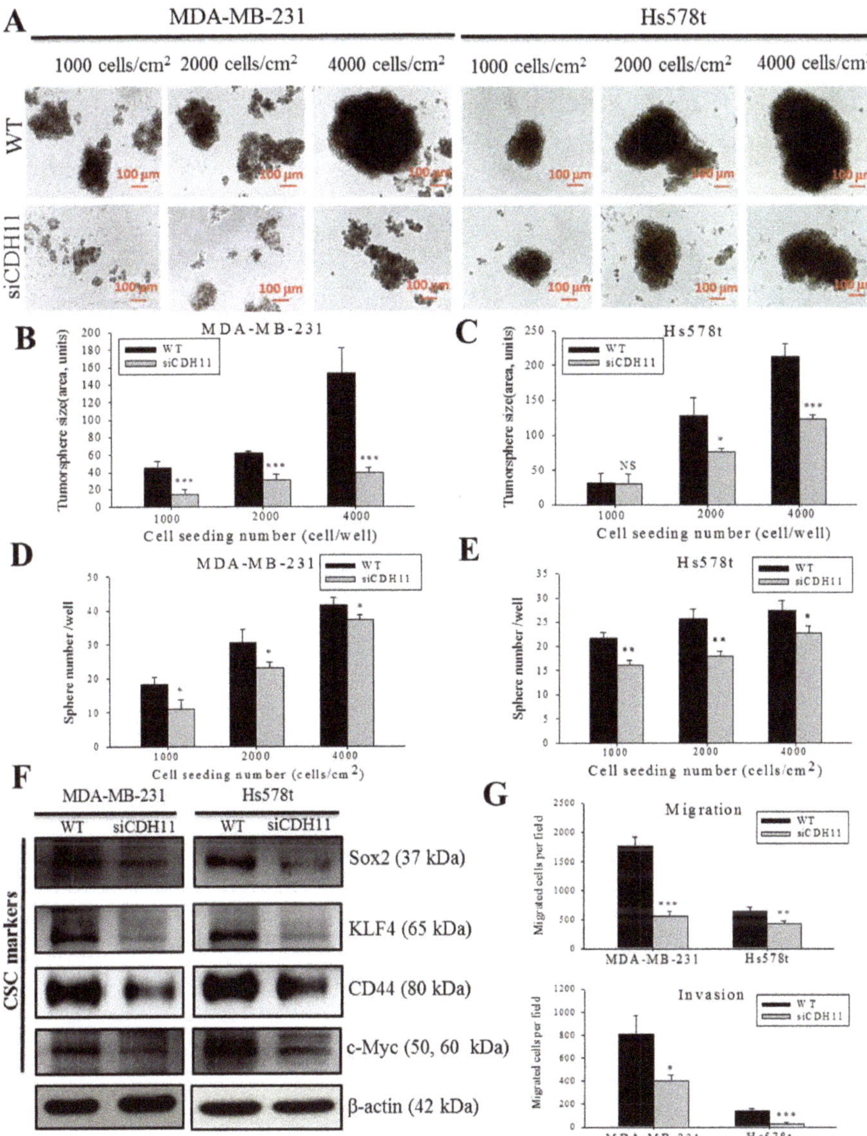

Figure 5. CDH11 inhibition suppresses the stem cell-like phenotype of TNBC cells. (**A**) Images showing tumorsphere size in wild-type cells and their siCDH11 counterparts of varying cell density. Histograms comparing the tumorsphere sizes of wild-type or si-CDH11 (**B**) MDA-MB-231 and (**C**) Hs578t. Histograms comparing the number of tumorspheres formed by wild-type or si-CDH11 (**D**) MDA-MB-231 and (**E**) Hs578t cells, based on ImageJ quantification. Only tumor spheres $\geq$40 µm in size were counted. (**F**) The effect of siCDH11 on the expression Sox2, KLF4, CD44, and c-Myc proteins in MDA-MB-231 and Hs578t cells. (**G**) Graphical representation of the migration (*upper panel)* and invasion *(lower panel)* of MDA-MB-231 and Hs578t cells. Results represent mean $\pm$ SD of 3 independent assays in triplicate. * $p < 0.05$, ** $p < 0.01$, *** $p < 0.001$; WT, wild type; NS, not significant; Scale bar: 100 µm.

3.4. CDH11 Inhibition Markedly Attenuate the Migration And invasion of TNBC Cells

Since the CSCs-like phenotype is often associated with enhanced oncogenicity and metastatic traits [24], we further examined whether and how alteration in CDH11 expression affects the characteristic aggressiveness of TNBC cells, using WT and siCDH11 MDA-MB-231 and Hs578t cells. We demonstrated that the siCDH11 cells exhibited significantly reduced migration and invasion ability; fewer migrated siCDH11 MDA-MB-231 and Hs579t cells were observed than for the WT cells (Figure 5G). Similarly, siCDH11 resulted in significantly less invaded cells, compared to the WT cells (Figure 5G and Figure S1B). Since, cell migration and invasion are objective measures of the aggressiveness of cancerous cells; these results further corroborate the vital role of CDH11 in the enhanced metastatic phenotype and aggressiveness of TNBC cells.

3.5. Silencing CDH11 Significantly Attenuates Tumorigenicity and Tumor Growth of TNBC Cells, In Vivo

Having previously demonstrated that silencing CDH11 suppresses the stem cell-like phenotype of TNBC cells in vitro, to determine the probable inhibitory effect of silencing CDH11 on tumor formation and growth in vivo, we generated tumor xenograft models derived from NOD/SCID mice ($n = 5$/group) inoculated with 2×10^6 WT or siCDH11 MDA-MB-231 or Hs578t cells subcutaneously in the hind-flank and observed for 27 days (Figure 6A). We demonstrated that the mice inoculated with siCDH11 MDA-MB-231 or Hs578T developed very significantly smaller tumors, compared to the WT group (MDA-MB-231: ~6.58-fold smaller, $p = 0.008$; Hs578t: ~5.75-fold smaller, $p = 0.003$ on day 27) (Figure 6B,C). siRNA effect in the siCDH11 group was maintained by intra-tumoral injection with 10 μmol/L of 'siCDH11-atelocollagen' complex (Figure S2). Of translational relevance, in parallel mice assays, survival analyses show that over 45 days, mice bearing siCDH11 MDA-MB-231 or Hs578T cells-derived tumors enjoyed 20–50% ($p = 0.027$) or 16–50% ($p = 0.011$) survival advantage, respectively, compared to their WT-inoculated litter-mate counterparts (Figure 6D,E). Taken together, these findings suggest that CDH11 play an important role in tumorigenicity and tumor growth of TNBC cells in vivo. As depicted in the schematic abstract (Figure 6F), these data showing siCDH11-associated impaired tumorigenesis and increased survival substantiate CDH11-targeting as a potential therapeutic strategy for patients with TNBC.

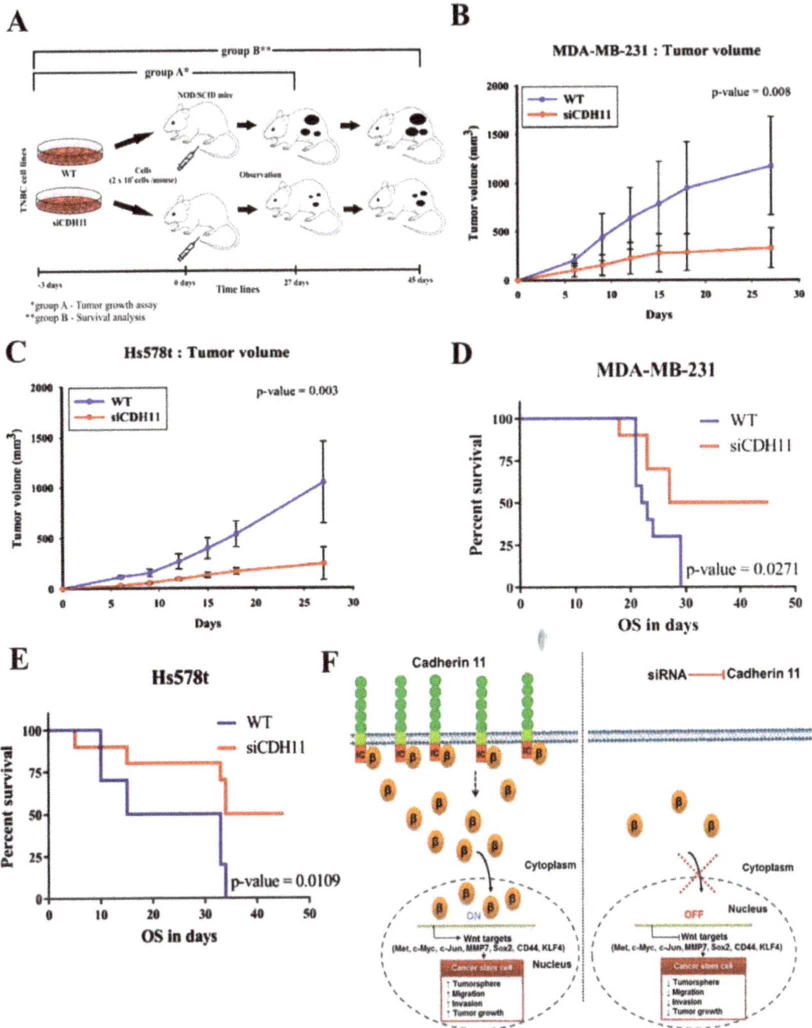

Figure 6. Silencing CDH11 significantly attenuates tumorigenicity of TNBC cells, in vivo. (**A**) Schema showing xenografts derived from NOD/SCID mice injected with 2×10^6 WT or siCDH11 MDA-MB-231 or Hs578t cells subcutaneously in the hind-flank and observed for 27 days (group A, $n = 5$/group) or 45 days (group B, $n = 5$/group). Graph showing difference in tumor volume over time for the WT or siCDH11 (**B**) MDA-MB-231 or (**C**) Hs578T tumor-bearing mice. Intergroup *p*-values were determined by 2-way ANOVA. Kaplan-Meier survival curves for mice bearing WT or siCDH11 (**D**) MDA-MB-231 or (**E**) Hs578T cell-derived tumors. WT, wild type. (**F**) Schematic abstract showing that CDH11 inhibition suppresses the TNBC CSCs-like and metastatic phenotypes through WNT signalling regulation. In the presence of WNT signal, cytoplasmic β-catenin is released from CDH11, accumulates in the perinuclear region, then translocates to the nucleus where it activates downstream CSCs-regulating target genes, including Met, c-Myc and Sox2 by binding to TCF/LEF. Once activated, these genes enhance self-renewal, migration, and invasion abilities of TNBC cells. In the absence of CDH11, membranar, cytoplasmic and nuclear β-catenin pool is diminished, WNT signalling is obtused, and the canonical WNT signalling pathway downstream target genes remains inactivated.

4. Discussion

Cadherin 11 is a transmembrane protein with documented conflicting roles in several types of cancers [25–28]. Recently, it was suggested that CDH11 acts as a tumor suppressor gene as demonstrated by its methylated and silenced status in malignant tissues such as in the nasopharyngeal, esophageal, gastric, hepatocellular carcinoma, colon, and breast carcinoma [26]. Similarly, in vitro and in vivo studies of murine retinoblastoma, cdh11 acted as a tumor suppressor gene through promotion of tumor cell death [29]. Conversely, in ovarian cancer patients, CDH11 exhibited no evidence of tumor-suppressive or oncogenic function, nor bore any prognostic / predictive relevance [27]. However, cumulative evidence supports the role of cdh11 as an oncogene, including that reported in the present study. Higher CDH11 mRNA expression level has been shown in malignant breast tumor than in non-malignant breast tumor or normal breast tissues [14]. Another study showed that increased expression of CDH11 is characteristic of enhanced invasiveness in breast cancer cell lines, in vitro [30]. In the present study, we demonstrated that high CDH11 expression positively correlates with poor prognosis in the more aggressive breast cancers subtypes, namely BL and TNBC subtypes (Figure 1). Thus, we posit that the high expression of CDH11 in patients with BL breast cancer and TNBC indicates that CDH11 plays an important role in these breast cancer molecular subtypes as an oncogene. This is consistent with the findings from prostate and renal cancers, wherein tissue samples from the bone metastatic site exhibited higher expression levels of the CDH11 protein than observed in primary tumor or adjacent normal tissue [15,16].

Based on breast cancer big data analyses, we showed that CDH11 expression positively correlates with expression of WNT signalling components such as β-catenin, WNT2, and TCF2, and that tissues from TNBC patients show concurrent high expression of CDH11 and β-catenin proteins; thus, we demonstrate that CDH11 positively modulates β-catenin expression and activates the canonical WNT signalling pathway (Figure 2). These findings are consistent with contemporary knowledge that β-catenin is a core mediator of the canonical WNT signalling pathway. In the presence of WNT molecule, β-catenin translocates from the cytoplasm to the nucleus, induces downstream target genes, subsequently activating the canonical WNT pathway [9]; however, with absent or deactivated WNT molecule, cytoplasmic β-catenin binds to the APC-Axin-CK1-GSK3β complex, otherwise called destruction complex, leading to β-catenin phosphorylation [11]. The phosphorylated β-catenin is degraded in the proteasome, thus, reducing β-catenin bioavailability [11]. Cadherin acts as a reservoir for calcium-dependent adhesion-competent β-catenin when WNT signalling is inactivated [13].

Validating our hypothesis that CDH11 may promote β-catenin nuclear translocation and WNT activation, we demonstrated that CDH11 modulates β-catenin expression levels and that CDH11/β-catenin signalling axis plays a regulatory role for the canonical WNT signalling pathway in TNBC (Figure 3), as concomitant decreased CDH11 and β-catenin in the TNBC cell was observed, with siCDH11 cells exhibiting reduced CDH11/β-catenin nuclear co-localization compared to the control WT cells (Figure 4). This is consistent with studies suggesting that the transcriptional suppression of β-catenin was associated with inhibition of the WNT signalling activity in MCF-4 and A549 cell lines [31]. Based on our immunofluorescence and IHC staining, CDH11 localizes in the membrane and cytoplasm. In rheumatoid arthritis, matrix metalloproteinases (MMPs)- and γ-secretase- induced cleavage induced CDH11 activity, through its release from the cytoplasmic domain and subsequent nuclear translocation [32,33]. This in-part explains our observed nuclear co-localization of CDH11 and β-catenin and subsequent WNT signalling activation.

Aside possessing the ability to self-renew and differentiate into heterogeneous tumor cells, CSCs are implicated in the drug-resistance, relapse, distant metastasis and poor prognosis in TNBC patients [34,35]. We demonstrated that CDH11 knockdown significantly reduces the size of formed mammospheres, with concomitant downregulation of CSCs markers, Sox2, KLF4, CD44, and c-Myc (Figure 5), which is consistent with previous findings showing that inhibition of the canonical WNT signalling decreased Sox2 [36], CD44 [24], and c-Myc, as well as suppressed tumorsphere formation and inhibited tumor formation in xenograft tumor mice models [34,37]. Taken together, our results highlight a modulatory role for the CDH11/β-catenin signalling axis in TNBC CSCs-like.

Having established the functional relevance of the CDH11/β-catenin signalling axis in TNBC CSCs-like phenotype, we further confirmed that this axis modulates TNBC metastatic phenotype as demonstrated by markedly suppressed migration and invasion ability of CDH-deficient TNBC cells (Figure 5). This finding is consistent with the previously documented role of CDH11 in the promotion of migration and invasion in colorectal and prostate cancer [38,39]. Finally, substantiating findings from our in vitro assays, our in vivo studies highlight the putative potential of CDH11-targeting to suppress tumorigenesis, and inhibit tumor growth in TNBC xenograft models, while concurrently improving survival time (Figure 6 and Figure S2).

5. Conclusions

In summary, our findings highlight the critical roles CDH11 plays in TNBC. We demonstrate for the first time, to the best of our knowledge, that by targeting β-catenin, CDH11 regulates the canonical WNT signalling pathway, inhibits the CSCs-like and metastatic phenotypes of TNBC cells, and represents a novel therapeutic approach in TNBC treatment. We herein provide a basis for further exploration of CDH11 as a putative candidate for targeted therapy in triple negative breast cancer.

Supplementary Materials: The following are available online at http://www.mdpi.com/2077-0383/8/2/148/s1, Figure S1: CDH11 inhibition suppresses the stem cell-like and metastatic phenotypes of TNBC cells. Figure S2: siRNA effect in the siCDH11 group was maintained by intra-tumoral 'siCDH11-atelocollagen' complex injection.

Author Contributions: P.B.S., O.A.B.: Project conception, Collation and/or assembly of data, Data analysis and interpretation, Bioinformatics, Manuscript writing and revision; J.-H.C.: Project conception, Experimental design, Collation and/or assembly of data, Data analysis and interpretation; T.A., S.M.H.: Data assembly and analysis; C.-T.Y., T.-Y.C.: Experimental design, Data analysis and interpretation, Provision of useful research insight and essential reagents, Administrative oversight; All authors read and approved the final submitted version.

Funding: This work was supported by National Science Council of Taiwan: Tsu-Yi Chao (MOST103-2325-B-038-002 and MOST105-2314-B038-080), and grant from Taipei Medical University -National Taiwan University of Science and Technology Joint Research Program (TMU-NTUST-103-03) to Chi-Tai Yeh.

Acknowledgments: The authors thank the research assistants of Core Facility Center, Department of Medical Research, Taipei Medical University-Shuang Ho Hospital, especially Iat-Hang Fong and Sam Huang for their technical assistance with cell-based assays.

Conflicts of Interest: The authors declare no conflict of interest.

Abbreviations

CDH11	cadherin 11
CSC	cancer stem cell
TNBC	triple negative breast cancer
WNT	wingless-type MMTV integration site
siRNA	small interfering RNA
NOD/SCID	non-obese diabetic/severe combined immunodeficiency

References

1. Rakha, E.A.; Chan, S. Metastatic Triple-negative Breast Cancer. *Clin. Oncol.* **2011**, *23*, 587–600. [CrossRef] [PubMed]

2. Balko, J.M.; Schwarz, L.J.; Luo, N.; Estrada, M.V.; Giltnane, J.M.; Davila-Gonzalez, D.; Wang, K.; Sanchez, V.; Dean, P.T.; Combs, S.E.; et al. Triple-negative breast cancers with amplification of JAK2 at the 9p24 locus demonstrate JAK2-specific dependence. *Sci. Transl. Med.* **2016**, *8*, 334ra53. [CrossRef] [PubMed]

3. Largillier, R.; Ferrero, J.M.; Doyen, J.; Barriere, J.; Namer, M.; Mari, V.; Courdi, A.; Hannoun-Levi, J.M.; Ettore, F.; Birtwisle-Peyrottes, I.; et al. Prognostic factors in 1038 women with metastatic breast cancer. *Ann. Oncol.* **2008**, *19*, 2012–2019. [CrossRef] [PubMed]

4. Al-Hajj, M.; Wicha, M.S.; Benito-Hernandez, A.; Morrison, S.J.; Clarke, M.F. Prospective identification of tumorigenic breast cancer cells. *Proc. Natl. Acad. Sci. USA* **2003**, *100*, 3983–3988. [CrossRef] [PubMed]

5. Koury, J.; Zhong, L.; Hao, J. Targeting Signalling Pathways in Cancer Stem Cells for Cancer Treatment. *Stem Cells Int.* **2017**, *2017*. [CrossRef] [PubMed]

6. Abdullah, L.N.; Chow, E.K.-H. Mechanisms of chemoresistance in cancer stem cells. *Clin. Transl. Med.* **2013**, *2*, 3. [CrossRef] [PubMed]

7. Ling, L.; Nurcombe, V.; Cool, S.M. Wnt signalling controls the fate of mesenchymal stem cells. *Gene* **2009**, *433*, 1–7. [CrossRef]

8. Pohl, S.-G.; Brook, N.; Agostino, M.; Arfuso, F.; Kumar, A.P.; Dharmarajan, A. Wnt signalling in triple-negative breast cancer. *Oncogenesis* **2017**, *6*, e310. [CrossRef]

9. Zhan, T.; Rindtorff, N.; Boutros, M. Wnt signalling in cancer. *Oncogene* **2017**, *36*, 1461–1473. [CrossRef]

10. Xu, J.; Prosperi, J.R.; Choudhury, N.; Olopade, O.I.; Goss, K.H. Beta-Catenin Is Required for the Tumorigenic Behavior of Triple-Negative Breast Cancer Cells. *PLoS ONE* **2015**, *10*, 1–11.

11. Hendriksen, J.; Jansen, M.; Brown, C.M.; van der Velde, H.; van Ham, M.; Galjart, N.; Offerhaus, G.J.; Fagotto, F.; Fornerod, M. Plasma membrane recruitment of dephosphorylated -catenin upon activation of the Wnt pathway. *J. Cell Sci.* **2008**, *121*, 1793–1802. [CrossRef] [PubMed]

12. Orsulic, S.; Huber, O.; Kemler, R. E-cadherin binding prevents β-catenin nuclear localization and β-catenin/LEF-1-mediated transactivation. *J. Cell Sci.* **1999**, *112*, 1237–1245. [PubMed]

13. Howard, S.; Deroo, T.; Fujita, Y.; Itasaki, N. A positive role of cadherin in wnt/β-catenin signalling during epithelial-mesenchymal transition. *PLoS ONE* **2011**, *6*, e23899. [CrossRef] [PubMed]

14. Pohlodek, K.; Tan, Y.Y.; Singer, C.F.; Gschwantler-Kaulich, D. Cadherin-11 expression is upregulated in invasive human breast cancer. *Oncol. Lett.* **2016**, *12*, 4393–4398. [CrossRef] [PubMed]

15. Satcher, R.L.; Pan, T.; Cheng, C.J.; Lee, Y.C.; Lin, S.C.; Yu, G.; Li, X.; Hoang, A.G.; Tamboli, P.; Jonasch, E.; et al. Cadherin-11 in Renal Cell Carcinoma Bone Metastasis. *PLoS ONE* **2014**, *9*, e89880. [CrossRef]

16. Chu, K.; Cheng, C.-J.; Ye, X.; Lee, Y.-C.; Zurita, A.J.; Chen, D.-T.; Yu-Lee, L.-Y.; Zhang, S.; Yeh, E.T.; Hu, M.C.-T.; et al. Cadherin-11 Promotes the Metastasis of Prostate Cancer Cells to Bone. *Mol. Cancer Res.* **2008**, *6*, 1259–1267. [CrossRef]

17. Anders, C.K.; Carey, L.A.; Frazier, D.P.; Kendig, R.D. Biology, Metastatic Patterns and Treatment of Patients with Triple-Negtive Breast Cancer. *Clin. Breast Cancer* **2010**, *9*, S73–S81. [CrossRef]

18. Altaner, C.; Altanerova, V. Stem cell based glioblastoma gene therapy. *Neoplasma* **2012**, *59*, 622–630. [CrossRef]

19. Uhlen, M.; Fagerberg, L.; Hallstrom, B.M.; Lindskog, C.; Oksvold, P.; Mardinoglu, A.; Sivertsson, A.; Kampf, C.; Sjostedt, E.; Asplund, A.; et al. Tissue-based map of the human proteome. *Science* **2015**, *347*, 1260419. [CrossRef]

20. Assefnia, S.; Dakshanamurthy, S.; Guidry Auvil, J.M.; Hampel, C.; Anastasiadis, P.Z.; Kallakury, B.; Uren, A.; Foley, D.W.; Brown, M.L.; Shapiro, L.; et al. Cadherin-11 in poor prognosis malignancies and rheumatoid arthritis: Common target, common therapies. *Oncotarget* **2014**, *5*, 1458–1474. [CrossRef]

21. Fedchenko, N.; Reifenrath, J. Different approaches for interpretation and reporting of immunohistochemistry analysis results in the bone tissue—A review. *Diagn. Pathol.* **2014**, *9*, 221. [CrossRef] [PubMed]

22. Yadav, B.S.; Chanana, P.; Jhamb, S. Biomarkers in triple negative breast cancer: A review. *World J. Clin. Oncol.* **2015**, *6*, 252–263. [CrossRef] [PubMed]

23. Prat, A.; Adamo, B.; Cheang, M.C.U.; Anders, C.K.; Carey, L.A.; Perou, C.M. Molecular Characterization of Basal-Like and Non-Basal-Like Triple-Negative Breast Cancer. *Oncologist* **2013**, *18*, 123–133. [CrossRef] [PubMed]

24. Jang, G.B.; Kim, J.Y.; Cho, S.D.; Park, K.S.; Jung, J.Y.; Lee, H.Y.; Hong, I.S.; Nam, J.S. Blockade of Wnt/β-catenin signalling suppresses breast cancer metastasis by inhibiting CSC-like phenotype. *Sci. Rep.* **2015**, *5*, 12465. [CrossRef] [PubMed]

25. Birtolo, C.; Pham, H.; Morvaridi, S.; Chheda, C.; Go, V.L.W.; Ptasznik, A.; Edderkaoui, M.; Weisman, M.H.; Noss, E.; Brenner, M.B.; et al. Cadherin-11 Is a Cell Surface Marker Up-Regulated in Activated Pancreatic Stellate Cells and Is Involved in Pancreatic Cancer Cell Migration. *Am. J. Pathol.* **2017**, *187*, 146–155. [CrossRef] [PubMed]

26. Li, L.; Ying, J.; Li, H.; Zhang, Y.; Shu, X.; Fan, Y.; Tan, J.; Cao, Y.; Tsao, S.W.; Srivastava, G.; et al. The human cadherin 11 is a pro-apoptotic tumor suppressor modulating cell stemness through Wnt/β-catenin signalling and silenced in common carcinomas. *Oncogene* **2012**, *31*, 3901–3912. [CrossRef] [PubMed]

27. VON Bülow, C.; Oliveira-Ferrer, L.; Loning, T.; Trillsch, F.; Manher, S.; Milde-Langosch, K. Cadherin-11 mRNA and protein expression in ovarian tumors of different malignancy: No evidence of oncogenic or tumor-suppressive function. *Mol. Clin. Oncol.* **2015**, *3*, 1067–1072. [CrossRef]

28. Zhu, Q.; Wang, Z.; Zhou, L.; Ren, Y.; Gong, Y.; Qin, W.; Bai, L.; Hu, J.; Wang, T. The role of cadherin-11 in microcystin-LR-induced migration and invasion in colorectal carcinoma cells. *Oncol. Lett.* **2018**, *15*, 1417–1422. [CrossRef]

29. Marchong, M.N.; Yurkowski, C.; Ma, C.; Spencer, C.; Pajovic, S.; Gallie, B.L. Cdh11 acts as a tumor suppressor in a murine retinoblastoma model by facilitating tumor cell death. *PLoS Genet.* **2010**, *6*, e1000923. [CrossRef]

30. Pishvaian, M.J.; Feltes, C.M.; Thompson, P.; Bussemakers, M.J.; Schalken, J.A.; Byers, S.W. Cadherin-11 is expressed in invasive breast cancer cell lines. *Cancer Res.* **1999**, *59*, 947–952.

31. Ding, F.; Wang, M.; Du, Y.; Du, S.; Zhu, Z.; Yan, Z. BHX Inhibits the Wnt Signalling Pathway by Suppressing β-catenin Transcription in the Nucleus. *Sci. Rep.* **2016**, *6*, 1–9. [CrossRef] [PubMed]

32. Noss, E.H.; Watts, G.F.M.; Zocco, D.; Keller, T.L.; Whitman, M.; Blobel, C.P.; Lee, D.M.; Brenner, M.B. Evidence for cadherin-11 cleavage in the synovium and partial characterization of its mechanism. *Arthritis Res. Ther.* **2015**, *17*, 1–12. [CrossRef] [PubMed]

33. Mullooly, M.; McGowan, P.M.; Kennedy, S.A.; Madden, S.F.; Crown, J.; O'Donovan, N.; Duffy, M.J. ADAM10: A new player in breast cancer progression? *Br. J. Cancer* **2015**, *113*, 945–951. [CrossRef] [PubMed]

34. Yang, A.; Qin, S.; Schulte, B.A.; Ethier, S.P.; Tew, K.D.; Wang, G.Y. MYC inhibition depletes cancer stem-like cells in triple-negative breast cancer. *Cancer Res.* **2017**, *77*, 6641–6650. [CrossRef] [PubMed]

35. Idowu, M.O.; Kmieciak, M.; Dumur, C.; Burton, R.S.; Grimes, M.M.; Powers, C.N.; Manjili, M.H. CD44(+)/CD24(−/low) cancer stem/progenitor cells are more abundant in triple-negative invasive breast carcinoma phenotype and are associated with poor outcome. *Hum. Pathol.* **2012**, *43*, 364–373. [CrossRef]

36. Van Raay, T.J.; Moore, K.B.; Iordanova, I.; Steele, M.; Jamrich, M.; Harris, W.A.; Vetter, M.L. Frizzled 5 signalling governs the neural potential of progenitors in the developing Xenopus retina. *Neuron* **2005**, *46*, 23–36. [CrossRef] [PubMed]

37. Leis, O.; Eguiara, A.; Lopez-Arribillaga, E.; Alberdi, M.J.; Hernandez-Garcia, S.; Elorriaga, K.; Pandiella, A.; Rezola, R.; Martin, A.G. Sox2 expression in breast tumours and activation in breast cancer stem cells. *Oncogene* **2012**, *31*, 1354–1365. [CrossRef] [PubMed]

38. Feltes, C.M.; Kudo, A.; Blaschuk, O.; Byers, S.W. An alternatively spliced cadherin-11 enhances human breast cancer cell invasion. *Cancer Res.* **2002**, *62*, 6688–6697. [PubMed]

39. Tomita, K.; Van Bokhoven, A.; Van Leenders, G.J.L.H.; Ruijter, E.T.G.; Jansen, C.F.J.; Bussemakers, M.J.G.; Schalken, J.A. Cadherin switching in human prostate cancer progression. *Cancer Res.* **2000**, *60*, 3650–3654. [PubMed]

Journal of
Clinical Medicine

Review

Different Shades of L1CAM in the Pathophysiology of Cancer Stem Cells

Marco Giordano and Ugo Cavallaro *

Unit of Gynaecological Oncology Research, European Institute of Oncology IRCSS, 20128 Milan, Italy;
marco.giordano@ieo.it
* Correspondence: ugo.cavallaro@ieo.it

Received: 2 April 2020; Accepted: 13 May 2020; Published: 16 May 2020

Abstract: L1 cell adhesion molecule (L1CAM) is aberrantly expressed in several tumor types where it is causally linked to malignancy and therapy resistance, acting also as a poor prognosis factor. Accordingly, several approaches have been developed to interfere with L1CAM function or to deliver cytotoxic agents to L1CAM-expressing tumors. Metastatic dissemination, tumor relapse and drug resistance can be fueled by a subpopulation of neoplastic cells endowed with peculiar biological properties that include self-renewal, efficient DNA repair, drug efflux machineries, quiescence, and immune evasion. These cells, known as cancer stem cells (CSC) or tumor-initiating cells, represent, therefore, an ideal target for tumor eradication. However, the molecular and functional traits of CSC have been unveiled only to a limited extent. In this context, it appears that L1CAM is expressed in the CSC compartment of certain tumors, where it plays a causal role in stemness itself and/or in biological processes intimately associated with CSC (e.g., epithelial-mesenchymal transition (EMT) and chemoresistance). This review summarizes the role of L1CAM in cancer focusing on its functional contribution to CSC pathophysiology. We also discuss the clinical usefulness of therapeutic strategies aimed at targeting L1CAM in the context of anti-CSC treatments.

Keywords: L1CAM; cancer stem cells; chemoresistance; epithelial-mesenchymal transition; cancer therapy; cell adhesion molecule

---

1. Background

The L1 cell adhesion molecule (L1CAM, also known as CD171) was described for the first time by Schachner et al. in the central nervous system (CNS) [1]. In that context, L1CAM has been primarily implicated in the development and plasticity of the nervous system, where it plays a pivotal role in neuronal migration and differentiation, neurite outgrowth, axon guidance, fasciculation of axons and dendrites, myelination, and synaptogenesis [2–4]. Accordingly, the knockout of the murine gene results in profound neurological disorders [5,6], and mutations in human L1CAM are causally related to a spectrum of CNS defects that are collectively defined as L1 syndrome [7].

Following the discovery of L1CAM, other closely related cell adhesion molecules (CAMs) have been described, defining an L1 subfamily of which L1CAM is the archetype. In vertebrates, the L1 subfamily comprises four different members which share an analogous structural organization: Close Homolog of L1 (CHL1), Neuronal Cell Adhesion Molecule (NrCAM), Neurofascin and L1CAM itself [8]. All these proteins, in turn, belong to the Immunoglobulin superfamily of CAMs (Ig-CAMs), which owes its name to the presence of Ig-like domains in the extracellular portion of these proteins.

2. Molecular Characteristics of L1CAM

2.1. Structural Determinants of L1CAM Interactions

L1CAM is a single-pass membrane glycoprotein with a molecular weight of 200–220 kDa which exhibits three different portions: an extracellular domain, a transmembrane domain and a highly conserved cytoplasmic domain (Figure 1a) [9].

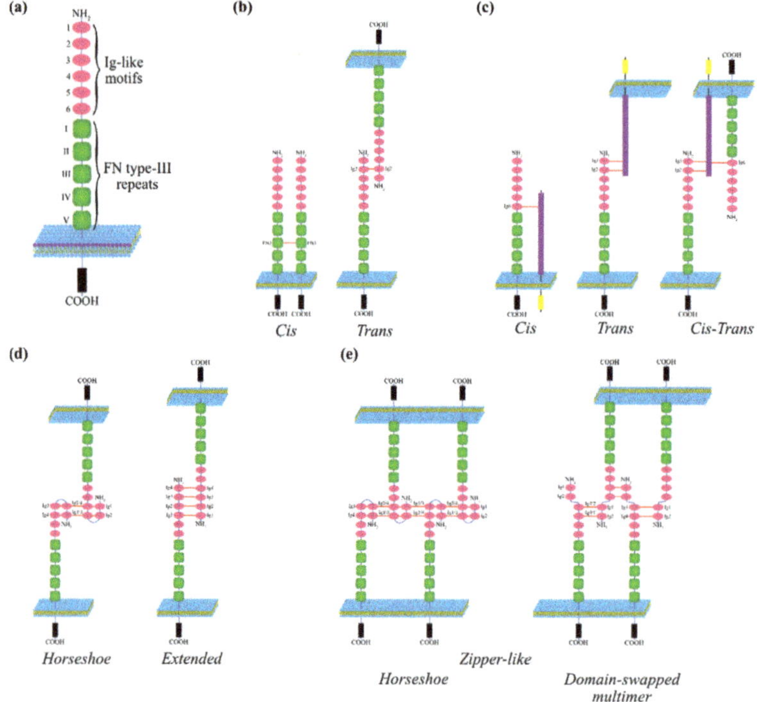

Figure 1. L1 cell adhesion molecule (L1CAM) structures and interactions. (**a**) Schematic structure of L1CAM, with the ectodomain comprising six Ig-like motifs (Ig1-Ig6, magenta ellipses) and five Fibronectin type-III repeats (FN1-FN5, green squares), the transmembrane (blue line) and the intracellular domain (black rectangle) (**b**) L1CAM homophilic interactions involve the FN3 repeat of two consecutive molecules (*cis*) that form hydrogen bonds (red line). When the molecules are exposed on two different cells, the involved residues can span between all the Ig-like motifs and FN2 or FN3 repeats (*trans*). For space limitations and clarity, only some homophilic bonds are displayed. See the main text for more details. (**c**) L1CAM heterophilic interactions involve almost all extracellular domains, with the specific domain depending on the partner. Only some examples of binding are shown here; more details are available in the main text. A generic partner is depicted as a transmembrane protein (violet rectangle: extracellular domain; yellow rectangle: intracellular domain). The two types of heterophilic interaction can be concomitant (*cis-trans*). (**d**) L1CAM can acquire different conformations during its interactions. The horseshoe structure is characterized by the bond between Ig-1 with Ig-4 and Ig-2 with Ig-3 that leads to a curvature of the molecule. Two horseshoe structures are connected by Ig-1 and Ig-2 repeats of adjacent molecules. The extended conformation involves bonds between the first four Ig structures of two neighbor L1CAM molecules. (**e**) More than two L1CAM molecules might be connected acquiring a very complex conformation (zipper-like). The domain-swapped multimer comprises interactions between Ig1 with Ig-4 and Ig2 with Ig-3. For space limitations and clarity, only interactions with full-length L1CAM are displayed. See text for more details.

The ectodomain comprises six N-terminal Ig-like motifs (Ig1–Ig6) followed by five fibronectin type III repeats (FN1–FN5) [4,8]. These structural elements and their dynamic arrangements are crucial to establish and drive the multiple interactions of L1CAM. Indeed, L1CAM exerts its function through inter-molecular interactions that can be either homophilic (i.e., L1CAM-L1CAM) or heterophilic with different partners. L1CAM can engage in *cis*-interactions, binding to another protein at the surface of the same cell (Figure 1b,c), *trans*-interactions with a protein localized on an adjacent cell (Figure 1b,c), or *cis/trans*-interactions where both events can occur simultaneously (Figure 1c) [10]. When these interactions take place, the proteins involved may assume different structural conformations in order to facilitate their binding that are illustrated in Figure 1. Su et al. predicted that L1CAM can acquire a horseshoe quaternary structure (Figure 1d). Another possible conformation that L1CAM acquires is the so-called extended quaternary structure (Figure 1d). Interestingly, these four domains are critical for neurite outgrowth and L1CAM adhesive properties, even though additional sequences within all the six Ig domains are required for an optimal activity [11–13]. Cryo-electron tomography studies also revealed a more complex mode of homophilic interaction whereby horseshoes from L1CAM proteins on opposing membranes meet as *trans* pairs, forming a lattice that is stabilized by protein-carbohydrate and carbohydrate-carbohydrate interactions (Figure 1e) [14]. The existence of such horseshoe-dependent structures is still controversial as some researchers attribute this conformation to a complicated mixture of other quaternary structures at a higher level of complexity rather than to a structure *per se* [14].

2.2. L1CAM Interactions

L1CAM is devoid of enzymatic activity and, therefore, needs molecular effectors for transducing intracellular signals and regulating the multiple processes in which it is involved. In this context, L1CAM often couples with other cell-surface molecules that, instead, have the capability to activate a downstream signaling. The proteins involved in a functional and/or physical interaction with L1CAM belong to different classes including other Ig-CAMs (such as NCAM), proteoglycans (e.g., neurocan), integrins, extra cellular matrix proteins (laminin), co-receptors (neuropilin-1), cytoskeletal proteins (ankyrin), and Receptor Tyrosine Kinases (RTKs) such as Fibroblast Growth Factor (FGF) and Epidermal Growth Factor (EGF) receptors. The main features of these interactions are summarized in Table 1.

Table 1. Homophilic and heterophilic L1CAM interactors. ND = not defined. *not conclusively demonstrated but inferred from the data provided in the corresponding reference.

Interactors	L1CAM Motif Involved	Type of Interaction	References
L1CAM	Ig1-6, FN2-3	*cis, trans*	[11,15–17]
NCAM	Ig5	*cis*	[18,19]
Neurocan	Ig1	ND	[20]
Integrins	Ig6, FN3	*cis*	[17,21,22]
Laminin	Ig6*	*trans*	[23]
Neuropilin	Ig1	*cis, trans*	[10,24]
Ankyrin	Cytoplasmic domain		[25,26]
FGFR	FN3, Ig1, Ig2	*cis, trans*	[27]
EGFR	FN5, Ig3*	*cis, trans*	[28]

2.2.1. L1CAM Interacting Partners and Functional Implications

L1CAM and NCAM interact with each other in *cis* [18,19,29]. This interaction allows L1CAM to bind other L1CAM molecules in *trans* and, therefore, has been termed "assisted homophilic binding". This binding has synergistic effects on L1CAM-mediated cell aggregation and adhesion in neuroblastoma cells [19,29]. L1CAM can also interact with neurocan [20,30]. The Ig6 motif of L1CAM contains the highly conserved aminoacidic sequence Arg-Gly-Asp (RGD) that is crucial for the *cis* interaction of L1CAM with $\alpha_v\beta_3$, $\alpha_v\beta_1$, $\alpha_5\beta_1$, $\alpha_v\beta_5$, $\alpha_{IIB}\beta_3$ integrins [21,22,31]. The FN3 repeat is also involved in the binding of L1CAM with integrins, in particular with $\alpha_5\beta_1$, $\alpha_v\beta_3$, $\alpha_9\beta_1$ [17]. The interaction with laminin occurs, although not exclusively, via the human natural killer-1 (HNK-1) carbohydrate [23]. The binding between L1CAM and Neuropilin-1 (NRP1 or NP-1) requires the small aminoacidic motif FASNKL [10,24]. Castellani and collaborators showed that the switching of semaphorin-3A (Sema3A)-induced axonal repulsion into attraction depends on *cis* vs. *trans* interactions of L1CAM with NP-1, respectively. In this scenario, L1CAM and NP-1 could be considered as co-receptors for Sema3A. Hence, this *cis* interaction is required as part of the Sema3A receptor complex and is necessary for the switching mechanism [10]. The cytoskeletal protein ankyrin is a prominent intracellular partner of L1CAM, and their interaction occurs through the highly conserved amino acid motifs LADY and FIGQY [25,26]. In neurons, L1CAM interaction with ankyrin is critical for the synaptic targeting of retinal axons and it also induces, in co-operation with EphrinB signaling, axon branch attraction in vivo [32]. The first RTK proposed to interact with L1CAM (and other adhesion molecules) is the fibroblast growth factor receptor (FGFR) [33]. Among the four members of the FGFR family, the direct interaction with L1CAM so far has been demonstrated only for FGFR1 [27,34]. L1CAM can also bind all the members of the EGFR family (erbB1-erbB4) [35]. When L1CAM *trans* interacts with EGFR, the binding is very weak and is not sufficient *per se* for EGFR autophosphorylation, even though a tyrosine kinase activity was detected at cell contact sites in *D. melanogaster* [36]. This implies that *cis* interactions between the two types of molecules may be required to enhance L1CAM-induced activation of EGFR. To note, a *cis* contact with erbB3 has been described in vivo [35].

2.2.2. The Regulation of L1CAM Interactions via Phosphorylation of the Cytoplasmic Tail

A key feature of L1CAM that can modulate its interactions is the phosphorylation of its cytoplasmic domain. Three kinases have been implicated in this process: ERK2, c-SRC and PKCα. The first to be identified was ERK2, a key player in the mitogen-activated protein kinase (MAPK) signaling, that interacts physically with L1CAM. Schaefer et al. have found that ERK2 is able to phosphorylate L1CAM at Ser^{1204} and Ser^{1248} n vitro, and suggested that this phosphorylation regulates the binding of L1CAM to ankyrin [37]. L1CAM and ERK2 crosstalk is bidirectional, since L1CAM stimulates ERK2 activation and MAPK signaling, most likely via FGFR. L1CAM-mediated ERK2 activation has been proposed to occur upon L1CAM endocytosis since both L1CAM and ERK2 are present in endocytic vesicles [37]. The second kinase implicated in L1CAM phosphorylation is the nonreceptor tyrosine kinase c-SRC. The latter phosphorylates the Tyr^{1176} residue within the YRSLE motif [38] that is required for endocytosis of L1CAM via clathrin-coated pits [39]. The authors have shown that Tyr^{1176} phosphorylation prevents L1CAM binding to clathrin-associated AP-2 complex, suggesting a model whereby transient dephosphorylation of the YRSLE motif allows L1CAM endocytosis. Finally, protein kinase C-alpha (PKCα) phosphorylates the Thr^{1172} residue within the L1CAM cytoplasmic domain, thus regulating important properties of pancreatic adenocarcinoma cells such as motility [40,41]. Notably, Thr^{1172} phosphorylation promotes conformational changes that, in turn, influence the interactions of L1CAM even within its extracellular portion. For example, the binding to integrins is profoundly affected by the phosphorylation status of Thr^{1172} [40,41]. It is worth mentioning, however, that these studies relied principally on the use of recombinant proteins, and therefore need further validation in more physiological settings.

In summary, L1CAM engages in interactions with a myriad of molecules that impinge on several signal transduction pathways which, in turn, orchestrate fundamental aspects of cell physiology.

2.3. Proteolytic Processing of L1CAM

By analogy to other Ig-CAMs, L1CAM can be cleaved by several proteinases, a process that regulates both the cell-autonomous and non-cell-autonomous signaling of the protein [42]. These proteolytic events and the generated L1CAM fragments are summarized in Figure 2.

A cleavage site in the FN3 domain of L1CAM is recognized by the serine proteases PC5A, plasmin and trypsin [43–45] (Figure 2a). The proteolytic action of these enzymes generates two fragments, one of about 140-kDa and one of 80-kDa. Lutz and collaborators demonstrated the production of alternative fragments of 135-kDa (soluble) and 70-kDa that comprise the intracellular and transmembrane domains and part of the extracellular domain (Figure 2a). The nuclear translocation of the 70-kDa fragment, which involves sumoylation of the residue Lys^{1172}, is mediated by importin, results from trafficking via endosomes and depends on the presence of a nuclear localization signal encompassing Lys^{1147}. However, the authors did not investigate the possible role of nuclear L1CAM fragment in gene transcription [46]. Besides the serine protease-mediated processing, the ectodomain of L1CAM can be cleaved closer to the membrane, thus producing a bigger fragment of about 180-200 kDa (Figure 2b). The proteases responsible for such process are Neuropsin, β-secretase 1 (BACE1) and the metalloproteases ADAM10 and ADAM17 [47–51]. Notably, the two types of processing may occur sequentially (Figure 2c). First, a fragment of 140-kDa is produced upon cleaving inside the FN3 domain. The remaining portion of the ectodomain is further processed to produce a soluble fragment of 50-kDa [43]. Whatever protease accounts for the iuxtamembrane cleavage of L1CAM, in addition to the above mentioned 180-200 kDa portion of the ectodomain, a 30-32 kDa fragment is always produced. This, in turn, is further processed by γ-secretase resulting in a smaller fragment of 28 kDa that is released from the membrane, translocates into the nucleus and triggers gene transcription [51]. However, soluble L1CAM can also derive from events other than proteolytic cleavage. Indeed, Angiolini et al. have recently described a novel, endothelial-specific isoform of L1CAM, that results from a peculiar alternative splicing of its pre-mRNA. In particular, the splicing factor NOVA2 induces the skipping of the exon that encodes the transmembrane domain, generating a soluble L1CAM that retains both the extracellular and the cytoplasmic domains and is endowed with angiogenic activity [52].

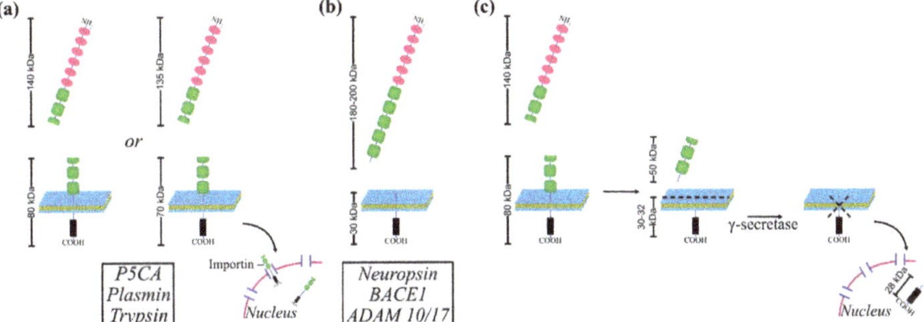

Figure 2. Processing of L1CAM. (**a**) Full-length L1CAM undergoes a cleavage inside the FN3 repeat by serine proteases PC5A, plasmin or trypsin. Thus, a soluble fragment of about 135-140 kDa (top) and another of 75-80 kDa are produced. The former is released while the remaining fragment, which comprises part of the ectodomain, the transmembrane domain and the cytoplasmic domain, is internalized and imported into the nucleus via importin. (**b**) L1CAM is cleaved closer to the membrane by neuropsin, BACE1 or ADAM10/17 proteases generating a soluble fragment of about 180–200 kDa (top) and one of 30 kDa. (**c**) The processes described in (**a,b**) can occur sequentially. First, L1CAM is cleaved inside the FN3 repeat; a second cut occurs close to the membrane producing a fragment of 50 kDa and another of 30-32 kDa, which remains anchored to the membrane. The latter is further processed by γ-secretase and is released from the plasma membrane, eventually translocating into the nucleus.

3. Clinical Relevance of L1CAM in Cancer Diagnosis and Prognosis

The definition of L1CAM's role in cell motility and plasticity within the nervous system prompted many groups to investigate whether the protein exerts an analogous role in different contexts. Numerous studies focused on the function of L1CAM in tumor-related processes. Indeed, L1CAM has emerged as a causal factor in tumor invasion and metastasis. Such a role is extensively described in a number of excellent reviews [42,53,54] and will not be discussed here. We will mainly focus on the clinical relevance of L1CAM in cancer patients, and in particular on the correlation of its levels with the prognosis, the diagnosis and other clinical parameters in certain cancer types (Table 2).

Table 2. L1CAM clinical relevance in cancer.

Cancer Type	Prognostic Value	Clinico-Pathological Parameters Correlating with L1CAM	References
Endometrial cancer	Negative (OS, DFS)	High grade, lymph node metastasis, tumor relapse	[55–60]
Ovarian cancer	Negative (OS, DFS)	Low tumor resectability, lymph node metastasis, chemoresistance,	[61–64]
Melanoma	Negative (DFS)	Metastasis	[65]
Breast cancer	Negative (DFS)	Larger tumor size, lymph node involvement, higher histologic grade, advanced TNM stage	[66]
Gastric cancer	Negative (OS, DFS)	Distant metastasis	[67]
Colon cancer	Negative (OS)	Advanced cancer stage, distant metastasis and tumor recurrence	[68]
Pancreatic cancer	Negative (OS)	Lymph node involvement, vascular invasion, perineural invasion and higher degree of pain	[69]
Non-small cell lung cancer	Negative (PFS)	None	[70]
Kidney cancer	Negative (OS)	ND	[71]

The potential of L1CAM as a prognostic factor has received great attention from investigators working on gynecological tumors. Among these, the tumor in which the prognostic role of L1CAM has been investigated most extensively is endometrial cancer.

Bosse et al. [55] evaluated L1CAM levels in a large retrospective cohort of early stage endometrial cancer. They have found that patients with >10% tumor cells positive for L1CAM had a remarkably shorter overall survival than those below that threshold. Smogeli et al. [56] found that, in the subgroup of patients who did not receive adjuvant chemotherapy, L1CAM expression was significantly associated with shorter disease-free survival and with risk of relapse in univariate analysis. In a different study that analysed biopsies from 1134 endometrial cancer patients, L1CAM high expression predicted poor disease-specific survival – defined as time from surgery to death – both in the entire cohort and among low-risk patients, who normally receive no or limited adjuvant treatment [57]. Moreover, high expression of L1CAM correlated significantly with the occurrence of lymph node metastases, both in the whole patient population and in the low-risk subgroup. In addition, the authors evaluated the serum level of soluble L1CAM (sL1CAM) in preoperative samples from a subgroup of 372 patients with endometrial cancer. A higher amount of sL1CAM was detected in cancer patients with respect to healthy controls, in line with previous studies [61]. High levels of sL1CAM predicted poor disease-specific survival in both the entire cohort and in the low-risk group. sL1CAM levels were also predictive for lymph node metastasis in the entire cohort. However, once adjusted for age, FIGO stage, histologic type and grade, circulating sL1CAM levels failed to exhibit independent prognostic power.

L1CAM has also been assessed as a biomarker for preoperative risk factor stratification of endometrial carcinoma. However, it did not significantly improve risk stratification when compared to classical factors [72], implying that L1CAM should not be used to plan preoperative treatment. Another study showed that L1CAM is highly expressed in recurrent respect to non-recurrent endometrial cancer and correlates with lower disease-free survival [58].

A multicentric study named ENITEC, which involved 10 different European centers, assessed L1CAM expression in 1199 cases that included early and late-stage endometriod endometrial cancers (EECs) as well as non-endometriod endometrial cancers (NEECs) [59]. In early-stage EECs, L1CAM expression was associated with grade 3 histology and lymphovascular space invasion (LVSI) while in late-stage EECs its expression correlated also with the presence of nodal disease. Moreover, L1CAM levels were associated with shorter disease-free and overall survival in both early and late-stage EECs. Very recently, L1CAM expression was also found to correlate with distant recurrence in early-stage endometrial cancer patients with negative peritoneal cytology, a subgroup that normally has a relatively good prognosis [60].

In the context of gynecological oncology, the clinical utility of L1CAM has also been extensively studied in ovarian cancer (OC). In this tumor type, L1CAM shows a wide range of expression, from a small fraction of cancer cells to high levels in most of the tumor mass (Figure 3a–c).

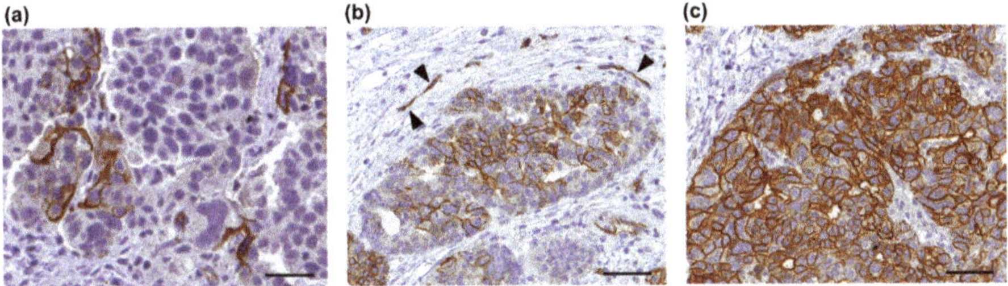

Figure 3. L1CAM levels in ovarian cancer. Three examples of high-grade serous ovarian cancer with different levels of L1CAM are shown. L1CAM is visualized by immunohistochemistry on formalin-fixed paraffin-embedded specimens. The protein is localized on the cancer cell membrane with levels that range from (**a**) a minority of cells to (**c**) most of the tumor mass. The arrowheads in (**b**) show ovarian cancer-associated vessels that are L1CAM-positive as previously described [56]. Scale bars: (**a**), 25 μm; (**b**,**c**) 50 μm.

Fogel et al. reported the expression of L1CAM in poorly differentiated OC, where it acts as a biomarker of worse prognosis [61]. Other groups have then confirmed the association of L1CAM with bad outcome [42,62], although some studies have questioned the prognostic value of L1CAM in OC [73]. Of note, L1CAM was consistently detected at the invasive front of OC [61,62,64], supporting its role in tumor invasion.

In type-I ovarian cancer, often confined to the ovary and resulting in good prognosis, L1CAM was assayed in patients with either endometriod or clear cell histotype [63]. The authors showed that L1CAM levels correlated with poor disease-specific overall survival and disease-free survival in endometrioid, but not in clear cell ovarian carcinomas. Moreover, overall survival was worse in early-stage patients with high L1CAM levels, making L1CAM a potential stratification marker for a high-risk subgroup among these putative good-prognosis patients. Although L1CAM alone did not result in multivariate analysis as an independent prognostic factor for overall disease survival, its positivity was associated with incomplete response to primary therapy in endometroid ovarian cancer but, once again, not in the clear cell histotype [63]. Finaly, Altevogt and collaborators measured the levels of both membrane and soluble L1CAM in a cohort of high grade serous ovarian cancer patients [64]. They found that soluble L1CAM isolated from patients' ascitic fluid correlated with poor outcome in terms of PFS and showed a trend toward prolonged OS.

The correlation of L1CAM expression with poor prognosis is by no means limited to the gynecological cancers described above. Indeed, L1CAM has been defined as a negative prognostic factor in melanoma [65], breast cancer [66], gastric cancer [67], colon cancer [68], pancreatic cancer [69], non-small cell lung cancer [70], kidney cancer [71], etc., as elegantly reviewed elsewhere [42,53,54].

In summary, the clinical utility of L1CAM as a diagnostic and prognostic factor has emerged from many studies in different types of cancers, albeit with some controversial observations, which is in line with its involvement in malignancy-associated features of tumor cells [42,53,54].

4. L1CAM Mechanism of Action in Stemness and in Stem-Related Processes

A common fate of all cancer patients, especially for those who have been diagnosed at late stage, is that they undergo chemotherapy and/or radiotherapy. However, even in case of a satisfactory response of the primary tumor to the treatments, many patients experience a disease relapse, and the scenario is even worsened by the acquisition of chemoresistance which makes the recurrent tumor refractory to standard therapies. Tumor relapse and drug resistance are commonly attributed to a subset of cells that, taking advantage of an armory of biological weapons, evade chemotherapy with diverse resistance mechanisms. Such cell subpopulation appears related to cancer stem cells (CSC) [74]. This hypothesis is supported by certain characteristics of CSC such as: a low proliferative potential that makes them insensitive to chemotherapy which usually targets actively dividing cells; increased levels of molecular pumps that efflux drugs out of the cell; adaptation to inflammation; efficient DNA repair; and altered apoptotic mechanisms [75].

CSC have been the subject of intense debate and controversy over the last two decades. Some researchers refer to them as a transient state of tumor cells rather than a discrete subpopulation [76], a concept strictly related to tumor cell plasticity; others are even questioning their existence [77,78]. According to many investigators, much of the controversy and skepticism around CSC derives from a semantic issue, namely what criteria must be considered to define bona fide CSC [79]. As a matter of fact, a subset of cancer cells with distinct tumor-initiating ability does evade the conventional therapies and fuels tumor recurrence [80,81]. These features of CSC account for the common use of the term cancer-initiating cells (CIC) as a synonym [82]. Collectively, there is a body of experimental and clinical evidence that supports the existence and the pro-malignant function of cancer cells with stem-like features [79].

4.1. L1CAM Function in CSC and Its Contribution to Cancer Stemness-Associated Processes

L1CAM has been linked to many CSC-related processes, yet only a few studies have demonstrated its direct involvement in stemness. Tumor recurrence and metastasis is accompanied and fueled by specific biological processes, such as the acquisition of chemoresistance and EMT, that have been found to be intimately associated with cancer stemness [83]. Hence, it is not surprising that, due to its causal role in CSC biology, L1CAM has been also implicated in these processes in cancer, even though these studies were not always conducted on CSC subpopulations.

4.1.1. Glioblastoma

Similarly to what occurred during its discovery, the first connection of L1CAM with CSC was found in the central nervous system [84]. Through a loss-of-function approach, the authors showed that silencing L1CAM in glioblastoma stem cells (GSC), defined by CD133 expression, reduced their growth and survival both in vitro and in vivo. They also described the underlying mechanism that entails the loss of Olig2, caused by L1CAM silencing, and the upregulation of cyclin-dependent kinase inhibitor p21$^{WAF1/CIP1}$, which acts as a tumor suppressor [84]. In another study, Cheng and coworkers showed that L1CAM regulates DNA damage responses and radiosensitivity of CD133$^+$ or CD15$^+$ GSC through the nuclear translocation of its intracellular domain [85]. In particular, L1CAM expression is induced by DNA damage, and is critical for the activation of ataxia telangiectasia mutated (ATM) and of the downstream checkpoint proteins Chk2, Rad17, and Chk1, which repair the DNA as part of the DNA damage response signaling pathway [86]. Mechanistically, the portion of L1CAM involved in this process is the highly conserved cytoplasmic domain that acts as signal transducer upon DNA damage, regulating NBS1 expression through c-Myc to enhance DNA damage checkpoint activation. The findings discussed above might implicate L1CAM as a potential marker of glioblastoma CSC. Indeed, L1CAM has been frequently employed to identify stem-like population in glioblastoma [84,85,87,88]. Nevertheless, CSC-related functions, such as tumor initiation and self-renewal abilities, have not been defined yet by comparing L1CAM-positive and L1CAM-negative subpopulations of glioma cells [89]. Thus, the role of L1CAM as a GSC marker remains to be conclusively demonstrated.

Glioblastoma provides also a prototypical example of L1CAM's role in chemoresistance [90]. Held-Feindt et al. started from the earlier observation that TGF-β1 is a potent inducer of L1CAM expression in tumor cells [91,92]. Notably, these authors employed clinically relevant models of primary tumor cells and cultivated them as suspension neurospheres as a tool to enrich for glioblastoma stem-like cells. They showed that L1CAM promoted chemoresistance to temozolomide, which was mediated by TGF-β1 and led to the down-regulation of caspase-8 in both stem-like and bulk glioblastoma cells. These data raised the question of whether chemoresistance is restricted to CSC compartment only or rather it can also be ascribed to non-stem subpopulations. However, it is appropriate to point out that the cells used in the study derived from the differentiation of stem-like cells. It is conceivable, in this case, that such differentiated cells might have retained chemoresistance as an imprinting from their stem state.

4.1.2. Colorectal Cancer

The identification of L1CAM as a target of β-catenin/TCF signaling in colorectal cancer (CRC) [93,94] paved the way to the definition of its role in CRC stem cells. Gavert et al. [95] observed initially that, while the ectopic expression of L1CAM enhanced CRC metastasis, likely via the activation of NF-κB signaling, L1CAM was not co-expressed with the CRC stem cell markers EpCAM, CD133 and CD44. However, L1CAM was subsequently reported to be expressed in LGR5[+] CRC stem cells, where it promoted metastatic dissemination through the induction of the clusterin gene (*CLU*), an event independent from NF-κB signaling [96]. In particular, L1CAM enhanced the transactivation of *CLU* by the transcription factor STAT-1. The link between L1CAM and STAT-1, however, remains to be defined at the molecular level. The role of L1CAM in CRC stemness has also been investigated by Basu et al. who focused on the Wnt target gene and transcription factor Achete scute-like 2 (ASCL2), a key regulator of stemness that is exclusively expressed in LGR5[+] intestinal stem cells [97,98]. Basu et al. implicated ASCL2 as an L1CAM effector in CRC progression [99]. Indeed, L1CAM expression in human CRC cells dramatically increased the expression of ASCL2 which, in turn, was required for L1CAM-induced CRC cell proliferation, motility and tumorigenesis. Finally, L1CAM and ASCL2 were found to co-localize in human CRC tissue, suggesting a possible cooperation in conferring a more invasive phenotype to CRC cells. A very recent study has provided compelling evidence that high L1CAM expression marks a subpopulation of CRC cells with tumor propagation, metastasis-initiating and chemoresistance features [100]. Intriguingly, L1CAM[high] cells partially overlapped with LGR5[+] CRC stem-like cells. In this context, L1CAM expression was not only a biomarker but also a prerequisite for metastasis initiation and chemoresistance. Ganesh et al. also provided mechanistic insights into the upregulation of L1CAM by showing that the loss of E-cadherin-dependent epithelial integrity releases L1CAM from the transcriptional repression operated by REST, a factor that prevents L1CAM expression in non-neuronal tissues [100]. Very recently Fang et al. correlated L1CAM with pERK 1/2 levels in CRC lymph node metastasis [101]. The authors showed that L1CAM expression in tissue samples increased from poorly differentiated through well differentiated CRC reaching the highest levels in metastatic CRC tissue. The same behavior was observed for pERK 1/2. Future work should clarify whether the positive correlation between L1CAM and pERK levels reflects a functional link and whether L1CAM-regulated ERK signaling contributes to CRC dissemination.

4.1.3. Pancreatic Cancer

L1CAM-associated chemoresistance has also been proposed in pancreatic cancer. Lund and collaborators generated a pancreatic carcinoma cell line resistant to the chemotherapeutic 5-Fluorouracil (5-FU) and, upon transcriptomic profiling, identified L1CAM interaction pathway as one of the top-ranking hits among 319 upregulated genes [102]. Silencing L1CAM resulted in decreased invasiveness of the 5-FU resistant cell line. Driven by their microarray data and by previous observation in pancreatic cancer [92], the authors knocked down Slug and β-catenin in chemoresistant cells and found that the former, but not the latter, modulates L1CAM protein levels. Yet whether a Slug/L1CAM axis accounts for chemoresistance in pancreatic carcinoma remains to be investigated.

4.1.4. Gynecological Cancers

Among tumors of the gynecological tract, the role of L1CAM in cancer stemness has been studied in ovarian and endometrial carcinoma. A study conducted on ovarian cancer cell lines revealed that L1CAM, in combination with CD133, marks a subpopulation of ovarian CSCs [103]. In light of the known correlation of L1CAM with ovarian cancer aggressiveness [61,62], these findings, once confirmed in patient-derived tissue, might implicate L1CAM/CD133-positive CSC in the malignant properties of this tumor type. Furthermore, the repertoire of ovarian CSC-associated markers is still a highly debated and controversial issue [82], and L1CAM may offer a new tool for the unequivocal identification of this elusive cell population. The study on L1CAM+/CD133+ ovarian CSCs showed also, through the genetic manipulation of ovarian cancer cell lines, that L1CAM is causally involved in CSC-associated radioresistance as well as in self-renewal and tumor initiation. Further research, however, should aim at elucidating the underlying molecular mechanisms.

In ovarian cancer cells, the L1CAM gene was found to be under the regulation of TWIST1, a transcription factor that is causally linked to increased tumorigenicity as well as resistance to cisplatin [104]. Indeed, L1CAM was upregulated upon the forced expression of TWIST1 and, more important, it was required for TWIST1-induced chemoresistance. Mechanistically, upon cisplatin treatment, L1CAM silencing partially prevented Akt activation, which was a key player in cisplatin resistance of ovarian cancer cells. These data, therefore, pointed to a TWIST1/L1CAM/Akt signaling pathway that drives chemoresistance in ovarian cancer.

As discussed earlier, L1CAM correlates with malignancy in endometrial cancer. In agreement with those observations, recent studies have established a link between L1CAM and endometrial cancer stemness [105]. The authors observed first that L1CAM promotes epithelial-mesenchymal transition (EMT) as exemplified by the concomitant downregulation of E-cadherin and induction of vimentin. Given the well-established association between EMT and cancer stemness [83], the authors then investigated CSC-related features. They found that L1CAM-expressing cells exhibit resistance to *anoikis* and higher clonogenic potential in non-adherent conditions, two peculiar features of CSC [82]. Moreover, L1CAM expression was accompanied by the upregulation of Musashi-1 and CD133, both considered to be CSC markers in endometrial cancer [106].

4.1.5. Retinoblastoma

Recently, L1CAM has been reported to be both sufficient and necessary for conferring chemoresistance to retinoblastoma, the most common intraocular cancer in children [107]. Proteins related to apoptosis and multi-drug resistance (MDR) are frequently involved in the resistance of cancer cells and CSCs to chemotherapy [108–110]. Along this line, L1CAM depletion in retinoblastoma cells resulted in a marked increase of the pro-apoptotic proteins which cleaved caspase-3 and cytochrome C, whereas the anti-apoptotic proteins Bcl-2, Bcl-xL, and pro-caspase-3, were reduced. Moreover, the drug efflux pumps ABCA1, ABCB1, ABCC2, and ABCG2 were significantly reduced in L1CAM-depleted cells whereas L1CAM overexpression increased their levels.

4.1.6. L1CAM Impact on Stemness-Related Features of Tumor Microenvironment

The contribution of L1CAM to tumor development can also occur via the modulation of stemness-related properties within the tumor microenvironment, which in turn modulates cancer cell behavior in a non-autonomous manner. For example, in tumor endothelium L1CAM promotes endothelial-to-mesenchymal transition (EndMT), a phenomenon reminiscent of EMT [111]. In this context, L1CAM induces the expression, among many other genes, of KLF4 and CD44, both well-known as stemness-associated factors. EndMT generates cells that retain the properties of multipotent stem cells and can differentiate into several cell types (e.g., fibroblast, pericyte, bone, etc.) [112]. Furthermore, EndMT produces key cellular components of the tumor microenvironment such as cancer-associated fibroblasts that support tumor progression [113]. This might open a novel scenario whereby L1CAM could also orchestrate stemness in the tumor microenvironment, adding a further layer of complexity to its role in cancer progression.

4.1.7. L1CAM Expression Obtained by Omics Data Unveiled its Involvement in CSC Processes

By comparing cancer stem cell genetic profile with their non neoplastic counterpart, L1CAM emerged very frequently altered. These data endorse the crucial role of the adhesion molecule in cancer stem cells.

Nakata et al. demonstrated that LGR5 gene is associated with stem features in glioblastoma [114]. A transcriptomic analysis of LGR5-silenced glioblastoma stem cells showed that L1CAM is downregulated and therefore is regulated by LGR5, although the molecular mechanism remains to be elucidated. In another study, Okawa et al. profiled and compared the proteome and secretome of glioblastoma multiforme stem cells (GNS) and normal neural stem cells [115]. The authors found that several CSC markers were enriched in the glioblastoma stem cell population, and L1CAM was one of them. Interestingly, L1CAM was found aberrantly expressed both in the secretome and in the cell-associated proteome of GNS, suggesting that its function is specifically involved in the stem cell compartment of glioblastoma.

Finally, Gemei et al. analyzed the genetic profile of the 3AB-OS osteosarcoma cell line that is an immortalized CSC line and represents the stem component of MG63 parental cell line used for genetic comparison [116]. L1CAM was among the genes whose expression increased in the c3AB-OS cell line with respect to the differentiated MG63.

A transcriptomics profile combined with high-throughput flow cytometry was conducted on self-renewing (i.e., stem-like) and non-self-renewing cells from the sonic hedgehog (SHH) subgroup of medulloblastoma. In this case, unlike the studies reported above, L1CAM was found specifically downregulated in CSCs at both gene and protein levels [117]. This suggests that the expression of L1CAM as a CSC-associated marker is tumor type-dependent, and in certain tumors L1CAM can even mark selectively the non-stem cancer cell population.

In conclusion, several studies have provided compelling evidence that implicates L1CAM, directly or indirectly, in the regulation of CSC pathophysiology. Yet none of them have conclusively indicated or demonstrated L1CAM to be a CSC (or CIC) marker. Therefore, the jury is still out on the use of L1CAM for the isolation, identification and characterization of the CSC compartment in clinically relevant settings.

4.2. Beyond Cancer: L1CAM in Normal Stem Cells

Besides the role of L1CAM in cancer stemness, it is worth mentioning that a few studies have implicated the molecule also in normal stem cells. Son and collaborators [118] demonstrated that the stem cell markers octamer-binding transcription factor 4 (Oct4), Nanog, sex-determining region Y–box 2 (Sox2), forkhead box protein D3 (FoxD3), and SSEA-3 were downregulated in L1CAM-depleted human embryonic stem cells (hESC). Conversely, the same genes resulted in being upregulated in L1CAM-overexpressing hESC. L1CAM-depleted hESC displayed also an increased expression of lineage markers (i.e., Olig2, GFAP, Pax6 as ectoderm marker; CD31, T-Branch, LEF1 as mesoderm marker; FOXA2, GATA4, SOX17 as endoderm marker). Moreover, the authors showed that L1CAM is essential for hESC pluripotency. In particular, when normal H9 cells were differentiated into embryoid bodies (EBs) most lineage markers did not increase in L1CAM-depleted cells whereas they were upregulated in control cells. Mechanistically, FGFR1 appeared to be involved in this process, although the exact mechanism remains to be elucidated. L1CAM formed a complex with FGFR1 and a reduced activation of FGFR1, ERK and AKT was observed in L1CAM-depleted cells [118]. Another study, performed on murine ESC (mESC), showed that L1CAM is required for neuronal differentiation of mESC and depends on the fucosyltransferase FUT9 and sialyltransferase ST3Gal4 through a signaling pathway that involves the activation of phospholipase C-gamma (PLCγ) [119]. It remains unclear whether the discrepancy between the impact of L1 in hESC, where it sustains stemness, and mESC, where it promotes differentiation, reflects a species-specific role of L1 in ESC or rather depends on different experimental conditions between the two studies.

Further complexity is added by a recent study on L1CAM in human neural progenitors (hNP). In this context, the ectopic expression of L1CAM extracellular domain altered differentiation and motility in hNP without affecting cell proliferation [120]. L1CAM-expressing hNP lost their progenitor status and became committed to differentiation. Interestingly, these authors co-cultured hNP with chick embryo brain cells to assess phenotype alterations in a more complex microenvironment. In these culture conditions L1CAM ectodomain again promoted the loss of the progenitor phenotype and induced the differentiation of hNP towards astrocytes. Taken together with the stemness-sustaining role of L1CAM in hESC (see above), these findings on promoting commitment in hNP might imply that the outcome of L1CAM expression is cell context-dependent.

5. L1CAM as a Therapeutic Target

The expression pattern of L1CAM in cancer and its functional role in CSC point to this molecule as a viable target for novel therapeutic strategies. While no attempts have been reported yet to test the druggability of L1CAM in CSC, several studies have supported the potential of L1CAM-targeted treatments in different tumor types, as summarized in Table 3. The three main strategies that have been designed for this purpose are illustrated in Figure 4 and described below.

Table 3. L1CAM-targeted therapeutic approaches in different cancers.

Cancer Type	Therapeutic Strategy	Antibody Clone	Effect	References
Ovarian cancer	Antibody alone	chCE7, L1-11A	↓ Proliferation ↓ Migration	[121]
Ovarian cancer	Antibody alone	CE7	↓ Proliferation	[62]
Ovarian cancer	Antibody alone	L1-9.3	↓ Tumor growth ↑ Survival	[122]
Melanoma Pancreatic cancer	Antibody alone	L1-9.3	↓ Tumor growth EMT induction	[123]
Cholangiocarcinoma	Antibody alone	cA10-A3	↓ Tumor growth	[124]
Pancreatic cancer	Antibody + gemcitabine	L1-14.10, L1-9.3	↓ Tumor growth ↑ Apoptosis	[125]
Ovarian cancer	Antibody + paclitaxel	L1-14.10, L1-9.3	↓ Tumor growth ↑ Apoptosis	[125]
Cholangiocarcinoma	Antibody + gemcitabine	Ab417	↓ Tumor growth	[126]
Cholangiocarcinoma	Antibody + cisplatin	Ab417	↓ Tumor growth	[126]
Ovarian cancer	Radioimmunoconjugate	^{177}Lu-DOTA-chCE7	↑ Survival	[127]
Ovarian cancer	Radioimmunoconjugate	^{161}Tb-chCE7	↓ Tumor growth	[128]
Neuroblastoma	Radioimmunoconjugate	^{131}I-chCE7	↓ Tumor growth	[129]
Cholangiocarcinoma	Radioimmunoconjugate	^{177}Lu-NOTA-cA10-A3	↓ Tumor growth ↑ Apoptosis ↓ Proliferation	[130]
Ovarian cancer	Radioimmunoconjugate + paclitaxel	^{177}Lu-DOTA-chCE7	↑ Survival Tumor growth delay	[131]
Ovarian cancer	Radioimmunoconjugate + protein kinase inhibitor	^{177}Lu-DOTA-chCE7	↓ Proliferation ↓ Tumor growth	[132]
Neuroblastoma	CAR-T cell	CE7R	Various responses	[133]
Ovarian cancer	CAR-T cell	CE7R	↓ Tumor growth ↓ Ascites	[134]
Retinoblastoma	CAR-T cell	CE7R	Cytotoxicity	[135]

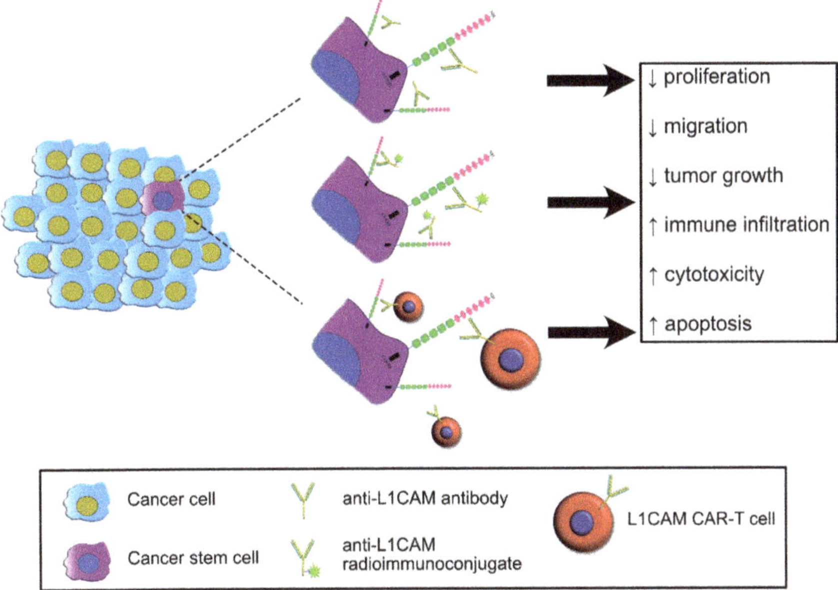

Figure 4. L1CAM is a viable therapeutic target in cancer. Three strategies for targeting L1CAM in cancer cells are depicted. Such strategies are based on neutralizing antibodies, radioimmunoconjugates or chimeric antigen receptor-redirected T (CAR-T) cells. These treatments, alone, or in combination with chemotherapy, result in the reduction of tumorigenicity due to increased apoptosis and cytotoxic effects, possibly accompanied by increased immune infiltration into the tumor site. Based on the expression and function of L1CAM in cancer stem cells, these L1CAM-targeted approaches may prove effective anti-CSC strategies, as illustrated in the figure. See the main text for more details.

5.1. Monoclonal Antibodies

In many cases, L1CAM-targeted approaches have relied on neutralizing antibodies (Figure 4). The latter, for example, exhibited a remarkable potential as antitumor agents in preclinical models of ovarian cancer. Arlt et al. induced antibody-mediated reduction of proliferation and migration in vitro as well as tumor growth in vivo using two independent anti-L1CAM monoclonal antibodies (chCE7 and L1-11A) [121]. We have also reported a reduction of proliferation in IGROV1 cells upon anti-L1CAM monoclonal antibody CE7 treatment [62]. Wolterink and coworkers generated novel monoclonal L1CAM antibodies [122]. The selected clone L1-9.3 inhibited SKOV3ip tumor growth, increasing mouse survival. The transcriptome analysis of treated mice revealed also that L1-9.3 treatment could interfere with apoptotic and tumor growth pathways. To note, the treatment with L1-9.3 was accompanied by massive monocyte infiltration, and monocyte depletion via clodronate liposomes abolished the therapeutic effect of the antibody. This role of monocytes in the response to L1CAM-mediated immunotherapy would be consistent with a crosstalk between L1CAM and the tumor immune microenvironment. Such a hypothesis was further supported in preclinical models of pancreatic cancer. Sebens and collaborators [136] demonstrated that L1CAM induced an immune suppressive phenotype in malignant pancreatic ductal adenocarcinoma (PDAC). In particular, T-regs, but not T-effs, were more prone in migrating on L1CAM expressing H6c7 and Panc1 cells. Moreover, in such microenvironment T-effs reduced their proliferation and inhibited autologous T cell proliferation. L1CAM promoted the establishment of a microenvironment that could favor immune escape and might contribute to tumor progression and chemoresistance [136].

The therapeutic effects of anti-L1CAM antibodies were also assessed in melanoma and pancreatic carcinoma. Doberstein et al. evaluated the therapeutic efficacy of anti-L1CAM treatment on the syngeneic tumor models RET (melanoma) and Panc02 (pancreatic adenocarcinoma), both genetically manipulated to express L1CAM [123]. They also employed the clone L1-9.3 and confirmed both the antibody-mediated reduction of tumor growth and the involvement of immune effector mechanisms in its anti-tumor effect. In addition, the authors showed that anti-L1CAM treatment induced EMT via EGFR phosphorylation. Another study reported cytotoxicity and anti-tumor activity of L1CAM chimeric antibody cA10-A3 in a mouse model of intrahepatic cholangiocarcinoma [124]. Of note, despite L1CAM expression in the nervous system and other body districts, no signs of behavioral changes or other signs of toxicity were reported with the use of anti-L1CAM antibodies for all the studies presented here. This supports the potential clinical utility of such tools as antitumor strategy.

As outlined in Section 4.1, L1CAM is causally related to chemoresistance in various cancer types, which suggests that interfering with L1CAM function may enhance the response to other treatments. Indeed, anti-L1CAM antibodies improved the efficacy of chemotherapeutic drugs in preclinical models of ovarian and pancreatic cancer [137]. The authors employed two different anti-L1CAM antibodies: L1-14.10 and L1-9.3. Pancreatic adenocarcinoma Colo357 and ovarian cancer SKOV3ip cell lines were treated with both antibodies in combination with either gemcitabine or paclitaxel, respectively. The addition of anti-L1CAM inhibited tumor growth much more than chemotherapy alone and increased apoptosis. Lower levels of NF-κB were observed upon combo treatment along with a reduction in vascular endothelial growth factor (VEGF) production and CD31-positive vessels. To note, increased monocyte infiltration was also observed in this study upon combination treatment. Along the same line, Cho et al. have demonstrated that treating intrahepatic cholangiocarcinoma (Choi-CK) xenograft mouse model with gemcitabine or cisplatin in combination with anti-L1CAM antibody Ab417 inhibited tumor growth [126]. The combination of L1CAM antibodies with conventional chemotherapy or other targeted approaches remains an area of research that may have relevant implications and deserves further investigation.

5.2. Radioimmunoconjugates

Besides these strategies for functional inactivation, the expression pattern of L1CAM in cancer suggests that anti-L1CAM antibodies may also be harnessed to deliver cytotoxic agents to tumor cells. It is rather surprising, in this regard, that no reports are available in the literature about antibody-drug conjugates based on L1CAM. In fact, L1CAM undergoes endocytosis [39], which is enhanced by antibody binding [138,139], supporting the hypothesis that antibody-drug conjugates would represent suitable tools against L1CAM-expressing cancer cells.

Anti-L1CAM antibodies, instead, have been widely used in preclinical models of ovarian carcinoma and other tumor types as radioimmunotherapy tools upon conjugation with different radioactive isotopes (Figure 4). Fischer et al. used a chimeric anti-L1CAM monoclonal antibody (chCE7) conjugated with 1,4,7,10-tetraazacyclododecane-N-N′-N′-N‴-tetra acetic acid (DOTA) and labeled with the low-energy β-emitter lutetium-177 (^{177}Lu) to treat human ovarian cancer-bearing mice. A single treatment with ^{177}Lu-DOTA-chCE7 was able to increase mice survival upon subcutaneous injection of SKOV3.ip1 [127]. Grünberg and coworkers showed that the treatment of nude mice bearing IGROV1 xenografts with terbium-161-labelled chCE7 increased radiotoxicity in respect to the radioimmunoconjugate alone [128]. In the neuroblastoma model SK-N-SH, the efficacy of Iodine-131 (^{131}I)-labelled chCE7 was compared with that of ^{131}I-metaiodobenzylguanidine (MIBG), currently used for treating recurrent or refractory neuroblastoma [129]. ^{131}I-MIBG was less effective than ^{131}I-chCE7 in reducing tumor volume although none of them abolished tumor growth. Both treatments elicited a transient response since, after the initial reduction, more pronounced for ^{131}I-chCE7, the tumor started to regrow reaching the starting volume. Interestingly, both molecules were also tested as imaging tools in seven patients with recurrent neuroblastoma. The two imaging approaches were complementary in targeting the tumor and therefore the authors proposed their use in a combined radioimmunotherapy [129]. Another group has recently conjugated the anti-L1CAM antibody cA10-A3 with 1,4,7-triazacyclononane-1,4,7-triacetic acid (NOTA) labelled with ^{177}Lu and evaluated its efficacy in cholangiocarcinoma [130]. The treatment with ^{177}Lu-NOTA-cA10-A3 of mice xenografted with L1CAM-overexpressing SCK-L1 cells reduced tumor volume by promoting cell apoptosis and reducing cell proliferation.

In addition, radio-labelled L1CAM antibodies have also been employed together with other drugs. Lindenblatt et al. coupled ^{177}Lu-DOTA-chCE7 with paclitaxel treatment. They reported a synergistic inhibitory effect on IGROV1 viability via cell cycle arrest in the radiosensitive G2/M phase [131]. The combined treatment prolonged survival and also delayed tumor latency in vivo. In another study, the same group combined ^{177}Lu-DOTA-chCE7 with protein kinase inhibitors (PKIs) to treat ovarian cancer [132]. Among five compounds (alisertib, MK1775, MK2206, saracatinib, temsirolimus), they selected MK1775 for further characterization due to its higher efficacy. The combination of MK1775 with ^{177}Lu-DOTA-chCE7, administering either the PKI firstly or both compounds at the same time, synergistically reduced IGROV1 proliferation. The treatment induced DNA double strand breaks in IGROV1 cells, as reflected by histone H2A.X phosphorylation at Ser-139 (γH2A.X), resulting in tumor cell apoptosis. γH2A.X foci were also found in SKOV3ip xenografts upon treatment with either MK1775 alone or in combination with the radioimmunoconjugate. The combined treatment reduced tumor volume in mouse xenografts with respect to the PKI alone, even if the treatment with only ^{177}Lu-DOTA-chCE7 was sufficient by itself to reduce tumor growth to a similar extent [132]. A comparison of these findings with those obtained by combining ^{177}Lu-DOTA-chCE7 with paclitaxel (see above) will help to inform future therapeutic strategies. Overall, these data support the feasibility of combination treatments based on L1CAM radioimmunoconjugates as efficacious strategies.

Besides radioimmunotherapy, the conjugation of L1CAM antibodies with various radioisotopes has generated reagents that found interesting applications in tumor imaging. Indeed, the successful detection of neoplastic lesions has been reported not only in preclinical cancer models [130,140,141] but also in patients as described above [129]. Due to the expression and function of L1CAM in cancer stem cells for some tumors, the possibility of imaging this cell subpopulation might be of help not only for prognostic purposes but also to monitor the evolution of the disease.

5.3. CAR-T Cells

L1CAM has also been investigated as a target for cellular therapies based on the adoptive transfer of chimeric antigen receptor-redirected T (CAR-T) cells (Figure 4). This approach dates back to 1993 and was first published by the immunologist Zelig Eshhar [142]. Since its first application, CAR-T cell therapy proved to be an intriguing approach and today has become of paramount importance for liquid cancers where it offers a successful therapeutic strategy [143,144]. It is based on the ex vivo isolation of tumor-reactive T lymphocytes engineered to express a chimeric antigen to redirect against the tumor. However, such a strategy quickly became of interest for its potential also in solid tumors where it was pioneered by Jensen et al. in neuroblastoma patients [133]. The authors generated CAR-T cells co-expressing an L1CAM-specific chimeric receptor (CE7R) and the fusion gene encoding the selection-suicide enzyme hygromycin phosphotransferase–thymidine kinase (HyTK). These CAR-T cells were then tested in a phase-I clinical trial for children with recurrent/refractory neuroblastoma. Treated patients did not show any sign of overt toxicity. Despite that the study's primary endpoint was limited to safety of the treatment, the authors also monitored patients' response. Among the six patients who underwent adoptive CAR-T cell therapy, one patient had first a stable disease and then a partial response, one patient displayed a complete response and another patient had stable disease. However, among the six patients only one experienced a prolonged survival. Subsequent studies also supported CAR-T therapy feasibility in other neoplasms. Hong et al. evaluated the applicability of L1CAM-based CAR-T cells in ovarian cancer [134]. The authors showed that CE7R T-cells were able to kill a panel of human ovarian cancer cell lines. Moreover, CE7R T-cells also recognized and killed ascites-derived primary cells. In SKOV3 xenograft mice, the treatment significantly inhibited tumor growth and reduced ascites production compared to control mice. Andersch et al. employed CAR-T cells co-targeting both L1CAM and the GD2 ganglioside in retinoblastoma cell lines, and reported that this approach led to elimination of tumor cells in vitro [135]. Interestingly, they also described an escape mechanism whereby retinoblastoma cells downregulate the expression of both antigens after CAR-T treatment. Further validation of this approach using in vivo preclinical models will undoubtedly provide important insights into its clinical utility.

In summary, various studies conducted on different tumor types have shown the remarkable potential of L1CAM-targeting strategies for the design of innovative antitumor therapies.

6. Potential Implications and Clinical Perspectives of L1CAM in Cancer and CSC

Many lines of experimental and clinical evidence support the potential relevance of L1CAM in the management of tumor patients. For example, as outlined in Section 2.2.2, the ectodomain of L1CAM is released into the extracellular space upon proteolytic cleavage. This raises the possibility that shed L1CAM becomes detectable in blood and other body fluids. Indeed, several studies found a correlation between soluble L1CAM levels in liquid biopsies from cancer patients and different clinical parameters [57,61,144,145]. Along these lines, we have recently found a novel soluble isoform of L1CAM, endowed with angiogenic functions, that lacks the transmembrane domain and is released in the extracellular space. The levels of this isoform resulted in being associated with vessel density in ovarian cancer specimens suggesting an implication also in tumor angiogenesis [52]. Thus, L1CAM is a promising non-invasive biomarker with diagnostic and prognostic value.

The expression pattern of L1CAM in many different cancer types as well as its pivotal role in diverse tumor-related cellular processes, including cancer stemness, makes it a suitable target for antitumor therapies. Indeed, L1CAM-targeted therapies resulted in multiple advantages, among which are counteracting tumor cell invasion, EMT and metastasis initiation, and pro-malignant interactions between tumor cells and their microenvironment. Based on the studies discussed in the previous sections, this list should also include the inhibition of CSC function.

Yet, with the exception of a phase-I trial with CAR-T cells in neuroblastoma patients ([133], see also Section 5), to date no L1CAM-based treatments have reached clinical use, although there are biotech companies that are actively pursuing this goal. Future efforts in this direction should take into account various aspects, including, for example, the potential side effects of L1CAM-targeted treatments. The protein, indeed, is abundantly expressed in the nervous system as well as in other cell types of various organs (e.g., the hematopoietic system and various epithelia). While no studies on preclinical models have reported signs of overt toxicity of anti-L1CAM agents, additional early-phase clinical trials are warranted to validate these findings in humans.

Finally, by analogy to several targeted treatments, it is unlikely that any L1CAM-based strategy would give satisfactory results as a monotherapy. We believe that L1CAM-targeted treatments should be considered in the context of combination therapies. This view is supported, for example, by the reported role of L1CAM in conferring chemoresistance to cancer cells (see Section 4.2), which implies that neutralizing L1CAM would restore sensitivity to conventional antitumor drugs. Furthermore, emerging evidence (mentioned in Section 5.1) also points to a crosstalk of L1CAM with the tumor immune microenvironment which contributes to immune evasion. Hence, one can speculate that L1CAM inactivation synergizes with immunotherapy in overcoming the ability of many tumors to escape the immune attack.

So far, L1CAM has not been investigated as a target in the context of therapies directed against CSC. Yet the expression pattern and the role of the molecule in this subpopulation of tumor cells provide the rationale for assessing the impact of L1CAM-targeted treatments on CSC function. Indeed, L1CAM may represent an Achilles heel for such a cell population. As outlined above, a few strategies have been designed to either interfere with L1CAM activity or deliver antitumor agents to L1CAM-expressing cells. Thus, it will be of interest to test whether such strategies impact on tumor stemness and on CSC-related properties such as self-renewal, chemoresistance and cancer initiation. Given the role of L1CAM in CSC, it is intriguing to envisage L1CAM-based strategies to target this cancer cell subpopulation. However, based on the functional contribution of CSC to tumor metastasis and relapse, the therapeutic window becomes a key factor in order to achieve better efficacy. One can expect, in this regard, a lower effect in the neoadjuvant treatment of a primary tumor as compared to the prevention of metastatic dissemination or tumor recurrence.

Future research will clarify if and to what extent L1CAM targeting represents a suitable strategy for innovative antitumor therapies, focusing in particular on the opportunity of defeating CSC-driven metastasis, relapse and drug resistance, thus contributing to tumor eradication.

Funding: The work in the authors' laboratory over the last few years was funded by Associazione Italiana Ricerca sul Cancro (AIRC), Worldwide Cancer Research, Fondazione Istituto Europeo di Oncologia-Centro Cardiologico Monzino, the Italian Ministry of Health, Cariplo Foundation, and Ovarian Cancer Research Alliance.

Acknowledgments: We apologize to all those colleagues whose important work could not be discussed due to space limitations. Marco Giordano is supported by a fellowship from Fondazione Istituto Europeo di Oncologia-Centro Cardiologico Monzino. We thank Francesca Angiolini for the immunostaining of ovarian cancer samples and the members of the Cavallaro lab for critical reading of the manuscript.

Conflicts of Interest: The authors declare no conflict of interest. The funders had no role in the design of the study; in the collection, analyses, or interpretation of data; in the writing of the manuscript; or in the decision to publish the results.

References

1. Rathjen, F.G.; Schachner, M. Immunocytological and biochemical characterization of a new neuronal cell surface component (L1 antigen) which is involved in cell adhesion. *EMBO J.* **1984**, *3*, 1–10. [CrossRef] [PubMed]

2. Brümmendorf, T.; Kenwrick, S.; Rathjen, F.G. Neural cell recognition molecule L1: From cell biology to human hereditary brain malformations. *Curr. Opin. Neurobiol.* **1998**, *8*, 87–97. [CrossRef]

3. Patzke, C.; Acuna, C.; Giam, L.R.; Wernig, M.; Südhof, T.C. Conditional deletion of L1CAM in human neurons impairs both axonal and dendritic arborization and action potential generation. *J. Exp. Med.* **2016**, *213*, 499–515. [CrossRef] [PubMed]

4. Schmid, R.S.; Maness, P.F. L1 and NCAM adhesion molecules as signaling coreceptors in neuronal migration and process outgrowth. *Curr. Opin. Neurobiol.* **2008**, *18*, 245–250. [CrossRef] [PubMed]

5. Fransen, E.; D'Hooge, R.; Van Camp, G.; Verhoye, M.; Sijbers, J.; Reyniers, E.; Soriano, P.; Kamiguchi, H.; Willemsen, R.; Koekkoek, S.K.E.; et al. L1 knockout mice show dilated ventricles, vermis hypoplasia and impaired exploration patterns. *Hum. Mol. Genet.* **1998**, *7*, 999–1009. [CrossRef] [PubMed]

6. Dahme, M.; Bartsch, U.; Martini, R.; Anliker, B.; Schachner, M.; Mantei, N. Disruption of the mouse L1 gene leads to malformations of the nervous system. *Nat. Genet.* **1997**, *17*, 346–349. [CrossRef] [PubMed]

7. Fransen, E.; Van Camp, G.; Vits, L.; Willems, P.J. L1-associated diseases: Clinical geneticists divide, molecular geneticists unite. *Hum. Mol. Genet.* **1997**, *6*, 1625–1632. [CrossRef]

8. Maness, P.F.; Schachner, M. Neural recognition molecules of the immunoglobulin superfamily: Signaling transducers of axon guidance and neuronal migration. *Nat. Neurosci.* **2007**, *10*, 19–26. [CrossRef]

9. Moos, M.; Tacke, R.; Scherer, H.; Teplow, D.; Früh, K.; Schachner, M. Neural adhesion molecule L1 as a member of the immunoglobulin superfamily with binding domains similar to fibronectin. *Nature* **1988**, *334*, 701–703. [CrossRef]

10. Castellani, V.; De Angelis, E.; Kenwrick, S.; Rougon, G. Cis and trans interactions of L1 with neuropilin-1 control axonal responses to semaphorin 3A. *EMBO J.* **2002**, *21*, 6348–6357. [CrossRef]

11. Haspel, J.; Friedlander, D.R.; Ivgy-May, N.; Chickramane, S.; Roonprapunt, C.; Chen, S.; Schachner, M.; Grumet, M. Fast track—Critical and optimal ig domains for promotion of neurite outgrowth by L1/Ng-CAM. *J. Neurobiol.* **2000**, *42*, 287–302. [CrossRef]

12. De Angelis, E. Disease-associated mutations in L1 CAM interfere with ligand interactions and cell-surface expression. *Hum. Mol. Genet.* **2002**, *11*, 1–12. [CrossRef] [PubMed]

13. Gouveia, R.M.; Gomes, C.M.; Sousa, M.; Alves, P.M.; Costa, J. Kinetic analysis of L1 homophilic interaction: Role of the first four immunoglobulin domains and implications on binding mechanism. *J. Biol. Chem.* **2008**, *283*, 28038–28047. [CrossRef] [PubMed]

14. He, Y.; Jensen, G.J.; Bjorkman, P.J. Cryo-Electron Tomography of Homophilic Adhesion Mediated by the Neural Cell Adhesion Molecule L1. *Structure* **2009**, *17*, 460–471. [CrossRef] [PubMed]

15. Zhao, X.; Siu, C.H. Colocalization of the homophilic binding site and the neuritogenic activity of the cell adhesion molecule L1 to its second Ig-like domain. *J. Biol. Chem.* **1995**, *270*, 29413–29421. [CrossRef] [PubMed]

16. De Angelis, E.; MacFarlane, J.; Du, J.S.; Yeo, G.; Hicks, R.; Rathjen, F.G.; Kenwrick, S.; Brümmendorf, T. Pathological missense mutations of neural cell adhesion molecule L1 affect homophilic and heterophilic binding activities. *EMBO J.* **1999**, *18*, 4744–4753. [CrossRef]

17. Silletti, S.; Mei, F.; Sheppard, D.; Montgomery, A.M.P. Plasmin-sensitive dibasic sequences in the third fibronectin-like domain of L1-cell adhesion molecule (CAM) facilitate homomultimerization and concomitant integrin recruitment. *J. Cell Biol.* **2000**, *149*, 1485–1501. [CrossRef]

18. Heller, M.; Von der Ohe, M.; Kleene, R.; Mohajeri, M.H.; Schachner, M. The immunoglobulin-superfamily molecule basigin is a binding protein for oligomannosidic carbohydrates: An anti-idiotypic approach. *J. Neurochem.* **2003**, *84*, 557–565. [CrossRef]

19. Kadmon, G.; Kowitz, A.; Altevogt, P.; Schachner, M. The neural cell adhesion molecule N-CAM enhances L1-dependent cell-cell interactions. *J. Cell Biol.* **1990**, *110*, 193–208. [CrossRef] [PubMed]

20. Oleszewski, M.; Beer, S.; Katich, S.; Geiger, C.; Zeller, Y.; Rauch, U.; Altevogt, P. Integrin and neurocan binding to L1 involves distinct Ig domains. *J. Biol. Chem.* **1999**, *274*, 24602–24610. [CrossRef] [PubMed]

21. Montgomery, A.M.P.; Becker, J.C.; Siu, C.H.; Lemmon, V.P.; Cheresh, D.A.; Pancook, J.D.; Zhao, X.; Reisfeld, R.A. Human neural cell adhesion molecule L1 and rat homologue NILE are ligands for integrin $\alpha v \beta 3$. *J. Cell Biol.* **1996**, *132*, 475–485. [CrossRef] [PubMed]

22. Felding-Habermann, B.; Silletti, S.; Mei, F.; Siu, C.H.; Yip, P.M.; Brooks, P.C.; Cheresh, D.A.; O'Toole, T.E.; Ginsberg, M.H.; Montgomery, A.M.P. A single immunoglobulin-like domain of the human neural cell adhesion molecule L1 supports adhesion by multiple vascular and platelet integrins. *J. Cell Biol.* **1997**, *139*, 1567–1581. [CrossRef]

23. Hall, H.; Carbonetto, S.; Schachner, M. L1/HNK-1 Carbohydrate- and $\beta 1$ Integrin-Dependent Neural Cell Adhesion to Laminin-1. *J. Neurochem.* **2002**, *68*, 544–553. [CrossRef] [PubMed]

24. Stoeck, A.; Schlich, S.; Issa, Y.; Gschwend, V.; Wenger, T.; Herr, I.; Marmé, A.; Bourbie, S.; Altevogt, P.; Gutwein, P. L1 on ovarian carcinoma cells is a binding partner for Neuropilin-1 on mesothelial cells. *Cancer Lett.* **2006**, *239*, 212–226. [CrossRef] [PubMed]

25. Davis, J.Q.; Bennett, V. Ankyrin binding activity shared by the neurofascin/L1/NrCAM family of nervous system cell adhesion molecules. *J. Biol. Chem.* **1994**, *269*, 27163–27166. [PubMed]

26. Hortsch, M.; Nagaraj, K.; Godenschwege, T.A. The interaction between L1-type proteins and ankyrins—A master switch for L1-type cam function. *Cell. Mol. Biol. Lett.* **2009**, *14*, 57–69. [CrossRef]

27. Kulahin, N.; Li, S.; Hinsby, A.; Kiselyov, V.; Berezin, V.; Bock, E. Fibronectin type III (FN3) modules of the neuronal cell adhesion molecule L1 interact directly with the fibroblast growth factor (FGF) receptor. *Mol. Cell. Neurosci.* **2008**, *37*, 528–536. [CrossRef]

28. Nagaraj, K.; Kristiansen, L.V.; Skrzynski, A.; Castiella, C.; Garcia-Alonso, L.; Hortsch, M. Pathogenic human L1-CAM mutations reduce the adhesion-dependent activation of EGFR. *Hum. Mol. Genet.* **2009**, *18*, 3822–3831. [CrossRef]

29. Horstkorte, R.; Schachner, M.; Magyar, J.P.; Vorherr, T.; Schmitz, B. The fourth immunoglobulin-like domain of NCAM contains a carbohydrate recognition domain for oligomannosidic glycans implicated in association with L1 and neurite outgrowth. *J. Cell Biol.* **1993**, *121*, 1409–1422. [CrossRef]

30. Oleszewski, M.; Gutwein, P.; Von Der Lieth, W.; Rauch, U.; Altevogt, P. Characterization of the L1-neurocan-binding site. Implications for L1-L1 homophilic binding. *J. Biol. Chem.* **2000**, *275*, 34478–34485. [CrossRef]

31. Blaess, S.; Kammerer, R.A.; Hall, H. Structural analysis of the sixth immunoglobulin-like domain of mouse neural cell adhesion molecule L1 and its interactions with alpha(v)beta3, alpha(IIb)beta3, and alpha5beta1 integrins. *J. Neurochem.* **1998**, *71*, 2615–2625. [CrossRef] [PubMed]

32. Buhusi, M.; Schlatter, M.C.; Demyanenko, G.P.; Thresher, R.; Maness, P.F. L1 interaction with ankyrin regulates mediolateral topography in the retinocollicular projection. *J. Neurosci.* **2008**, *28*, 177–188. [CrossRef] [PubMed]

33. Williams, E.J.; Furness, J.; Walsh, F.S.; Doherty, P. Activation of the FGF receptor underlies neurite outgrowth stimulated by L1, N-CAM, and N-cadherin. *Neuron* **1994**, *13*, 583–594. [CrossRef]

34. Kulahin, N.; Li, S.; Kiselyov, V.; Bock, E.; Berezin, V. Identification of neural cell adhesion molecule L1-derived neuritogenic ligands of the fibroblast growth factor receptor. *J. Neurosci. Res.* **2009**, *87*, 1806–1812. [CrossRef]

35. Donier, E.; Gomez-Sanchez, J.A.; Grijota-Martinez, C.; Lakomá, J.; Baars, S.; Garcia-Alonso, L.; Cabedo, H. L1CAM binds ErbB receptors through Ig-like domains coupling cell adhesion and neuregulin signalling. *PLoS ONE* **2012**, *7*, e40674. [CrossRef]

36. Islam, R.; Kristiansen, L.V.; Romani, S.; Garcia-Alonso, L.; Hortsch, M. Activation of EGF Receptor Kinase by L1-mediated Homophilic Cell Interactions. *Mol. Biol. Cell* **2004**, *15*, 2003–2012. [CrossRef]

37. Schaefer, A.W.; Kamiguchi, H.; Wong, E.V.; Beach, C.M.; Landreth, G.; Lemmon, V. Activation of the MAPK signal cascade by the neural cell adhesion molecule L1 requires L1 internalization. *J. Biol. Chem.* **1999**, *274*, 37965–37973. [CrossRef]

38. Schaefer, A.W.; Kamei, Y.; Kamiguchi, H.; Wong, E.V.; Rapoport, I.; Kirchhausen, T.; Beach, C.M.; Landreth, G.; Lemmon, S.K.; Lemmon, V. L1 endocytosis is controlled by a phosphorylation-dephosphorylation cycle stimulated by outside-in signaling by L1. *J. Cell Biol.* **2002**, *157*, 1223–1232. [CrossRef]

39. Kamiguchi, H.; Long, K.E.; Pendergast, M.; Schaefer, A.W.; Rapoport, I.; Kirchhausen, T.; Lemmon, V. The neural cell adhesion molecule L1 interacts with the AP-2 adaptor and is endocytosed via the clathrin-mediated pathway. *J. Neurosci.* **1998**, *18*, 5311–5321. [CrossRef]

40. Chen, M.M.; Lee, C.Y.; Leland, H.A.; Lin, G.Y.; Montgomery, A.M.; Silletti, S. Inside-out regulation of L1 conformation, integrin binding, proteolysis, and concomitant cell migration. *Mol. Biol. Cell* **2010**, *21*, 1671–1685. [CrossRef]

41. Chen, M.M.; Leland, H.A.; Lee, C.Y.; Silletti, S. Tyrosine and serine phosphorylation regulate the conformation and subsequent threonine phosphorylation of the L1 cytoplasmic domain. *Biochem. Biophys. Res. Commun.* **2009**, *389*, 257–264. [CrossRef] [PubMed]

42. Kiefel, H.; Bondong, S.; Hazin, J.; Ridinger, J.; Schirmer, U.; Riedle, S.; Altevogt, P. L1CAM: A major driver for tumor cell invasion and motility. *Cell Adhes. Migr.* **2012**, *6*, 374–384. [CrossRef] [PubMed]

43. Kalus, I.; Schnegelsberg, B.; Seidah, N.G.; Kleene, R.; Schachner, M. The proprotein convertase PC5A and a metalloprotease are involved in the proteolytic processing of the neural adhesion molecule L1. *J. Biol. Chem.* **2003**, *278*, 10381–10388. [CrossRef]

44. Nayeem, N.; Silletti, S.; Yang, X.M.; Lemmon, V.P.; Reisfeld, R.A.; Stallcup, W.B.; Montgomery, A.M.P. A potential role for the plasmin(ogen) system in the posttranslational cleavage of the neural cell adhesion molecule L1. *J. Cell Sci.* **1999**, *112*, 4739–4749. [PubMed]

45. Sadoul, K.; Sadoul, R.; Faissner, A.; Schachner, M. Biochemical Characterization of Different Molecular Forms of the Neural Cell Adhesion Molecule L1. *J. Neurochem.* **1988**, *50*, 510–521. [CrossRef] [PubMed]

46. Lutz, D.; Wolters-Eisfeld, G.; Joshi, G.; Djogo, N.; Jakovcevski, I.; Schachner, M.; Kleene, R. Generation and nuclear translocation of sumoylated transmembrane fragment of cell adhesion molecule L1. *J. Biol. Chem.* **2012**, *287*, 17161–17175. [CrossRef]

47. Matsumoto-Miyai, K.; Ninomiya, A.; Yamasaki, H.; Tamura, H.; Nakamura, Y.; Shiosaka, S. NMDA-dependent proteolysis of presynaptic adhesion molecule L1 in the hippocampus by neuropsin. *J. Neurosci.* **2003**, *23*, 7727–7736. [CrossRef]

48. Zou, Y.; Uddin, M.M.; Padmanabhan, S.; Zhu, Y.; Bu, P.; Vancura, A.; Vancurova, I. The proto-oncogene Bcl3 induces immune checkpoint PD-L1 expression, mediating proliferation of ovarian cancer cells. *J. Biol. Chem.* **2018**, *293*, 15483–15496. [CrossRef]

49. Maretzky, T.; Schulte, M.; Ludwig, A.; Rose-John, S.; Blobel, C.; Hartmann, D.; Altevogt, P.; Saftig, P.; Reiss, K. L1 Is Sequentially Processed by Two Differently Activated Metalloproteases and Presenilin/ -Secretase and Regulates Neural Cell Adhesion, Cell Migration, and Neurite Outgrowth. *Mol. Cell. Biol.* **2005**, *25*, 9040–9053. [CrossRef]

50. Gutwein, P.; Mechtersheimer, S.; Riedle, S.; Stoeck, A.; Gast, D.; Joumaa, S.; Zentgraf, H.; Fogel, M.; Altevogt, D.P. ADAM10-mediated cleavage of L1 adhesion molecule at the cell surface and in released membrane vesicles. *FASEB J.* **2003**, *17*, 292–294. [CrossRef]

51. Riedle, S.; Kiefel, H.; Gast, D.; Bondong, S.; Wolterink, S.; Gutwein, P.; Altevogt, P. Nuclear translocation and signalling of L1-CAM in human carcinoma cells requires ADAM10 and presenilin/γ-secretase activity. *Biochem. J.* **2009**, *420*, 391–402. [CrossRef] [PubMed]

52. Angiolini, F.; Belloni, E.; Giordano, M.; Campioni, M.; Forneris, F.; Paronetto, M.P.; Lupia, M.; Brandas, C.; Pradella, D.; Di Matteo, A.; et al. A novel L1CAM isoform with angiogenic activity generated by NOVA2-mediated alternative splicing. *Elife* **2019**, *8*, e44305. [CrossRef] [PubMed]

53. Altevogt, P.; Doberstein, K.; Fogel, M. L1CAM in human cancer. *Int. J. Cancer* **2016**, *138*, 1565–1576. [CrossRef] [PubMed]

54. Gavert, N.; Ben-Shmuel, A.; Raveh, S.; Ben-Ze'ev, A. L1-CAM in cancerous tissues. *Expert Opin. Biol. Ther.* **2008**, *8*, 1749–1757. [CrossRef] [PubMed]

55. Bosse, T.; Nout, R.A.; Stelloo, E.; Dreef, E.; Nijman, H.W.; Jürgenliemk-Schulz, I.M.; Jobsen, J.J.; Creutzberg, C.L.; Smit, V.T.H.B.M. L1 cell adhesion molecule is a strong predictor for distant recurrence and overall survival in early stage endometrial cancer: Pooled PORTEC trial results. *Eur. J. Cancer* **2014**, *50*, 2602–2610. [CrossRef] [PubMed]

56. Smogeli, E.; Davidson, B.; Cvancarova, M.; Holth, A.; Katz, B.; Risberg, B.; Kristensen, G.; Lindemann, K. L1CAM as a prognostic marker in stage I endometrial cancer: A validation study. *BMC Cancer* **2016**, *16*, 596. [CrossRef]

57. Tangen, I.L.; Kopperud, R.K.; Visser, N.C.M.; Staff, A.C.; Tingulstad, S.; Marcickiewicz, J.; Amant, F.; Bjørge, L.; Pijnenborg, J.M.A.; Salvesen, H.B.; et al. Expression of L1CAM in curettage or high L1CAM level in preoperative blood samples predicts lymph node metastases and poor outcome in endometrial cancer patients. *Br. J. Cancer* **2017**, *117*, 840–847. [CrossRef]

58. Corrado, G.; Laquintana, V.; Loria, R.; Carosi, M.; De Salvo, L.; Sperduti, I.; Zampa, A.; Cicchillitti, L.; Piaggio, G.; Cutillo, G.; et al. Endometrial cancer prognosis correlates with the expression of L1CAM and miR34a biomarkers. *J. Exp. Clin. Cancer Res.* **2018**, *37*, 139. [CrossRef] [PubMed]

59. Van Der Putten, L.J.M.; Visser, N.C.M.; Van De Vijver, K.; Santacana, M.; Bronsert, P.; Bulten, J.; Hirschfeld, M.; Colas, E.; Gil-Moreno, A.; Garcia, A.; et al. L1CAM expression in endometrial carcinomas: An ENITEC collaboration study. *Br. J. Cancer* **2016**, *115*, 716–724. [CrossRef]

60. Vizza, E.; Mancini, E.; Laquintana, V.; Loria, R.; Carosi, M.; Baiocco, E.; Cicchillitti, L.; Piaggio, G.; Patrizi, L.; Sperduti, I.; et al. The prognostic significance of positive peritoneal cytology in endometrial cancer and its correlations with L1-CAM biomarker. *Surg. Oncol.* **2019**, *28*, 151–157. [CrossRef]

61. Fogel, M.; Gutwein, P.; Mechtersheimer, S.; Riedle, S.; Stoeck, A.; Smirnov, A.; Edler, L.; Ben-Arie, A.; Huszar, M.; Altevogt, P. L1 expression as a predictor of progression and survival in patients with uterine and ovarian carcinomas. *Lancet* **2003**, *362*, 869–875. [CrossRef]

62. Zecchini, S.; Bianchi, M.; Colombo, N.; Fasani, R.; Goisis, G.; Casadio, C.; Viale, G.; Liu, J.; Herlyn, M.; Godwin, A.K.; et al. The differential role of L1 in ovarian carcinoma and normal ovarian surface epithelium. *Cancer Res.* **2008**, *68*, 1110–1118. [CrossRef]

63. Soovares, P.; Pasanen, A.; Bützow, R.; Lassus, H. L1CAM expression associates with poor outcome in endometrioid, but not in clear cell ovarian carcinoma. *Gynecol. Oncol.* **2017**, *146*, 615–622. [CrossRef] [PubMed]

64. Bondong, S.; Kiefel, H.; Hielscher, T.; Zeimet, A.G.; Zeillinger, R.; Pils, D.; Schuster, E.; Castillo-Tong, D.C.; Cadron, I.; Vergote, I.; et al. Prognostic significance of L1CAM in ovarian cancer and its role in constitutive NF-κB activation. *Ann. Oncol.* **2012**, *23*, 1795–1802. [CrossRef] [PubMed]

65. Thies, A.; Schachner, M.; Moll, I.; Berger, J.; Schulze, H.J.; Brunner, G.; Schumacher, U. Overexpression of the cell adhesion molecule L1 is associated with metastasis in cutaneous malignant melanoma. *Eur. J. Cancer* **2002**, *38*, 1708–1716. [CrossRef]

66. Wu, J.D.; Hong, C.Q.; Huang, W.H.; Wei, X.L.; Zhang, F.; Zhuang, Y.X.; Zhang, Y.Q.; Zhang, G.J. L1 Cell Adhesion Molecule and Its Soluble Form sL1 Exhibit Poor Prognosis in Primary Breast Cancer Patients. *Clin. Breast Cancer* **2018**, *18*, e851–e861. [CrossRef]

67. Ichikawa, T.; Okugawa, Y.; Toiyama, Y.; Tanaka, K.; Yin, C.; Kitajima, T.; Kondo, S.; Shimura, T.; Ohi, M.; Araki, T.; et al. Clinical significance and biological role of L1 cell adhesion molecule in gastric cancer. *Br. J. Cancer* **2019**, *121*, 1058–1068. [CrossRef]

68. Boo, Y.J.; Park, J.M.; Kim, J.; Chae, Y.S.; Min, B.W.; Um, J.W.; Moon, H.Y. L1 expression as a marker for poor prognosis, tumor progression, and short survival in patients with colorectal cancer. *Ann. Surg. Oncol.* **2007**, *14*, 1703–1711. [CrossRef]

69. Ben, Q.W.; Wang, J.C.; Liu, J.; Zhu, Y.; Yuan, F.; Yao, W.Y.; Yuan, Y.Z. Positive expression of L1-CAM is associated with perineural invasion and poor outcome in pancreatic ductal adenocarcinoma. *Ann. Surg. Oncol.* **2010**, *17*, 2213–2221. [CrossRef]

70. Yu, H.; Zhou, P.; Li, D.; Li, W. L1CAM-positive expression is associated with poorer survival outcomes in resected non-small cell lung cancer patients. *Int. J. Clin. Exp. Pathol.* **2019**, *12*, 2665–2671.

71. Doberstein, K.; Wieland, A.; Lee, S.B.B.; Blaheta, R.A.A.; Wedel, S.; Moch, H.; Schraml, P.; Pfeilschifter, J.; Kristiansen, G.; Gutwein, P. L1-CAM expression in ccRCC correlates with shorter patients survival times and confers chemoresistance in renal cell carcinoma cells. *Carcinogenesis* **2011**, *32*, 262–270. [CrossRef] [PubMed]

72. Pasanen, A.; Loukovaara, M.; Tuomi, T.; Bützow, R. Preoperative risk stratification of endometrial carcinoma: L1CAM as a biomarker. *Int. J. Gynecol. Cancer* **2017**, *27*, 1318–1324. [CrossRef] [PubMed]

73. Aktas, B.; Kasimir-Bauer, S.; Wimberger, P.; Kimmig, R.; Heubner, M. Utility of mesothelin, L1CAM and afamin as biomarkers in primary ovarian cancer. *Anticancer Res.* **2013**, *33*, 329–336. [PubMed]

74. Bajaj, J.; Diaz, E.; Reya, T. Stem cells in cancer initiation and progression. *J. Cell Biol.* **2020**, *219*, e201911053. [CrossRef] [PubMed]

75. Turdo, A.; Veschi, V.; Gaggianesi, M.; Chinnici, A.; Bianca, P.; Todaro, M.; Stassi, G. Meeting the challenge of targeting cancer stem cells. *Front. Cell Dev. Biol.* **2019**, *7*, 16. [CrossRef] [PubMed]

76. Hermann, P.C.; Sainz, B. Pancreatic cancer stem cells: A state or an entity? *Semin. Cancer Biol.* **2018**, *53*, 223–231. [CrossRef] [PubMed]

77. Rowan, K. Are cancer stem cells real? After four decades, debate still simmers. *J. Natl. Cancer Inst.* **2009**, *101*, 546–547. [CrossRef]

78. Jordan, C.T. Cancer Stem Cells: Controversial or Just Misunderstood? *Cell Stem Cell* **2009**, *4*, 203–205. [CrossRef]

79. Pattabiraman, D.R.; Weinberg, R.A. Tackling the cancer stem cells-what challenges do they pose? *Nat. Rev. Drug Discov.* **2014**, *13*, 497–512. [CrossRef]

80. Ratajczak, M.Z.; Bujko, K.; Mack, A.; Kucia, M.; Ratajczak, J. Cancer from the perspective of stem cells and misappropriated tissue regeneration mechanisms. *Leukemia* **2018**, *32*, 2519–2526. [CrossRef]

81. Nimmakayala, R.K.; Batra, S.K.; Ponnusamy, M.P. Unraveling the journey of cancer stem cells from origin to metastasis. *Biochim. Biophys. Acta Rev. Cancer* **2019**, *1871*, 50–63. [CrossRef]

82. Lupia, M.; Cavallaro, U. Ovarian cancer stem cells: Still an elusive entity? *Mol. Cancer* **2017**, *16*, 64. [CrossRef] [PubMed]

83. Shibue, T.; Weinberg, R.A. EMT, CSCs, and drug resistance: The mechanistic link and clinical implications. *Nat. Rev. Clin. Oncol.* **2017**, *14*, 611–629. [CrossRef] [PubMed]

84. Bao, S.; Wu, Q.; Li, Z.; Sathornsumetee, S.; Wang, H.; McLendon, R.E.; Hjelmeland, A.B.; Rich, J.N. Targeting cancer stem cells through L1CAM suppresses glioma growth. *Cancer Res.* **2008**, *68*, 6043–6048. [CrossRef] [PubMed]

85. Cheng, L.; Wu, Q.; Huang, Z.; Guryanova, O.A.; Huang, Q.; Shou, W.; Rich, J.N.; Bao, S. L1CAM regulates DNA damage checkpoint response of glioblastoma stem cells through NBS1. *EMBO J.* **2011**, *30*, 800–813. [CrossRef]

86. Maréchal, A.; Zou, L. DNA damage sensing by the ATM and ATR kinases. *Cold Spring Harb. Perspect. Biol.* **2013**, *5*, a012716. [CrossRef]

87. Mills, B.N.; Albert, G.P.; Halterman, M.W. Expression Profiling of the MAP Kinase Phosphatase Family Reveals a Role for DUSP1 in the Glioblastoma Stem Cell Niche. *Cancer Microenviron.* **2017**, *10*, 57–68. [CrossRef]

88. Erhart, F.; Blauensteiner, B.; Zirkovits, G.; Printz, D.; Soukup, K.; Klingenbrunner, S.; Fischhuber, K.; Reitermaier, R.; Halfmann, A.; Lötsch, D.; et al. Gliomasphere marker combinatorics: Multidimensional flow cytometry detects CD44+/CD133+/ITGA6+/CD36+ signature. *J. Cell. Mol. Med.* **2019**, *23*, 281–292. [CrossRef]

89. Brescia, P.; Richichi, C.; Pelicci, G. Current strategies for identification of glioma stem cells: Adequate or unsatisfactory? *J. Oncol.* **2012**, *2012*, 376894. [CrossRef]

90. Held-Feindt, J.; Schmelz, S.; Hattermann, K.; Mentlein, R.; Mehdorn, H.M.; Sebens, S. The neural adhesion molecule L1CAM confers chemoresistance in human glioblastomas. *Neurochem. Int.* **2012**, *61*, 1183–1191. [CrossRef]

91. Geismann, C.; Morscheck, M.; Koch, D.; Bergmann, F.; Ungefroren, H.; Arlt, A.; Tsao, M.S.; Bachem, M.G.; Altevogt, P.; Sipos, B.; et al. Up-regulation of L1CAM in pancreatic duct cells is transforming growth factor β1- and slug-dependent: Role in malignant transformation of pancreatic cancer. *Cancer Res.* **2009**, *69*, 4517–4526. [CrossRef] [PubMed]

92. Geismann, C.; Arlt, A.; Bauer, I.; Pfeifer, M.; Schirmer, U.; Altevogt, P.; Müerköster, S.S.; Schäfer, H. Binding of the transcription factor Slug to the L1CAM promoter is essential for transforming growth factor-β1 (TGF-β)-induced L1CAM expression in human pancreatic ductal adenocarcinoma cells. *Int. J. Oncol.* **2011**, *38*, 257–266. [PubMed]

93. Gavert, N.; Conacci-Sorrell, M.; Gast, D.; Schneider, A.; Altevogt, P.; Brabletz, T.; Ben-Ze'Ev, A. L1, a novel target of β-catenin signaling, transforms cells and is expressed at the invasive front of colon cancers. *J. Cell Biol.* **2005**, *168*, 633–642. [CrossRef] [PubMed]

94. Conacci-Sorrell, M.E.; Ben-Yedidia, T.; Shtutman, M.; Feinstein, E.; Einat, P.; Ben-Ze'ev, A. Nr-CAM is a target gene of the β-catenin/LEF-1 pathway in melanoma and colon cancer and its expression enhances motility and confers tumorigenesis. *Genes Dev.* **2002**, *16*, 2058–2072. [CrossRef]

95. Gavert, N.; Vivanti, A.; Hazin, J.; Brabletz, T.; Ben-Ze'ev, A. L1-Mediated colon cancer cell metastasis does not require changes in EMT and cancer stem cell markers. *Mol. Cancer Res.* **2011**, *9*, 14–24. [CrossRef]

96. Shapiro, B.; Tocci, P.; Haase, G.; Gavert, N.; Ben-Ze'ev, A. Clusterin, a gene enriched in intestinal stem cells, is required for L1-mediated colon cancer metastasis. *Oncotarget* **2015**, *6*, 34389–34401. [CrossRef]

97. Van der Flier, L.G.; van Gijn, M.E.; Hatzis, P.; Kujala, P.; Haegebarth, A.; Stange, D.E.; Begthel, H.; van den Born, M.; Guryev, V.; Oving, I.; et al. Transcription Factor Achaete Scute-Like 2 Controls Intestinal Stem Cell Fate. *Cell* **2009**, *136*, 903–912. [CrossRef]

98. Schuijers, J.; Junker, J.P.; Mokry, M.; Hatzis, P.; Koo, B.K.; Sasselli, V.; Van Der Flier, L.G.; Cuppen, E.; Van Oudenaarden, A.; Clevers, H. Ascl2 acts as an R-spondin/wnt-responsive switch to control stemness in intestinal crypts. *Cell Stem Cell* **2015**, *16*, 158–170. [CrossRef]

99. Basu, S.; Gavert, N.; Brabletz, T.; Ben-Ze'ev, A. The intestinal stem cell regulating gene ASCL2 is required for L1-mediated colon cancer progression. *Cancer Lett.* **2018**, *424*, 9–18. [CrossRef]

100. Ganesh, K.; Basnet, H.; Kaygusuz, Y.; Laughney, A.M.; He, L.; Sharma, R.; O'Rourke, K.P.; Reuter, V.P.; Huang, Y.-H.; Turkekul, M.; et al. L1CAM defines the regenerative origin of metastasis-initiating cells in colorectal cancer. *Nat. Cancer* **2020**, *1*, 28–45. [CrossRef]

101. Fang, Q.-X.; Zheng, X.-C.; Zhao, H.-J. L1CAM is involved in lymph node metastasis via ERK1/2 signaling in colorectal cancer. *Am. J. Transl. Res.* **2020**, *12*, 837–846. [PubMed]

102. Lund, K.; Dembinski, J.L.; Solberg, N.; Urbanucci, A.; Mills, I.G.; Krauss, S. Slug-dependent upregulation of L1CAM is responsible for the increased invasion potential of pancreatic cancer cells following long-term 5-FU treatment. *PLoS ONE* **2015**, *10*, e0123684. [CrossRef] [PubMed]

103. Terraneo, N.; Jacob, F.; Peitzsch, C.; Dubrovska, A.; Krudewig, C.; Huang, Y.L.; Heinzelmann-Schwarz, V.; Schibli, R.; Béhé, M.; Grünberg, J. L1 cell adhesion molecule confers radioresistance to ovarian cancer and defines a new cancer stem cell population. *Cancers* **2020**, *12*, 217. [CrossRef]

104. Roberts, C.M.; Tran, M.A.; Pitruzzello, M.C.; Wen, W.; Loeza, J.; Dellinger, T.H.; Mor, G.; Glackin, C.A. TWIST1 drives cisplatin resistance and cell survival in an ovarian cancer model, via upregulation of GAS6, L1CAM, and Akt signalling. *Sci. Rep.* **2016**, *6*, 37652. [CrossRef] [PubMed]

105. Chen, J.; Gao, F.; Liu, N. L1CAM promotes epithelial to mesenchymal transition and formation of cancer initiating cells in human endometrial cancer. *Exp. Ther. Med.* **2018**, *15*, 2792–2797. [CrossRef] [PubMed]

106. Götte, M.; Wolf, M.; Staebler, A.; Buchweitz, O.; Kelsch, R.; Schüring, A.N.; Kiesel, L. Increased expression of the adult stem cell marker Musashi-1 in endometriosis and endometrial carcinoma. *J. Pathol.* **2008**, *215*, 317–329. [CrossRef] [PubMed]

107. Jo, D.H.; Lee, K.; Kim, J.H.; Jun, H.O.; Kim, Y.; Cho, Y.L.; Yu, Y.S.; Min, J.K.; Kim, J.H. L1 increases adhesion-mediated proliferation and chemoresistance of retinoblastoma. *Oncotarget* **2017**, *8*, 15441–15452. [CrossRef]

108. Fletcher, J.I.; Haber, M.; Henderson, M.J.; Norris, M.D. ABC transporters in cancer: More than just drug efflux pumps. *Nat. Rev. Cancer* **2010**, *10*, 147–156. [CrossRef]

109. Chang, G. Multidrug resistance ABC transporters. *FEBS Lett.* **2003**, *555*, 102–105. [CrossRef]

110. Garcia-Mayea, Y.; Mir, C.; Masson, F.; Paciucci, R.; LLeonart, M.E. Insights into new mechanisms and models of cancer stem cell multidrug resistance. *Semin. Cancer Biol.* **2020**, *60*, 166–180. [CrossRef]

111. Magrini, E.; Villa, A.; Angiolini, F.; Doni, A.; Mazzarol, G.; Rudini, N.; Maddaluno, L.; Komuta, M.; Topal, B.; Prenen, H.; et al. Endothelial deficiency of L1 reduces tumor angiogenesis and promotes vessel normalization. *J. Clin. Investig.* **2014**, *124*, 4335–4350. [CrossRef]

112. Medici, D. Endothelial-Mesenchymal Transition in Regenerative Medicine. *Stem Cells Int.* **2016**, *2016*, 6962801. [CrossRef] [PubMed]

113. Zeisberg, E.M.; Potenta, S.; Xie, L.; Zeisberg, M.; Kalluri, R. Discovery of endothelial to mesenchymal transition as a source for carcinoma-associated fibroblasts. *Cancer Res.* **2007**, *67*, 10123–10128. [CrossRef] [PubMed]

114. Nakata, S.; Campos, B.; Bageritz, J.; Lorenzo Bermejo, J.; Becker, N.; Engel, F.; Acker, T.; Momma, S.; Herold-Mende, C.; Lichter, P.; et al. LGR5 is a marker of poor prognosis in glioblastoma and is required for survival of brain cancer stem-like cells. *Brain Pathol.* **2013**, *23*, 60–72. [CrossRef] [PubMed]

115. Okawa, S.; Gagrica, S.; Blin, C.; Ender, C.; Pollard, S.M.; Krijgsveld, J. Proteome and Secretome Characterization of Glioblastoma-Derived Neural Stem Cells. *Stem Cells* **2017**, *35*, 967–980. [CrossRef]

116. Gemei, M.; Corbo, C.; D'Alessio, F.; Di Noto, R.; Vento, R.; Del Vecchio, L. Surface proteomic analysis of differentiated versus stem-like osteosarcoma human cells. *Proteomics* **2013**, *13*, 3293–3297. [CrossRef]

117. Liang, L.; Aiken, C.; McClelland, R.; Morrison, L.C.; Tatari, N.; Remke, M.; Ramaswamy, V.; Issaivanan, M.; Ryken, T.; Del Bigio, M.R.; et al. Characterization of novel biomarkers in selecting for subtype specific medulloblastoma phenotypes. *Oncotarget* **2015**, *6*, 38881–38900. [CrossRef]

118. Son, Y.S.; Seong, R.H.; Ryu, C.J.; Cho, Y.S.; Bae, K.H.; Chung, S.J.; Lee, B.; Min, J.K.; Hong, H.J. Brief report: L1 cell adhesion molecule, a novel surface molecule of human embryonic stem cells, is essential for self-renewal and pluripotency. *Stem Cells* **2011**, *29*, 2094–2099. [CrossRef]

119. Li, Y.; Huang, X.; An, Y.; Ren, F.; Yang, Z.Z.; Zhu, H.; Zhou, L.; He, X.; Schachner, M.; Xiao, Z.; et al. Cell recognition molecule L1 promotes embryonic stem cell differentiation through the regulation of cell surface glycosylation. *Biochem. Biophys. Res. Commun.* **2013**, *440*, 405–412. [CrossRef]

120. Pusey, M.A.; Pace, K.; Fascelli, M.; Linser, P.J.; Steindler, D.A.; Galileo, D.S. Ectopic expression of L1CAM ectodomain alters differentiation and motility, but not proliferation, of human neural progenitor cells. *Int. J. Dev. Neurosci.* **2019**, *78*, 49–64. [CrossRef]

121. Arlt, M.J.E.; Novak-Hofer, I.; Gast, D.; Gschwend, V.; Moldenhauer, G.; Grünberg, J.; Honer, M.; Schubiger, P.A.; Altevogt, P.; Krüger, A. Efficient inhibition of intra-peritoneal tumor growth and dissemination of human ovarian carcinoma cells in nude mice by anti-L1-cell adhesion molecule monoclonal antibody treatment. *Cancer Res.* **2006**, *66*, 936–943. [CrossRef] [PubMed]

122. Wolterink, S.; Moldenhauer, G.; Fogel, M.; Kiefel, H.; Pfeifer, M.; Lüttgau, S.; Gouveia, R.; Costa, J.; Endell, J.; Moebius, U.; et al. Therapeutic antibodies to human L1CAM: Functional characterization and application in a mouse model for ovarian carcinoma. *Cancer Res.* **2010**, *70*, 2504–2515. [CrossRef] [PubMed]

123. Doberstein, K.; Harter, P.N.; Haberkorn, U.; Bretz, N.P.; Arnold, B.; Carretero, R.; Moldenhauer, G.; Mittelbronn, M.; Altevogt, P. Antibody therapy to human L1CAM in a transgenic mouse model blocks local tumor growth but induces EMT. *Int. J. Cancer* **2015**, *136*, E326–E339. [CrossRef]

124. Lee, E.S.; Jeong, M.S.; Singh, R.; Jung, J.; Yoon, H.; Min, J.K.; Kim, K.H.; Hong, H.J. A chimeric antibody to L1 cell adhesion molecule shows therapeutic effect in an intrahepatic cholangiocarcinoma model. *Exp. Mol. Med.* **2012**, *44*, 293–302. [CrossRef] [PubMed]

125. Schäfer, H.; Dieckmann, C.; Korniienko, O.; Moldenhauer, G.; Kiefel, H.; Salnikov, A.; Krüger, A.; Altevogt, P.; Sebens, S. Combined treatment of L1CAM antibodies and cytostatic drugs improve the therapeutic response of pancreatic and ovarian carcinoma. *Cancer Lett.* **2012**, *319*, 66–82. [CrossRef] [PubMed]

126. Cho, S.; Lee, T.S.; Song, I.H.; Kim, A.R.; Lee, Y.J.; Kim, H.; Hwang, H.; Jeong, M.S.; Kang, S.G.; Hong, H.J. Combination of anti-L1 cell adhesion molecule antibody and gemcitabine or cisplatin improves the therapeutic response of intrahepatic cholangiocarcinoma. *PLoS ONE* **2017**, *12*, e0170078. [CrossRef] [PubMed]

127. Fischer, E.; Grünberg, J.; Cohrs, S.; Hohn, A.; Waldner-Knogler, K.; Jeger, S.; Zimmermann, K.; Novak-Hofer, I.; Schibli, R. L1-CAM-targeted antibody therapy and 177Lu-radioimmunotherapy of disseminated ovarian cancer. *Int. J. Cancer* **2012**, *130*, 2715–2721. [CrossRef]

128. Grünberg, J.; Lindenblatt, D.; Dorrer, H.; Cohrs, S.; Zhernosekov, K.; Köster, U.; Türler, A.; Fischer, E.; Schibli, R. Anti-L1CAM radioimmunotherapy is more effective with the radiolanthanide terbium-161 compared to lutetium-177 in an ovarian cancer model. *Eur. J. Nucl. Med. Mol. Imaging* **2014**, *41*, 1907–1915. [CrossRef]

129. Hoefnagel, C.A.; Rutgers, M.; Buitenhuis, C.K.M.; Smets, L.A.; De Kraker, J.; Meli, M.; Carrel, F.; Amstutz, H.; Schubiger, P.A.; Novak-Hofer, I. A comparison of targetting of neuroblastoma with mIBG and anti L1-CAM antibody mAb chCE7: Therapeutic efficacy in a neuroblastoma xenograft model and imaging of neuroblastoma patients. *Eur. J. Nucl. Med.* **2001**, *28*, 359–368. [CrossRef]

130. Song, I.H.; Jeong, M.S.; Hong, H.J.; Shin, J., II; Park, Y.S.; Woo, S.K.; Moon, B.S.; Kim, K., II; Lee, Y.J.; Kang, J.H.; et al. Development of a theranostic convergence bioradiopharmaceutical for immuno-PET based radioimmunotherapy of L1CAM in cholangiocarcinoma model. *Clin. Cancer Res.* **2019**, *25*, 6148–6159. [CrossRef]

131. Lindenblatt, D.; Fischer, E.; Cohrs, S.; Schibli, R.; Grünberg, J. Paclitaxel improved anti-L1CAM lutetium-177 radioimmunotherapy in an ovarian cancer xenograft model. *EJNMMI Res.* **2014**, *4*, 54. [CrossRef] [PubMed]

132. Lindenblatt, D.; Terraneo, N.; Pellegrini, G.; Cohrs, S.; Spycher, P.R.; Vukovic, D.; Béhé, M.; Schibli, R.; Grünberg, J. Combination of lutetium-177 labelled anti-L1CAM antibody chCE7 with the clinically relevant protein kinase inhibitor MK1775: A novel combination against human ovarian carcinoma. *BMC Cancer* **2018**, *18*, 922. [CrossRef] [PubMed]

133. Park, J.R.; DiGiusto, D.L.; Slovak, M.; Wright, C.; Naranjo, A.; Wagner, J.; Meechoovet, H.B.; Bautista, C.; Chang, W.C.; Ostberg, J.R.; et al. Adoptive transfer of chimeric antigen receptor re-directed cytolytic T lymphocyte clones in patients with neuroblastoma. *Mol. Ther.* **2007**, *15*, 825–833. [CrossRef] [PubMed]

134. Hong, H.; Brown, C.E.; Ostberg, J.R.; Priceman, S.J.; Chang, W.C.; Weng, L.; Lin, P.; Wakabayashi, M.T.; Jensen, M.C.; Forman, S.J. L1 cell adhesion molecule-specific chimeric antigen receptor-redirected Human T cells exhibit specific and efficient antitumor activity against human ovarian cancer in mice. *PLoS ONE* **2016**, *11*, e0146885. [CrossRef] [PubMed]

135. Andersch, L.; Radke, J.; Klaus, A.; Schwiebert, S.; Winkler, A.; Schumann, E.; Grunewald, L.; Zirngibl, F.; Flemmig, C.; Jensen, M.C.; et al. CD171- and GD2-specific CAR-T cells potently target retinoblastoma cells in preclinical in vitro testing. *BMC Cancer* **2019**, *19*, 895. [CrossRef] [PubMed]

136. Grage-Griebenow, E.; Jerg, E.; Gorys, A.; Wicklein, D.; Wesch, D.; Freitag-Wolf, S.; Goebel, L.; Vogel, I.; Becker, T.; Ebsen, M.; et al. L1CAM promotes enrichment of immunosuppressive T cells in human pancreatic cancer correlating with malignant progression. *Mol. Oncol.* **2014**, *8*, 982–997. [CrossRef]

137. Aikawa, Y. Rabex-5 protein regulates the endocytic trafficking pathway of ubiquitinated neural cell adhesion molecule L1. *J. Biol. Chem.* **2012**, *287*, 32312–32323. [CrossRef]

138. Panicker, A.K.; Buhusi, M.; Erickson, A.; Maness, P.F. Endocytosis of β1 integrins is an early event in migration promoted by the cell adhesion molecule L1. *Exp. Cell Res.* **2006**, *312*, 299–307. [CrossRef]

139. Grünberg, J.; Novak-Hofer, I.; Honer, M.; Zimmermann, K.; Knogler, K.; Bläuenstein, P.; Ametamey, S.; Maecke, H.R.; Schubiger, P.A. In vivo evaluation of177Lu- and 67/64Cu-labeled recombinant fragments of antibody chCE7 for radioimmunotherapy and PET imaging of L1-CAM-positive tumors. *Clin. Cancer Res.* **2005**, *11*, 5112–5120. [CrossRef]

140. Zimmermann, K.; Grünberg, J.; Honer, M.; Ametamey, S.; Schubiger, P.A.; Novak-Hofer, I. Targeting of renal carcinoma with 67/64Cu-labeled anti-L1-CAM antibody chCE7: Selection of copper ligands and PET imaging. *Nucl. Med. Biol.* **2003**, *30*, 417–427. [CrossRef]

141. Eshhar, Z.; Waks, T.; Gross, G.; Schindler, D.G. Specific activation and targeting of cytotoxic lymphocytes through chimeric single chains consisting of antibody-binding domains and the γ or ζ subunits of the immunoglobulin and T-cell receptors. *Proc. Natl. Acad. Sci. USA* **1993**, *90*, 720–724. [CrossRef] [PubMed]

142. Maude, S.L.; Frey, N.; Shaw, P.A.; Aplenc, R.; Barrett, D.M.; Bunin, N.J.; Chew, A.; Gonzalez, V.E.; Zheng, Z.; Lacey, S.F.; et al. Chimeric antigen receptor T cells for sustained remissions in leukemia. *N. Engl. J. Med.* **2014**, *371*, 1507–1517. [CrossRef]

143. Kochenderfer, J.N.; Dudley, M.E.; Kassim, S.H.; Somerville, R.P.T.; Carpenter, R.O.; Maryalice, S.S.; Yang, J.C.; Phan, G.Q.; Hughes, M.S.; Sherry, R.M.; et al. Chemotherapy-refractory diffuse large B-cell lymphoma and indolent B-cell malignancies can be effectively treated with autologous T cells expressing an anti-CD19 chimeric antigen receptor. *J. Clin. Oncol.* **2015**, *33*, 540–549. [CrossRef] [PubMed]

144. Zander, H.; Rawnaq, T.; von Wedemeyer, M.; Tachezy, M.; Kunkel, M.; Wolters, G.; Bockhorn, M.; Schachner, M.; Izbicki, J.R.; Kaifi, J. Circulating levels of cell adhesion molecule l1 as a prognostic marker in gastrointestinal stromal tumor patients. *BMC Cancer* **2011**, *11*, 189. [CrossRef] [PubMed]

145. Wachowiak, R.; Krause, M.; Mayer, S.; Peukert, N.; Suttkus, A.; Müller, W.C.; Lacher, M.; Meixensberger, J.; Nestler, U. Increased L1CAM (CD171) levels are associated with glioblastoma and metastatic brain tumors. *Medicine* **2018**, *97*, e12396. [CrossRef] [PubMed]

 Journal of
Clinical Medicine

Review

Wnt Signaling in Ovarian Cancer Stemness, EMT, and Therapy Resistance

Miriam Teeuwssen * and Riccardo Fodde *

Department of Pathology, Erasmus MC Cancer Institute, Erasmus University Medical Center,
3015 GD Rotterdam, The Netherlands
* Correspondence: m.teeuwssen@eramusmc.nl (M.T.); r.fodde@erasmusmc.nl (R.F.)

Received: 2 September 2019; Accepted: 3 October 2019; Published: 11 October 2019

Abstract: Ovarian cancers represent the deadliest among gynecologic malignancies and are characterized by a hierarchical structure with cancer stem cells (CSCs) endowed with self-renewal and the capacity to differentiate. The Wnt/β-catenin signaling pathway, known to regulate stemness in a broad spectrum of stem cell niches including the ovary, is thought to play an important role in ovarian cancer. Importantly, Wnt activity was shown to correlate with grade, epithelial to mesenchymal transition, chemotherapy resistance, and poor prognosis in ovarian cancer. This review will discuss the current knowledge of the role of Wnt signaling in ovarian cancer stemness, epithelial to mesenchymal transition (EMT), and therapy resistance. In addition, the alleged role of exosomes in the paracrine activation of Wnt signaling and pre-metastatic niche formation will be reviewed. Finally, novel potential treatment options based on Wnt inhibition will be highlighted.

Keywords: Ovarian cancer; Wnt signaling; cancer stem cells; tumor progression; therapy resistance; exosomes

1. Introduction

Epithelial ovarian cancer (EOC) represents the deadliest among gynecologic malignancies [1]. This is mainly due to the fact that up to 80% of ovarian cancer patients present with symptoms and are subsequently diagnosed only at late disease stages, i.e., when metastases have already spread to pelvic organs (stage II), the abdomen (stage III), or beyond the peritoneal cavity (stage IV) [2].

EOC is an extremely heterogeneous disease. Multiple (epi)genetic alterations at a broad spectrum of oncogenes and tumor suppressor genes have been observed in ovarian cancer leading to deregulation of signal transduction pathways whose functions ranges from DNA repair, cell proliferation, apoptosis, cell adhesion, and motility. Based on these molecular alterations, ovarian cancer has been subdivided in two major type I and type II classes of tumors [3]. Type I tumors are slow growing, mostly restricted to the ovary, and develop from well-established precursor lesions called "borderline" tumors. Type I tumors comprise of four different subtypes, namely low-grade serous, mucinous, clear cell, and endometrioid cancers. The histological composition of these four types resemble normal cells present in the fallopian tube and/or ovarian surface epithelium, endocervix, vagina, and endometrium, respectively, thus suggesting different cells of origin for the different histotypes [3]. Type I lesions frequently carry mutations in *KRAS*, *BRAF*, *PTEN*, and *CTNNB1* (β-catenin), and often show a relatively stable karyotype.

Type II ovarian cancers include high-grade serous (HGS) and undifferentiated carcinomas, the vast majority of which characterized by *TP53* alterations and pronounced genomic instability [3]. Of note, inherited and somatic *BRCA1* and *BRCA2* mutations are usually found in type II tumors. It is under debate whether HGS ovarian cancers originate from the fimbria of the fallopian tube or from the ovarian surface epithelium (OSE) [4].

Ovarian cancers are thought, because of their distinctive progression and recurrence patterns, to be characterized by a hierarchical structure with cancer stem cells (CSCs) endowed with self-renewal and the capacity to differentiate, which continuously fuel the growth of the tumor mass and coexist with more committed cell types [5,6]. Notably, the Wnt/β-catenin signaling pathway, known to regulate stemness in a broad spectrum of stem cell niches including the ovary, is thought to play an important role in ovarian cancer. First, 16–54% of endometrioid ovarian cancers are characterized by mutations in β-catenin or, though at a considerably less frequency, in other members of the Wnt cascade such as *APC*, *AXIN1*, and *AXIN2* [7,8]. Second, other histotypes, and in particular serous ovarian carcinomas where mutations in Wnt-related genes are relatively uncommon, are characterized by constitutive Wnt signaling activation as indicated by alterations in β-catenin subcellular localization (i.e., nuclear and cytoplastic vs. membrane-bound) [9–12]. Importantly, Wnt activity was shown to correlate with grade [12], epithelial to mesenchymal transition (EMT) [7], chemo-resistance [13], and poor prognosis [14] in patients with ovarian carcinomas.

Here, we will review the current knowledge of the role of Wnt signaling in ovarian cancer stemness, EMT, and therapy resistance. The alleged role of exosomes in the paracrine activation of Wnt signaling, and novel potential treatment options based on Wnt inhibition will also be highlighted.

2. The Wnt/β-Catenin Signaling Pathway

Stem cells are distinguished from other somatic cells by their ability to self-renew and to give rise to distinct differentiated cell types throughout their lifetime [6]. The canonical Wnt signaling program plays a central role in controlling the balance between stemness and differentiation in several adult stem cell niches [15], including the ovary [7]. Accordingly, aberrant Wnt signaling is associated with pathological conditions like cancer [15].

Wnt proteins comprise a group of evolutionary conserved, lipid-modified glycoproteins [16] that operate at both short and long distances in order to regulate programs involved in proliferation, differentiation and stemness [15,17]. In absence of canonical Wnt ligands, intracellular β-catenin levels are regulated by the formation of a multiprotein "destruction complex" encompassing protein phosphatase 2A (PP2a), glycogen synthase kinase 3 (GSK3β) and casein kinase 1α (CK1α), and the scaffold proteins adenomatous polyposis coli (APC), and AXIN1/2. The destruction complex binds and phosphorylates β-catenin at specific serine and threonine residues, thereby targeting it for ubiquitination and subsequent degradation by the proteasome (Figure 1a). Instead, in the presence of Wnt ligands, co-activation of the Frizzled and LRP5/6 (low-density lipoprotein receptor-related proteins) receptors prevents the formation of the destruction complex, thereby stabilizing intracellular β-catenin and eventually leading to its translocation from the cytoplasm to the nucleus. Here, β-catenin interacts with members of the T-cell specific transcription factor/lymphoid enhancer binding factor (TCF/LEF) family of transcription factors and modulates the expression of a broad spectrum of Wnt downstream target genes regulating stemness, proliferation, and differentiation [15] (Figure 1b).

An illustrative example of the relevance of a tightly controlled Wnt signal regulation is provided by the intestinal stem cell niche, i.e., the crypt of Lieberkühn. At the bottom of the crypt, where the highly proliferative intestinal stem cells (ISC) reside, Wnt signaling is highly active due to signals from the surrounding stromal compartment [18], as also shown by nuclear β-catenin localization in both ISCs and the intercalating Paneth cells. Moving up along the crypt-villus axis, Wnt becomes progressively less active, following a signaling gradient inversely proportional to the differentiation grade of the epithelial lining [19]. In accordance with the central role played by this Wnt gradient, loss of function mutations at the tumor suppressor gene *APC* or gain of function mutations in the β-catenin (*CTNNB1*) oncogene leading to ligand-independent (i.e., constitutive) Wnt activation represent the main initiating events in the vast majority of sporadic colon cancer cases. Hence, the disruption of the homeostatic equilibrium among stemness, differentiation, and proliferation along the crypt-villus axis brought about by constitutive Wnt activation is sufficient to trigger colon cancer development [20].

The functional relevance of the Wnt pathway in controlling stemness, proliferation, and differentiation in organ-specific adult stem cell niches other than the intestinal tract is reflected by the broad spectrum of cancer where its deregulation contributes to tumor initiation and/or progression.

Accordingly, there is ample evidence from the scientific literature supporting an important role for Wnt signaling in both the onset and progression of ovarian cancer [7].

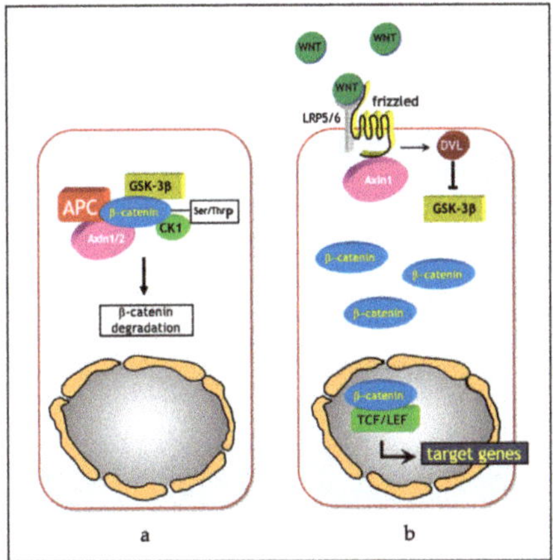

Figure 1. The Wnt/β-catenin signal transduction pathway in homeostasis. (**a**) In the absence of Wnt ligands, intracellular β-catenin levels are controlled by a destruction complex encompassing protein phosphatase 2A (PP2a), glycogen synthase kinase 3 (GSK3β) and casein kinase 1α (CK1α), adenomatous polyposis coli (APC), and AXIN1/2. This complex binds and phosphorylates β-catenin at serine and threonine residues, thereby targeting it for ubiquitination and proteolytic degradation by the proteasome. (**b**) In presence of Wnt, co-activation of the Frizzled and low-density lipoprotein receptor-related protein 5/6 (LRP5/6) (low-density lipoprotein receptor-related proteins) receptors prevents the formation of the destruction complex leading to the stabilization and consequent translocation of β-catenin from the cytoplasm to the nucleus. Here, β-catenin interacts with members of the T-cell specific transcription factor/lymphoid enhancer binding factor (TCF/LEF) family of transcription factors and modulates the expression of a broad spectrum of Wnt downstream target genes. DVL – disheveled. Adapted from [21].

3. Wnt Signaling in Ovarian Development and Tissue Homeostasis

Mammalian sex determination is a developmental process consisting of two distinct antagonistic genetic pathways allowing XX or XY undifferentiated gonads to differentiate into two different organs, namely the testis and the ovary [22]. The SRY-SOX9-FGF9 pathway supports testis development, while the RSPO1-Wnt-β-catenin-FOXL2 network promotes ovarian determination [22] (Figure 2).

Before sex determination, the undifferentiated gonad is composed of the coelomic epithelium, together with germ and mesenchymal cells. Here, both the Wnt signaling activators Wnt4 and R-spondin 1 (RPSO1) are important regulators of proliferation of the coelomic epithelium, as indicated by ablation of both *Rspo1* and *Wnt4* leading to reduced numbers of coelomic epithelial cells in XX and XY gonads and, consequently, to hypoplastic testis in XY mutant gonads [23].

During XY sex determination, the transcription factor sex-determining region Y (SRY) together with Splicing factor 1 (SF1) upregulate SRY-Box 9 (*SOX9*) gene expression. Subsequently, *SOX9* upregulation leads to the differentiation of coelomic epithelium into anti-Müllerian hormone producing Sertoli cells, thereby stimulating testis development [24]. Sertoli cells also secrete FGF9 (Fibroblast Growth Factor 9) thus inhibiting the pro-ovarian Wnt signaling pathway [25]. Furthermore, WT1 (Wilms Tumor 1) and ZNFR3 (Zinc Finger 3) also have been shown to downregulate Wnt signaling during male sexual differentiation [26,27]. Accordingly, genetic ablation of *Znrf3* leads to ectopic Wnt signaling in XY gonads and consequentially in the presentation of a female phenotype [27].

In females, both granulosa cells and ovarian surface epithelium (OSE) are derived from the coelomic epithelium. During fetal stages, *Rspo1* is expressed in the mesothelial lining of the coelomic cavity and within the fetal ovary [28], whereas *Wnt4* expression is localized to the gonad medulla and mesonephros between the gonad and the Müllerian duct [29]. Wnt4 and RSPO1 are essential for ovarian differentiation and oogenesis as they suppress *Sox9* expression, stimulate granulosa cell differentiation, and promote female sexual development by sustaining Müllerian duct differentiation [28,30,31]. Genetic ablation of *Wnt4, Rspo1,* or *Ctnnb1* in XX gonads lead to premature differentiation of granulosa cells in fetal stages and consequentially to the abrogation of ovary development at perinatal stages [32].

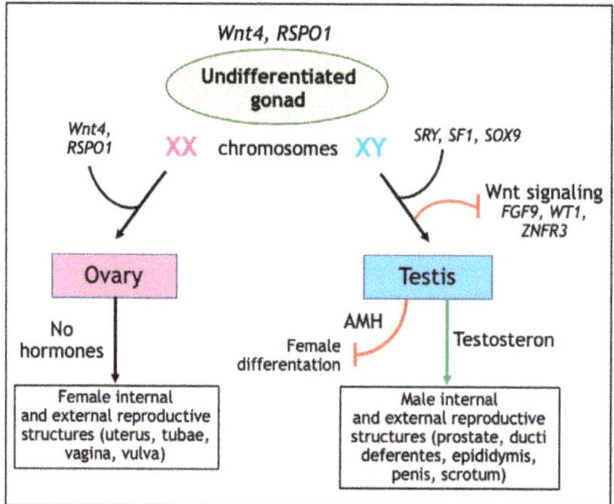

Figure 2. Schematic view of sex determination. In the undifferentiated gonad both Wnt4 and RSPO1 (R-spondin 1) are important regulators in particular for the proliferation of the coelomic epithelium. In XX gonads, expression of Wnt4, and RSPO1 leads to ovarian differentiation and oogenesis as they suppress *Sox9* expression, stimulate granulosa cell differentiation, and promote female sexual development by sustaining Müllerian duct differentiation. In XY gonads male reproductive organs are determined by the expression of sex-determining region Y (SRY) together with Splicing factor 1 (SF1) leading to upregulation of *Sox9* gene expression. In addition, Fibroblast Growth Factor 9 (FGF9), Wilms Tumor 1 (WT1), and Zinc Finger 3 (ZNFR3) inhibit the pro-ovarian Wnt signaling pathway. Also, anti-Müllerian hormone (AMH) prevents the development of the Müllerian duct into female reproductive organs.

Ng et al. (2014) [33] provided additional evidence highlighting the significance of the role played by Wnt during ovarian development and tissue maintenance and regeneration in adulthood. In this study, *Lgr5*, the marker of the above-mentioned and Wnt-driven intestinal stem cells, was shown to be broadly expressed during ovarian organogenesis, whereas it was restricted to the OSE in neonatal life. Using in vivo lineage tracing, $Lgr5^+$ cells were identified as stem or progenitor cells, able to contribute to the development of the OSE cell lineage, the mesovarian ligament, and the fimbriae. In adult ovaries, $Lgr5^+$ cells were restricted to the proliferative regions of the OSE and the mesovarian-fimbria junctional epithelium. In the OSE, $Lgr5^+$ cells are thought to preserve homeostasis and to underlie repair of the epithelial damage after ovulation [33]. Indeed, using a Wnt-reporter mouse model, the complete coelomic epithelium overlying the undifferentiated gonad displayed β-catenin/Tcf mediated LacZ expression gradually reduced to smaller populations during postnatal, pre-puberal, and adult life [34]. Of note, the $LacZ^+$ OSE cells were enriched in SP (side population) positive cells, a sub-population of stem-like cells identified by their capacity to efflux the dye Hoechst 33342 by ATP-binding cassette super-family G member 2 (ABCG2) transporter pumps [34], a clinically relevant feature acquired by chemotherapy resistant ovarian CSCs.

Apart from its role during embryonic development of the ovary, Wnt signaling was also shown to be an essential regulator of ovarian homeostasis, fertility, and tumorigenesis. Knock-out of *APC2*, a homologue of the *APC* tumor suppressor gene [35], resulted in the activation of ovarian Wnt signaling and in sub-fertility. The latter was due to disturbed follicular growth and the consequent reduced ovulation rate and corpora lutea formation [36]. Notably, aged $APC2^{-/-}$ mice developed granulosa cell tumors (GCT) with comparable histological features and molecular signatures to those of the corresponding human GCTs [36].

Overall, the central role played by Wnt in regulating the delicate balance between stemness, proliferation, and differentiation to ensure ovarian tissue homeostasis is reflected by its causal association with ovarian cancer onset and/or progression as discussed in the next section.

4. Wnt Signaling in Ovarian Cancer

As mentioned above, *CTNNB1* (β-catenin) mutations are found in 16–54% of endometrioid ovarian cancer cases. Likewise, genetic alterations in other members of the Wnt cascade, such as *APC*, *AXIN1*, and *AXIN2*, have also been detected in this specific ovarian cancer histotype [7,8]. In a conditional *APC* knock-out mouse model, it was shown that constitutive activation of Wnt/β-signaling in Müllerian duct-derived organs (i.e., fallopian tubes, uterus, cervix, and the upper two thirds of the vagina) results in the formation of endometrioid tumors in the oviduct, reminiscent of the corresponding histotype in man. Of note, in the same study the ovarian surface epithelium was unaffected, thus suggesting that the oviduct, rather than the OSE, encompasses the cell of origin of (endometrioid) ovarian cancer [37].

In addition to endometrioid ovarian carcinomas, mutations in *CTNNB1* are also found in rare cases of mucinous ovarian cancer [38]. Moreover, both *CTNNB1* and *APC* mutations have also been detected in non-epithelial microcystic stromal tumors (MSTs) of the ovary [39–41]. Accordingly, an increased incidence of MSTs has also been reported among patients affected by familial adenomatous polyposis (FAP) due to germline mutations in *APC* [40,41]

Yet, it should be clearly stated that endometrioid tumors represent a notable exception as mutations in Wnt-related genes are in general extremely rare in any other ovarian cancer histotype [7]. However, even in the absence of specific mutations, Wnt signaling has been reported to be frequently activated in the more common serous histotype as indicated by nuclear and cytoplasmic β-catenin subcellular localization [9–12]. In addition, expression profiling data have confirmed the frequent activation of Wnt signaling in ovarian cancer at large [42,43]. In particular, transcriptome analysis of ascites-derived ovarian cancer cells and tumor-associated macrophages (TAMs) has revealed that both canonical and non-canonical Wnt ligands (i.e., *WNT7A*, *WNT2A*, *WNT5A*, *WNT9A*) are expressed in tumor cells, whereas *LRP* and *FZD* are common to both tumor cells and TAMs [43].

Pangon et al. (2016) took advantage of the Cancer Genome Atlas (TCGA) to show that the oncogene *JRK* (jerky) is overexpressed in 15% of ovarian cancers in association with increased expression of canonical Wnt target genes [44]. JRK directly interacts with the β-catenin transcriptional complex, thereby stabilizing the β-catenin/TCF complex and ultimately resulting in increased β-catenin transcriptional activity and cell proliferation. Consistent with this, depletion of JRK in cancer cell lines repressed expression of β-catenin target genes and reduced cell proliferation [44].

More recently, noncoding RNAs (ncRNAs) have emerged as important post-translational regulators of Wnt-associated gene expression in ovarian cancer (Table 1). By using orthotopic mouse models of ovarian cancer, it was demonstrated that β-catenin plays a key role in the formation of metastasis by controlling the endoribonuclease Dicer, a key component of the microRNA (miR)-processing machinery. β-catenin directly targets Dicer, thereby downregulating multiple miRNAs including the miR-29 family known for its role as a negative EMT regulator. Silencing of β-catenin or overexpression of Dicer or mi-R29 in metastatic ovarian cancer cells reduced their migratory capacity, and attenuated metastasis formation upon β-catenin knockdown in orthotopic mouse models [45]. Of note, reduced expression of miR-29 is associated with ovarian cancer progression and strongly correlated with poor survival [46].

Several other miRs have been demonstrated to impact migration, invasion, and cancer progression via Wnt signaling in ovarian cancer [47–59]. Interestingly, miR-939 has been suggested to function as a tumor promotor by regulating Wnt signaling through direct suppression of the previously discussed *APC2* tumor suppressor [48].

Next to miRs, several long non-coding RNAs (lncRNAs) have been described to play a causative role in Wnt-associated cell proliferation, EMT, and chemotherapy resistance in ovarian cancer [60–65]. Table 1 summarizes the data relative to gene and non-coding RNA alterations leading to Wnt signaling activation in ovarian cancer.

Table 1. Gene and non-coding RNA alterations leading to Wnt signaling activation in ovarian cancers.

Gene/ncRNA	Ovarian Cancer Histotype[*]	Mechanism/Target	Reference
CTNNB1	Endometrioid.	Oncogenic activation.	[8,38,66–70]
CTNNB1	Mucinous.	Oncogenic activation.	[38]
CTNNB1	Microcystic Stromal Tumors (MST).	Oncogenic activation.	[39]
APC	Endometrioid.	Loss of tumor suppressor function.	[8]
APC	Microcystic Stromal Tumors (MST).	Loss of tumor suppressor function.	[40,41]
AXIN1	Endometrioid.	Loss of tumor suppressor function.	[8]
AXIN2	Endometrioid.	Loss of tumor suppressor function.	[8]
microRNA (miR)-10a	Granulosa cell tumor.	miR-10a targets *PTEN* and indirectly activates Wnt (and AKT) signaling. Oncogenic activation.	[54]
miR-15b	Epithelial ovarian cancer *.	miR-15b targets *WNT7A* 3'-untranslated region (3'-UTR) and thus inhibits Wnt signaling. Loss of tumor suppressor function.	[50]
miR-16	Epithelial ovarian cancer *.	miR-16 target(s) yet unknown; it inhibits Wnt signaling. Loss of tumor suppressor function.	[56]
miR-21	Epithelial ovarian cancer *.	miR-21 target(s) yet unknown; it activates Wnt signaling. Oncogenic activation.	[57]
miR-27a	Epithelial ovarian cancer *.	mir-27 targets the Wnt antagonist *FOXO1*. Oncogenic activation.	[55]
miR-29	Serous, mucinous, and clear cell ovarian cancer.	miR-29 target(s) yet unknown; it activates Wnt signaling. Oncogenic activation.	[45,46]
miR-92a-1	Epithelial ovarian cancer *.	miR-92a-1 targets the Wnt antagonist Dickkopf 1 (*DKK1*). Oncogenic activation.	[51]
miR-200c	Epithelial ovarian cancer *.	miR-200c target(s) yet unknown; it inhibits Wnt signaling. Loss of tumor suppressor function.	[47]
miR-214	Epithelial ovarian cancer *.	miR-214 target(s) yet unknown; it inhibits Wnt signaling. Loss of tumor suppressor function.	[53]
miR-219-5p	Epithelial ovarian cancer *.	miR-219-5p targets the EMT transcription factor *TWIST* and inhibits Wnt signaling. Loss of tumor suppressor function.	[52]
miR-654-5p	Epithelial ovarian cancer *.	miR-654-5p targets *CDCP1* and *PLAGL2*. Loss of tumor suppressor function.	[58]
miR-939	Epithelial ovarian cancer *.	miR-939 targets *APC2*. Loss of tumor suppressor function.	[48]
miR-1180	Epithelial ovarian cancer *.	miR-1180 targets *SFRP1*. Loss of tumor suppressor function.	[59]
miR-1207	Epithelial ovarian cancer *	miR-1207 targets *SFRP1*, *AXIN2*, and *ICAT*. Loss of tumor suppressor function.	[49]
HOTAIR [1]	Epithelial ovarian cancer *.	HOTAIR target(s) unknown; Wnt agonist. Oncogenic activation.	[60]
SNHG20 [2]	Epithelial ovarian cancer *.	SNHG20 target(s) unknown; Wnt agonist. Oncogenic activation.	[61]
HOXD-AS1 [3]	Epithelial ovarian cancer *.	HOXD-AS1 targets the Wnt antagonist miR-133a-3p. Oncogenic activation.	[62]
CCAT2 [4]	Epithelial ovarian cancer *.	Targets unknown; EMT and Wnt agonist. Oncogenic activation.	[63]
MALAT1 [5]	Epithelial ovarian cancer *.	Targets unknown; Wnt agonist. Oncogenic activation.	[64]
AWPPH [6]	Epithelial ovarian cancer *.	Targets unknown; Wnt agonist. Oncogenic activation.	[65]
HOXB-AS3 [7]	Serous ovarian cancer samples; other histotypes.	Targets unknown; Wnt agonist. Oncogenic activation.	[71]

*, histotype not characterized; [1], HOTAIR—HOX antisense intergenic RNA; [2], SNHG20—small nucleolar RNA host gene 20; [3], HOXD-AS1—HOXD cluster antisense RNA 1; [4], CCAT2—colon cancer-associated transcript 2; [5], MALAT1—metastasis associated lung adenocarcinoma 1; [6], AWPPH—associated with poor prognosis of hepatocellular carcinoma; [7], HOXB-AS3—HOXD cluster antisense RNA 3.

Apart from the above alterations in genes and non-coding RNAs, Wnt signaling activation in ovarian cancer might result from additional alternative epigenetic mechanisms, either cell-autonomous or induced by the tumor microenvironment. Epigenetic alterations leading to autocrine overexpression of Wnt ligands [72,73], receptors [74], and/or of other Wnt agonists like *FRAT1* [12] or *PYGO2* [75], or to the inhibition of antagonists such as the secreted frizzled receptors proteins (sFRP) and Dickkopf (DKK1) [14,76–78] have been reported in the literature. Likewise, paracrine secretion of Wnt-activating cues was observed from either the stroma surrounding the primary ovarian cancer, or from ascites in the case of late-stage disease. Several components of ovarian cancer ascites, known to be associated with shorter progression free survival [79,80], have been previously implicated in promoting Wnt signaling: leptin [81,82], urokinase-type plasminogen activator receptor (uPAR) [83], and macrophage migrating inhibitory factor (MIF) [84]. These soluble factors may act by activating the Wnt pathway in disseminated ovarian cancer cells present in ascites. Two additional ascites factors, namely osteoprotegerin (OPG) [85] and interleukin 8 (IL-8) [86], are in fact downstream Wnt targets and could serve as markers of Wnt signaling activity in ascites. In addition, β-1 integrin-mediated adhesion to the peritoneal mesothelium, a key step in the route to ovarian cancer metastasis, activates β-catenin signaling [87]. Of note, it has recently been shown that extracellular vesicles such as exosomes play a critical role in long-distance transmission of morphogens and in particular in Wnt signaling [88]. In the context of ovarian cancer ascites, exosomes may represent a stable source of paracrine Wnt signals [89]. The role of exosomes will be discussed at more length later on in this review.

5. Wnt Signaling in Ovarian Cancer Stem Cells, EMT, and Therapy Resistance

After diagnosis, tumor debulking surgery followed by carboplatin- and paclitaxel-based chemotherapy represent the standard first line therapy for high grade serous ovarian cancer patients. Although at this stage the primary response to chemotherapy is extremely efficient, most patients relapse and develop metastases locally and at distant organ sites [90]. This is mainly due to sub-populations of tumor cells likely to have acquired stem cell features (CSCs) through EMT and, consequently, the EMT-associated chemo-resistance [5]. In 2005, it was shown for the first time that the aggressiveness of human ovarian cancer results from alterations in stem and progenitor-like cells in the ovary [91]. Moreover, this study demonstrated that the small subpopulation of stem-like, tumor-propagating ovarian cancer cells were earmarked by expression of cluster of differentiation 44 (CD44) and other stem cell and EMT markers such as *KIT* (CD117), *SCF* (stem cell factor), *SLUG* (*SNAIL2*), and *VIM* (vimentin) [91]. After this initial report, several cell surface antigen markers have been identified which allow enrichment of ovarian CSCs from immortalized cell lines, primary tumors, and ascitic fluids: CD133, CD24, CD44, CD177, aldehyde dehydrogenase 1 (ALDH1), and SP [5]. Notably, active Wnt signaling has been shown to play a key role in the regulation and maintenance of ovarian cancer stemness [51,92,93].

Ovarian cancer follows a unique pattern of metastasis formation where, unlike many other cancer types, no anatomical barrier exists between the primary site and the abdominal cavity, thus greatly facilitating the dissemination of exfoliated malignant cells. In particular, disseminated ovarian cancer cells secrete vascular permeability factors and can block lymphatic drainage leading to accumulation of ascites fluid within the peritoneal cavity [94]. These malignant ascites provide a favorable tumor microenvironment (TME) enriched in secreted inflammatory cytokines [79], growth factors [95], and extracellular macromolecules (collagen, fibronectin, and laminin) [96]. In this environment, tumor cells form multicellular aggregates enriched in cancer stem/progenitor cells, the so-called 'spheroids', which eventually implant on the mesothelial lining of the peritoneum [97] (Figure 3). The attachment of these floating spheroids to the peritoneal lining and associated organs represents the major route for metastasis formation in ovarian cancer [98] where, as observed in other epithelial cancers, EMT was shown to play a key role [99]. Interestingly, although hematogenous spread is generally thought to play a relatively minor role in metastasis formation in ovarian cancer, it has recently been demonstrated in a parabiosis mouse model [100]. In this study, two mice, one of which intraperitoneal transplanted with ovarian cancer cells, were surgically connected to share blood supply. The development of ovarian cancer in the cancer-free animal likely results from hematogenous spread [100]. Likewise, circulating tumor cells have been identified in peripheral blood from ovarian cancer patients [101].

Overall, the naturally occurring spheroids in ascites are likely to underlie metastatic disease in ovarian cancer patients. In the next sections, we will discuss the current experimental evidence on the role of Wnt signaling in eliciting EMT and chemo-resistance in high grade serous ovarian cancer.

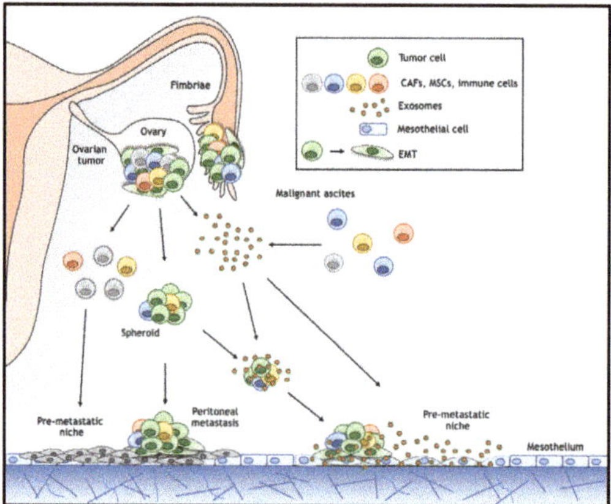

Figure 3. Model for peritoneal metastasis formation in ovarian cancer. Ovarian cancer follows a unique pattern of metastasis formation, where no anatomical barrier exists between the primary site and the abdominal cavity. Multicellular aggregates enriched in cancer stem/progenitor cells, the so-called spheroids, detach from the primary tumor and eventually implant on the mesothelial lining of the peritoneum. EMT was shown to play a key role facilitating the acquisition of stem-like features, anoikis resistance, and increased migration and invasion. The establishment of premetastatic niches composed of several cell populations, including tumor-associated neutrophils, is thought to be required for disseminating carcinoma cells to engraft at the distant site. Exosomes in ovarian cancer ascites have been proposed as a putative mechanism to facilitate long-range distance cell–cell communication thereby establishing both pre-metastatic niches in the peritoneal cavity and preserving stemness in disseminated cancer cells. CAFs: cancer associated fibroblasts; MSCs: mesenchymal stem cells.

5.1. Wnt Signaling and EMT in Ovarian Cancer

EMT is a reversible developmental program exploited by cancer cells to reversibly switch from an epithelial phenotype with apical-basal polarity and cell–cell adhesions, to a more motile mesenchymal state with spindle like morphology and front-back-end polarity [102]. Next to the motility and invasive features characteristic of the mesenchymal state, EMT is functionally associated with the acquirement of stem-like features, resistance to therapy, and immune suppression [103–105]. Last, the capacity of cells undergoing EMT to revert to an epithelial state by mesenchymal-to-epithelial transition (MET) is rate-limiting to allow the stem- and mesenchymal-like migrating CSCs to regain proliferative and epithelial features essential to colonize the metastatic site [102,106]. Various signaling pathways are involved in EMT, including transforming growth factor β (TGF-β), Notch, and Wnt/β-catenin. Activation of the Wnt/β-catenin pathway has been shown to be an important regulator of EMT in many different types of cancers [106–108], including ovarian cancer [109–112]. In this context, ovarian cancer cell lines with a high SNAIL to E-cadherin ratio, are characterized by enhanced CSC-like, motile, and therapy-resistant features when compared with epithelial ovarian cancer cell lines. Accordingly, *SNAIL* knockdown reversed the malignant properties and tumor burden of the more mesenchymal ovarian cancer cell lines in xenograft models [111]. *SNAIL* and other EMT transcription factors (EMT-TFs) have been shown to activate expression of the *GOLPH3* (Golgi phosphoprotein 3) gene, encoding for an oncoprotein frequently upregulated in ovarian cancer tissues and cell lines, through Wnt/β-catenin signaling activation [112]. Induction of EMT and the consequent acquisition of migratory and invasive cellular features downstream of Wnt activation have also been demonstrated in ovarian carcinomas where *IQGAP2*, a Wnt antagonist, is frequently silenced by DNA methylation [110]. Last, cyclin G2, an unconventional cyclin that opposes cell cycle progression and inhibits EMT, acts as a tumor suppressor in ovarian cancer by inhibiting Wnt/β-catenin signaling [109].

More recently, it has been suggested that, rather than being a binary process with fully opposing epithelial and mesenchymal phenotypes, EMT generates hybrid E/M cancer cells displaying both epithelial and mesenchymal characteristics [113,114]. Indeed, similar to the normal ovarian surface epithelial (OSE) cells previously shown to display both epithelial and mesenchymal characteristics and a remarkable phenotypic plasticity during post-ovulatory repair, double positive E-cadherin and vimentin cells have been observed in ovarian cancers [115]. Accordingly, different intermediate EMT states have been identified in ovarian cancer cell lines [116] and ascites-derived spheroids [117]. Here, ovarian cancer cells in hybrid E/M states were shown to exhibit stem-like features, anoikis resistance, and increased migration and invasion when compared with the fully epithelial and mesenchymal states [116–118].

Overall, it is yet unclear whether the hybrid E/M cells represent a 'metastable' cell population or are cells captured in a time frame during the transition between the epithelial to mesenchymal states. The elucidation of the complex network of intrinsic and extrinsic mechanisms underlying EMT during metastasis formation and the role of Wnt signaling therein represents an important future research challenge. In the next section, the current knowledge on the role of Wnt signaling in resistance to chemotherapy in ovarian cancer will be discussed.

5.2. Wnt Signaling and Therapy Resistance in Ovarian Cancer

As mentioned above, chemotherapy is extremely efficient in the first-line treatment of primary ovarian cancers although it inevitably leaves behind chemo-resistant CSCs likely to underlie relapse and metastasis in distant organ sites [90]. Wnt signaling has been associated with resistance to chemotherapy in different tumor types including ovarian cancer [119].

Chemo-resistance can be acquired through a broad spectrum of molecular and cellular mechanisms such as the upregulation of ATP-binding cassette (ABC) transporter pumps, the activation of EMT, and the exosome-mediated transport of molecules controlling a broad spectrum of pathways underlying drug resistance [120]. ABC transporters have indeed been shown to be expressed in ovarian cancer usually in association with cancer stemness and poor prognosis [121,122]. Notably, upregulation of the ABCG2 transporter pump and Wnt signaling activation downstream of cKIT mediate the onset of resistance to cisplatin and paclitaxel in ovarian CSCs [13]. In the same study, *ABCG2* expression and chemo-resistance to both cisplatin and paclitaxel could be reversed by β-catenin siRNA knockdown, once again highlighting the central role of Wnt signaling in these processes [13].

Another well-established mechanism underlying therapy resistance in ovarian cancer, as also mentioned in the previous section, is represented by EMT [99]. Su et al. (2010) showed that *SFRP5* (secreted frizzled-related protein 5), a well-known Wnt and EMT antagonist, is frequently downregulated in ovarian cancer by epigenetic silencing through promoter hypermethylation [123]. Accordingly, restoration of *SFRP5* expression inhibits Wnt signaling and EMT thus sensitizing ovarian cancer cells to chemotherapy. Activation of the EMT-TF *TWIST* and of *AKT2* signaling play key roles downstream of *SFRP5* silencing [123].

In addition to the above-mentioned cell-autonomous mechanisms, ascites also forms a unique tumor microenvironment likely to contribute to therapy resistance [124]. Malignant ascites provides a favorable tumor microenvironment consisting of cellular and non-cellular components, each likely to play a role in the development of resistance to carboplatin- and paclitaxel-based therapy. Among these, cancer associated fibroblasts (CAFs) represent an important component of ovarian cancer ascites [124]. CAFs are a subpopulation of fibroblasts capable of affecting tumor progression, dissemination, and therapy response through signaling to tumor cells and/or remodeling of the extracellular matrix (ECM) [125]. Recently, Ferrari et al. (2019) demonstrated that Dickkopf-3 (DKK3), the stromal expression of which is strongly associated with aggressive ovarian cancer, promotes CAFs' aggressive behavior by enhancing Yes-associated protein/transcriptional co-activator with PDZ-binding motif (YAP/TAZ) activity through Wnt/β-catenin signaling [126]. From a mechanistic perspective, DKK3 destabilizes the Wnt-antagonist Kremen, leading to increased LRP6 localization at the cell membrane. This in turn stabilizes YAP/TAZ and β-catenin levels leading to more global gene expression changes enhancing cancer stemness, malignant progression, and metastasis [126]. Other ascites cellular components such as macrophages have also been shown to take part in tumor progression and the development of therapy resistance. Ragahvan et al. (2019) showed that Wnt signaling participates in a bidirectional ovarian CSC-macrophage interaction [92]. By taking advantage of hetero-spheroids composed of macrophages and ovarian cancer cells in close contact with each other, it was shown that Wnt signaling, activated by secretion of the Wnt5b ligand from macrophages, led to an increase of the ovarian CSC compartment (ALDH$^+$) and to the enhancement of the immune-suppressive characteristics of the macrophages. Likewise, Wnt5b knockdown in macrophages resulted in a loss of the ALDH$^+$ ovarian CSC fraction. Most importantly, the hetero-spheroids were less sensitive to chemotherapeutics and were more invasive in in vitro assays [92]. Hence, macrophage-initiated Wnt activation is likely to play a central role in ovarian cancer stemness maintenance and in therapy resistance.

Notwithstanding more recent advances in chemotherapy (e.g., intraperitoneal delivery of cytotoxic drugs and the introduction of novel, more targeted agents such as bevacizumab and imatinib) [127–129], less than 30% of advanced ovarian cancer patients survive longer than five years after diagnosis [1]. Therefore, there is urgent need for novel therapeutic strategies based on improved understanding of the molecular and cellular mechanisms underlying dissemination and metastasis formation by ovarian cancer cells in the peritoneal cavity and their acquisition of dormant and chemo-resistant properties. Recently, the role played by extracellular vesicles and in particular by exosomes in tumor progression, dissemination, and resistance to therapy has opened new avenues in basic and translational cancer research. In the next section we will present and discuss the current knowledge on exosomes in ovarian cancer, especially in the context of intra-abdominal ascites and of long-range Wnt signaling activation.

6. Exosomes and Wnt Signaling in Ovarian Cancer Ascites

Malignant ascites provides a favorable tumor microenvironment and consists of a heterogeneous mixture of cells and secreted factors that modulate cancer cell behavior during tumor progression, metastasis formation, and acquirement of chemo-resistance. As mentioned, Wnt ligands are modified lipids and are therefore highly hydrophobic, thereby limiting their ability for extracellular diffusion [16]. Recently however, studies have shown that Wnts can be transported across tissues by exosomes [88,130]. In the following paragraphs we will highlight the current knowledge on the role played by exosomes in ovarian cancer ascites as a putative mechanism to activate Wnt signaling over long-range distances both in establishing pre-metastatic niches in the peritoneal cavity and in preserving stemness in disseminated cancer cells.

6.1. Exosomes

Exosomes are small extracellular vesicles ranging in diameter from 30 to 100 nm that are secreted by most eukaryotic cells. Secreted exosomes are important mediators in cell–cell communication as they carry molecules such as microRNAs, mRNAs, and both membrane-bound and secreted proteins [131]. Exosomes are thought to facilitate tumor survival and progression by stimulating angiogenesis and tumor growth, suppressing immune responses, remodeling of the extracellular matrix, promoting metastasis formation either directly and/or through the establishment of premetastatic niches [131] (Figure 3). Numerous studies have demonstrated the presence of exosomes in ovarian cancer cell line cultures, and in patient-derived serum and ascites [132–135]. Notably, it has been shown that active Wg (Wingless) and Wnt3a ligands are membrane-bound in exosomes from Drosophila and human cells, respectively [88]. Moreover, macrophage-derived and exosome-packaged Wnts are rate-limiting for the regenerative response of intestine intestinal stem cells after radiation [130]. In relation to cancer, fibroblast-derived exosomes carrying Wnt ligands increase cell migration and metastasis formation in breast cancer [136]. Hu et al. (2019) recently found that exosomes derived from stromal fibroblasts contain Wnt ligands capable of eliciting the de-differentiation of colon cancer cells into therapy resistant CSCs [137]. Alternatively, activation of Wnt signaling in target cells has been shown to occur by exosomes encompassing β-catenin in their cargo. Here, both 14-3-3 proteins and β-catenin were encompassed in the extracellular vesicles. 14-3-3 proteins bind to dishevelled segment polarity protein 2 (Dvl-2) and GSK3β thereby interfering with β-catenin phosphorylation and stimulating Wnt signaling [138].

Although to date no evidence has been presented supporting the presence of exosomes encompassing active Wnt ligands in ovarian cancer ascites, differential expression analysis of ovarian cancer exosomes compared with those from normal OSE cells indicate a potential involvement of miRNAs known to target the Wnt signal transduction pathway [139]. Moreover, recently it has been demonstrated that exosomes isolated from a highly invasive ovarian cancer cell line promote metastasis in vivo compared to exosomes from cells with low invasive capacity [140]. Quantitative proteomic analysis of tumor tissues of the mice treated with exosomes derived from these two different cell lines revealed a potential role for Wnt signaling in the role played by exosomes in tumor growth and metastasis in vivo [140]. Also, as discussed here below, ovarian cancer exosomes containing the Wnt target and transmembrane protein CD44 have been shown to participate in the formation of pre-metastatic niches [141].

6.2. Pre-Metastatic Niche

Ovarian carcinomas spread through the shedding of clusters of tumor cells from the primary lesion into the peritoneal cavity. In this context, the key event in metastatic seeding is the mesothelial adhesion of ovarian cancer cells in the intraperitoneal cavity. The establishment of premetastatic niches is thought to be required for disseminating cancer cells to engraft at the distant site [142]. Premetastatic niches comprise of a specialized and favorable micro-environment that facilitates colonization and promotes survival and outgrowth of disseminated tumor cells [142] (Figure 3). The relevance of the formation of pre-metastatic niches in ovarian cancer has been proposed by several studies [141,143]. Lee et al. (2019) demonstrated that inflammatory factors secreted by ovarian cancer cells mobilize neutrophils and stimulate them to create chromatin webs called 'neutrophil extracellular traps' (NETs) in the omentum in both tumor-bearing mice (before metastasis occurs) and in early-stage ovarian cancer patients. The NETs can sequentially capture ovarian cancer cells and thereby promote metastasis formation. Reversely, inhibiting NET formation abrogated omental colonization [143].

Next to NETs, ovarian cancer exosomes have also been shown to participate in the establishment of a pre-metastatic niche by alternative mechanisms. First, *MMP1* mRNA has been found in extracellular vesicles derived from ovarian cancer cell lines and ascites from ovarian cancer patients that promotes apoptotic cell death of the mesothelial cells, thus resulting in the destruction of the peritoneal barrier [144]. In addition, ovarian cancer cells' exosomes encompassing the cell-surface glycoprotein CD44 can transfer it to peritoneal mesothelial cells and induce their reprogramming by EMT activation. The modified mesothelium facilitates ovarian cancer invasion and metastasis formation [141]. Of note, CD44 is a major Wnt target gene in the intestinal epithelium [145] and is essential for Wnt induction during colon cancer progression [146], thus suggesting yet another functional link between Wnt signaling and ovarian cancer exosomes in pre-metastatic niche formation.

To interfere with the interaction between disseminated ovarian cancer cells and the exosome-receiving mesothelial cells, De la Fuente et al. (2015) developed a metastatic trap (M-Trap) [147]. By embedding exosomes purified from ovarian cancer patient ascites on a 3D scaffold, the authors showed that the M-Trap device was able to capture ovarian cancer cells in a mouse model of ovarian cancer. This led to a more focalized disease and an increase in survival rate [147]. These results lay the foundation for future clinical approaches to improve treatment of ovarian cancer patients with malignant ascites [147].

Overall, notwithstanding that treatment of advanced stage ovarian cancer patients still represents a major clinical challenge, recent advances in our understanding of the mechanisms underlying ovarian cancer ascites formation and the role they play in metastasis formation in the peritoneal cavity are of good auspices for the future. Exosomes in particular, may represent powerful tools in early diagnosis and treatment [135]. As for the latter, targeted exosome ablation or inhibition of exosome secretion may affect tumor progression or therapy resistance (Figure 4). In this scenario, the Wnt signaling pathway may also represent a relevant therapeutic target. In the next paragraph, treatment options based on targeting of Wnt signaling in ovarian cancer will be discussed.

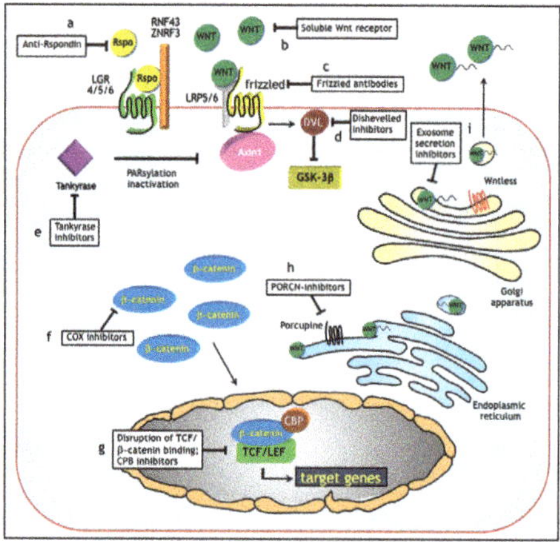

Figure 4. Therapeutic targets for the inhibition of Wnt signaling. (**a–c**) Wnt soluble receptors, anti-R-spondin antibodies, and antibodies directed against Frizzled receptors impair the ligand/receptor interaction and prevent downstream signaling. (**d**) Disheveled inhibitors block Wnt signaling by interfering with the Frizzled/Disheveled interaction. (**e**) Tankyrase activates Axin through PARsylation. Tankyrase inhibition increases Axin levels thus stimulating the formation of the β-catenin destruction complex and reducing the intracellular β-catenin pool. (**f**) cyclooxygenase (COX) inhibitors increase ubiquitination and proteasomal degradation of β-catenin. Next, COX2 inhibition leads to reduced levels of prostaglandin E2 (PGE2) known to positively affect Wnt signaling. (**g**) Disruption of its interaction with TCF inhibits β-catenin-mediated transcriptional activity. CREB-binding protein (CBP) inhibitors instead interfere with the interaction between TCF/LEF and CBP thereby reducing transcriptional activity. (**h**) *PORCN*-inhibitors hamper the palmitoylation of Wnt before its extracellular release. (**i**) Exosome secretion inhibitors reduce the transport of biomolecules like active Wnt ligands, RNAs and proteases that contribute to angiogenesis, tumor growth, immune response suppression, the remodeling and degradation of the extracellular matrix (ECM). Additional abbreviatons: RNF43 = RING finger 43; LGR4/5/6 = Leucine-rich repeat-containing G-protein coupled receptor 4/5/6; RSPO = R-spondin; ZRNF3 = Zinc RING finger 3; GSK3β = glycogen synthase kinase 3β; LRP5/6 = LDL Receptor Related Protein 5/6; TCF/LEF = T-cell specific transcription factor/lymphoid enhancer binding factor.

7. Targeting Wnt in Ovarian Cancer: Opportunities for Treatment?

During the last decade, the therapeutic response rate of ovarian cancer patients has improved through optimization of chemotherapy strategies, their intraperitoneal administration, and the introduction of targeted therapies [127–129]. However, despite these developments, the overall survival of ovarian cancer patients has not significantly improved [1]. Because of the role played in cancer stemness and in therapy resistance, the Wnt signaling pathway forms a candidate target for therapeutic intervention as different segment of this cascade are suitable for therapeutic targeting (Figure 4; Table 2).

Although R-spondins (RSPO) are unable to initiate Wnt signaling, they can, by binding to leucine-rich repeat-containing G-protein coupled receptors (LGR) enhance responses to low-dose Wnt proteins [148]. Functional RSPOs have been found in multiple human tumor types and anti-RSPO monoclonal antibodies shown to reduce the tumorigenicity of cancer cells in patient-derived tumor xenograft models of several malignancies including ovarian cancer [149]. Porcupine (PORCN) inhibitors form another relevant target to inhibit Wnt signaling. The acetyltransferase PORCN is responsible for post-translational modifications of Wnt proteins essential for the transport, secretion, and activity of the ligands. WNT974 is a selective PORCN inhibitor that has been shown to exert cytostatic effects on ascites-derived ovarian cancer cells as a consequence of Wnt signaling inhibition [150]. When combined with the conventional chemotherapeutic drug carboplatin, WNT974 administration led to increased cytotoxic effects and cell cycle arrest in ascites samples when compared with single drug treatments [150].

The FDA (Food and Drug Administration)-approved anti-helminth compound niclosamide represents yet another powerful Wnt inhibitor shown to repress ovarian CSC growth through downregulation of both the disheveled protein DVL2 and the surface receptor LRP6 [151]. Next to Wnt, niclosamide targets additional signaling pathways known to play a role in cancer stemness, including Notch, mTORC1, and Stat3 [152].

Besides the above mentioned Wnt targets and inhibitory compounds, inhibition of Wnt ligands secretion, inactivation of the extracellular portion of Frizzled receptors, and interference with the TCF/β-catenin complex represent additional and presently under investigation strategies [153] (Figure 4; Table 2).

Currently, different Wnt inhibitors are being evaluated in clinical trials for different cancer types including ovarian cancer. As a notable example, Ipafricept is a recombinant fusion protein that competes with the FZD8 receptor for its ligand thereby antagonizing Wnt signaling. Ipafricept reduces cancer stem cells, promotes differentiation, and synergizes with taxanes in ovarian cancer xenografts. More recently, a phase 1B trial was conducted with ipafricept in combination with carboplatin and paclitaxel in patients with recurrent platinum-sensitive ovarian cancer [154]. Unfortunately, although generally well-tolerated by patients, bone toxicity at efficacy doses limited ipafricept treatment [154]. Nonetheless, other Wnt inhibitors targeting PORCN and β-catenin are now being tested in clinical trials in different tumor types [155].

Table 2. Wnt inhibitors in ovarian cancer.

Molecular Targets	Inhibitors	Activity	Reference
Extracellular targeting	Anti-Rspondin	anti-RSPO monoclonal antibodies reduce tumorigenicity of cancer cells in patient-derived ovarian tumor xenograft models.	[149]
	Ipafricept (OMP54F28)	Recombinant fusion protein that competes with the membrane-bound Frizzled 8 (FZD8) receptor for its ligand; leads to tumor regression in combination with taxane in ovarian xenograft models; currently under clinical trial.	[154]
LRP6 inactivation	Salinomycin	Small molecule blocking Wnt induced LRP6 phosphorylation and induces its degradation; leads to repression of EMT in epithelial ovarian cancer.	[156,157]
Dishevelled	3289–8625	Small molecule disrupting the frizzled-disheveled interaction by targeting the PDZ domain; chemo-sensitizes paclitaxel-resistant ovarian cancer cells.	[158]
PORCN	WNT974	Small molecule inhibitors of Wnt acetyltransferase porcupine; increases cytostatic effects on ascites-derived ovarian cancer cells.	[150]
CK1α activation	Pyrvinium	Small molecule that selectively potentiates CK1α kinase activity leading to increased β-catenin phosphorylation; enhances sensitivity to chemotherapy of ovarian cancer cells.	[159,160]
Non-specific or overlapping targets	Niclosamide	Small molecule inhibitor promoting FZD1 endocytosis and suppressing LRP6 expression; inhibits growth and increases cell death in ovarian cancer.	[161–163]
	COX-inhibitors	Aspirin lowers the risk of ovarian cancer development; in case of ovarian cancer underlying mechanism yet unknown.	[164]

8. Conclusive Remarks

In conclusion, a considerable body of evidence supports the relevance of the role played by Wnt signaling in ovarian cancer stemness, progression to malignancy, and resistance to chemotherapy. Notwithstanding the potential and innovative therapeutic strategies currently in development to specifically target the Wnt pathway, plasticity of cancer cells still represents an escape mechanism leading to therapy resistance. Moreover, because of Wnt's essential role in tissue homeostasis and regeneration upon damage, its inhibition is likely to result in adverse events. Therefore, the identification and elucidation of the complex network of intrinsic and extrinsic mechanisms driving ovarian cancer progression and therapy resistance represent the major future research challenge in the translation of the fundamental understanding of metastasis and therapy.

Author Contributions: Writing-Original Draft Preparation, M.T. and R.F.; Writing-Review & Editing, M.T. and R.F.; Visualization, M.T.; Supervision, R.F.; Project Administration, M.T. and R.F.; Funding Acquisition, M.T. and R.F.

Funding: This research was funded by the Dutch Cancer Society (KWF), grant number EMCR 2015-7588.

Conflicts of Interest: The authors declare no conflict of interest.

References

1. Siegel, R.L.; Miller, K.D.; Jemal, A. Cancer statistics, 2018. *CA Cancer J. Clin.* **2018**, *68*, 7–30. [CrossRef] [PubMed]
2. Bast, R.C., Jr.; Hennessy, B.; Mills, G.B. The biology of ovarian cancer: New opportunities for translation. *Nat. Rev. Cancer* **2009**, *9*, 415–428. [CrossRef] [PubMed]
3. Kurman, R.J.; Shih Ie, M. Pathogenesis of ovarian cancer: Lessons from morphology and molecular biology and their clinical implications. *Int. J. Gynecol. Pathol.* **2008**, *27*, 151–160. [CrossRef] [PubMed]
4. Kurman, R.J.; Shih Ie, M. The origin and pathogenesis of epithelial ovarian cancer: A proposed unifying theory. *Am. J. Surg. Pathol.* **2010**, *34*, 433–443. [CrossRef] [PubMed]
5. Foster, R.; Buckanovich, R.J.; Rueda, B.R. Ovarian cancer stem cells: Working towards the root of stemness. *Cancer Lett.* **2013**, *338*, 147–157. [CrossRef]
6. Reya, T.; Morrison, S.J.; Clarke, M.F.; Weissman, I.L. Stem cells, cancer, and cancer stem cells. *Nature* **2001**, *414*, 105–111. [CrossRef] [PubMed]
7. Arend, R.C.; Londono-Joshi, A.I.; Straughn, J.M., Jr.; Buchsbaum, D.J. The Wnt/beta-catenin pathway in ovarian cancer: A review. *Gynecol. Oncol.* **2013**, *131*, 772–779. [CrossRef]
8. Wu, R.; Zhai, Y.; Fearon, E.R.; Cho, K.R. Diverse mechanisms of beta-catenin deregulation in ovarian endometrioid adenocarcinomas. *Cancer Res.* **2001**, *61*, 8247–8255.
9. Kildal, W.; Risberg, B.; Abeler, V.M.; Kristensen, G.B.; Sudbo, J.; Nesland, J.M.; Danielsen, H.E. beta-catenin expression, DNA ploidy and clinicopathological features in ovarian cancer: A study in 253 patients. *Eur. J. Cancer* **2005**, *41*, 1127–1134. [CrossRef]
10. Lee, C.M.; Shvartsman, H.; Deavers, M.T.; Wang, S.C.; Xia, W.; Schmandt, R.; Bodurka, D.C.; Atkinson, E.N.; Malpica, A.; Gershenson, D.M.; et al. beta-catenin nuclear localization is associated with grade in ovarian serous carcinoma. *Gynecol. Oncol.* **2003**, *88*, 363–368. [CrossRef]
11. Rask, K.; Nilsson, A.; Brannstrom, M.; Carlsson, P.; Hellberg, P.; Janson, P.O.; Hedin, L.; Sundfeldt, K. Wnt-signalling pathway in ovarian epithelial tumours: Increased expression of beta-catenin and GSK3beta. *Br. J. Cancer* **2003**, *89*, 1298–1304. [CrossRef] [PubMed]
12. Wang, Y.; Hewitt, S.M.; Liu, S.; Zhou, X.; Zhu, H.; Zhou, C.; Zhang, G.; Quan, L.; Bai, J.; Xu, N. Tissue microarray analysis of human FRAT1 expression and its correlation with the subcellular localisation of beta-catenin in ovarian tumours. *Br. J. Cancer* **2006**, *94*, 686–691. [CrossRef] [PubMed]
13. Chau, W.K.; Ip, C.K.; Mak, A.S.; Lai, H.C.; Wong, A.S. c-Kit mediates chemoresistance and tumor-initiating capacity of ovarian cancer cells through activation of Wnt/beta-catenin-ATP-binding cassette G2 signaling. *Oncogene* **2013**, *32*, 2767–2781. [CrossRef] [PubMed]
14. Jacob, F.; Ukegjini, K.; Nixdorf, S.; Ford, C.E.; Olivier, J.; Caduff, R.; Scurry, J.P.; Guertler, R.; Hornung, D.; Mueller, R.; et al. Loss of secreted frizzled-related protein 4 correlates with an aggressive phenotype and predicts poor outcome in ovarian cancer patients. *PLoS ONE* **2012**, *7*, e31885. [CrossRef] [PubMed]
15. Reya, T.; Clevers, H. Wnt signalling in stem cells and cancer. *Nature* **2005**, *434*, 843–850. [CrossRef] [PubMed]
16. Willert, K.; Brown, J.D.; Danenberg, E.; Duncan, A.W.; Weissman, I.L.; Reya, T.; Yates, J.R., 3rd; Nusse, R. Wnt proteins are lipid-modified and can act as stem cell growth factors. *Nature* **2003**, *423*, 448–452. [CrossRef] [PubMed]

17. Zecca, M.; Basler, K.; Struhl, G. Direct and long-range action of a wingless morphogen gradient. *Cell* **1996**, *87*, 833–844. [CrossRef]

18. Shoshkes-Carmel, M.; Wang, Y.J.; Wangensteen, K.J.; Toth, B.; Kondo, A.; Massasa, E.E.; Itzkovitz, S.; Kaestner, K.H. Subepithelial telocytes are an important source of Wnts that supports intestinal crypts. *Nature* **2018**, *557*, 242–246. [CrossRef]

19. Gregorieff, A.; Clevers, H. Wnt signaling in the intestinal epithelium: From endoderm to cancer. *Genes Dev.* **2005**, *19*, 877–890. [CrossRef]

20. Fodde, R.; Smits, R.; Clevers, H. APC, signal transduction and genetic instability in colorectal cancer. *Nat. Rev. Cancer* **2001**, *1*, 55–67. [CrossRef]

21. Fodde, R.; Brabletz, T. Wnt/beta-catenin signaling in cancer stemness and malignant behavior. *Curr. Opin. Cell Biol.* **2007**, *19*, 150–158. [CrossRef] [PubMed]

22. Biason-Lauber, A.; Chaboissier, M.C. Ovarian development and disease: The known and the unexpected. *Semin. Cell Dev. Biol.* **2015**, *45*, 59–67. [CrossRef] [PubMed]

23. Chassot, A.A.; Bradford, S.T.; Auguste, A.; Gregoire, E.P.; Pailhoux, E.; de Rooij, D.G.; Schedl, A.; Chaboissier, M.C. WNT4 and RSPO1 together are required for cell proliferation in the early mouse gonad. *Development* **2012**, *139*, 4461–4472. [CrossRef] [PubMed]

24. Sekido, R.; Lovell-Badge, R. Sex determination involves synergistic action of SRY and SF1 on a specific Sox9 enhancer. *Nature* **2008**, *453*, 930–934. [CrossRef] [PubMed]

25. Jameson, S.A.; Lin, Y.T.; Capel, B. Testis development requires the repression of Wnt4 by Fgf signaling. *Dev. Biol.* **2012**, *370*, 24–32. [CrossRef] [PubMed]

26. Chang, H.; Gao, F.; Guillou, F.; Taketo, M.M.; Huff, V.; Behringer, R.R. Wt1 negatively regulates beta-catenin signaling during testis development. *Development* **2008**, *135*, 1875–1885. [CrossRef]

27. Harris, A.; Siggers, P.; Corrochano, S.; Warr, N.; Sagar, D.; Grimes, D.T.; Suzuki, M.; Burdine, R.D.; Cong, F.; Koo, B.K.; et al. ZNRF3 functions in mammalian sex determination by inhibiting canonical WNT signaling. *Proc. Natl. Acad. Sci. USA* **2018**, *115*, 5474–5479. [CrossRef] [PubMed]

28. Parma, P.; Radi, O.; Vidal, V.; Chaboissier, M.C.; Dellambra, E.; Valentini, S.; Guerra, L.; Schedl, A.; Camerino, G. R-spondin1 is essential in sex determination, skin differentiation and malignancy. *Nat. Genet.* **2006**, *38*, 1304–1309. [CrossRef]

29. Kim, Y.; Kobayashi, A.; Sekido, R.; DiNapoli, L.; Brennan, J.; Chaboissier, M.C.; Poulat, F.; Behringer, R.R.; Lovell-Badge, R.; Capel, B. Fgf9 and Wnt4 act as antagonistic signals to regulate mammalian sex determination. *PLoS Biol.* **2006**, *4*, e187. [CrossRef]

30. Chassot, A.A.; Ranc, F.; Gregoire, E.P.; Roepers-Gajadien, H.L.; Taketo, M.M.; Camerino, G.; de Rooij, D.G.; Schedl, A.; Chaboissier, M.C. Activation of beta-catenin signaling by Rspo1 controls differentiation of the mammalian ovary. *Hum. Mol. Genet.* **2008**, *17*, 1264–1277. [CrossRef]

31. Vainio, S.; Heikkila, M.; Kispert, A.; Chin, N.; McMahon, A.P. Female development in mammals is regulated by Wnt-4 signalling. *Nature* **1999**, *397*, 405–409. [CrossRef] [PubMed]

32. Chassot, A.A.; Gillot, I.; Chaboissier, M.C. R-spondin1, WNT4, and the CTNNB1 signaling pathway: Strict control over ovarian differentiation. *Reproduction* **2014**, *148*, R97–R110. [CrossRef] [PubMed]

33. Ng, A.; Tan, S.; Singh, G.; Rizk, P.; Swathi, Y.; Tan, T.Z.; Huang, R.Y.; Leushacke, M.; Barker, N. Lgr5 marks stem/progenitor cells in ovary and tubal epithelia. *Nat. Cell. Biol.* **2014**, *16*, 745–757. [CrossRef] [PubMed]

34. Usongo, M.; Farookhi, R. beta-catenin/Tcf-signaling appears to establish the murine ovarian surface epithelium (OSE) and remains active in selected postnatal OSE cells. *BMC Dev. Biol.* **2012**, *12*, 17. [CrossRef] [PubMed]

35. Van Es, J.H.; Kirkpatrick, C.; van de Wetering, M.; Molenaar, M.; Miles, A.; Kuipers, J.; Destree, O.; Peifer, M.; Clevers, H. Identification of APC2, a homologue of the adenomatous polyposis coli tumour suppressor. *Curr. Biol.* **1999**, *9*, 105–108. [CrossRef]

36. Mohamed, N.E.; Hay, T.; Reed, K.R.; Smalley, M.J.; Clarke, A.R. APC2 is critical for ovarian WNT signalling control, fertility and tumour suppression. *BMC Cancer* **2019**, *19*, 677. [CrossRef]

37. Van der Horst, P.H.; van der Zee, M.; Heijmans-Antonissen, C.; Jia, Y.; DeMayo, F.J.; Lydon, J.P.; van Deurzen, C.H.; Ewing, P.C.; Burger, C.W.; Blok, L.J. A mouse model for endometrioid ovarian cancer arising from the distal oviduct. *Int. J. Cancer* **2014**, *135*, 1028–1037. [CrossRef]

38. Sagae, S.; Kobayashi, K.; Nishioka, Y.; Sugimura, M.; Ishioka, S.; Nagata, M.; Terasawa, K.; Tokino, T.; Kudo, R. Mutational analysis of beta-catenin gene in Japanese ovarian carcinomas: Frequent mutations in endometrioid carcinomas. *Jpn J. Cancer Res.* **1999**, *90*, 510–515. [CrossRef]

39. Lee, J.H.; Kim, H.S.; Cho, N.H.; Lee, J.Y.; Kim, S.; Kim, S.W.; Kim, Y.T.; Nam, E.J. Genetic analysis of ovarian microcystic stromal tumor. *Obstet. Gynecol. Sci.* **2016**, *59*, 157–162. [CrossRef]

40. Liu, C.; Gallagher, R.L.; Price, G.R.; Bolton, E.; Joy, C.; Harraway, J.; Venter, D.J.; Armes, J.E. Ovarian Microcystic Stromal Tumor: A Rare Clinical Manifestation of Familial Adenomatous Polyposis. *Int. J. Gynecol. Pathol.* **2016**, *35*, 561–565. [CrossRef]

41. Lee, S.H.; Koh, Y.W.; Roh, H.J.; Cha, H.J.; Kwon, Y.S. Ovarian microcystic stromal tumor: A novel extracolonic tumor in familial adenomatous polyposis. *Genes Chromosomes Cancer* **2015**, *54*, 353–360. [CrossRef] [PubMed]

42. Marchion, D.C.; Xiong, Y.; Chon, H.S.; Al Sawah, E.; Bou Zgheib, N.; Ramirez, I.J.; Abbasi, F.; Stickles, X.B.; Judson, P.L.; Hakam, A.; et al. Gene expression data reveal common pathways that characterize the unifocal nature of ovarian cancer. *Am. J. Obstet. Gynecol.* **2013**, *209*, 576.e1–576.e16. [CrossRef] [PubMed]

43. Reinartz, S.; Finkernagel, F.; Adhikary, T.; Rohnalter, V.; Schumann, T.; Schober, Y.; Nockher, W.A.; Nist, A.; Stiewe, T.; Jansen, J.M.; et al. A transcriptome-based global map of signaling pathways in the ovarian cancer microenvironment associated with clinical outcome. *Genome Biol.* **2016**, *17*, 108. [CrossRef] [PubMed]

44. Pangon, L.; Ng, I.; Giry-Laterriere, M.; Currey, N.; Morgan, A.; Benthani, F.; Tran, P.N.; Al-Sohaily, S.; Segelov, E.; Parker, B.L.; et al. JRK is a positive regulator of beta-catenin transcriptional activity commonly overexpressed in colon, breast and ovarian cancer. *Oncogene* **2016**, *35*, 2834–2841. [CrossRef] [PubMed]

45. To, S.K.Y.; Mak, A.S.C.; Eva Fung, Y.M.; Che, C.M.; Li, S.S.; Deng, W.; Ru, B.; Zhang, J.; Wong, A.S.T. beta-catenin downregulates Dicer to promote ovarian cancer metastasis. *Oncogene* **2017**, *36*, 5927–5938. [CrossRef] [PubMed]

46. Dai, F.; Zhang, Y.; Chen, Y. Involvement of miR-29b signaling in the sensitivity to chemotherapy in patients with ovarian carcinoma. *Hum. Pathol.* **2014**, *45*, 1285–1293. [CrossRef] [PubMed]

47. Zhang, Y.; Wang, J.; Wu, D.; Li, M.; Zhao, F.; Ren, M.; Cai, Y.; Dou, J. IL-21-secreting hUCMSCs combined with miR-200c inhibit tumor growth and metastasis via repression of Wnt/beta-catenin signaling and epithelial-mesenchymal transition in epithelial ovarian cancer. *Onco. Targets Ther.* **2018**, *11*, 2037–2050. [CrossRef]

48. Ying, X.; Li-ya, Q.; Feng, Z.; Yin, W.; Ji-hong, L. MiR-939 promotes the proliferation of human ovarian cancer cells by repressing APC2 expression. *Biomed. Pharmacother.* **2015**, *71*, 64–69. [CrossRef]

49. Wu, G.; Liu, A.; Zhu, J.; Lei, F.; Wu, S.; Zhang, X.; Ye, L.; Cao, L.; He, S. MiR-1207 overexpression promotes cancer stem cell-like traits in ovarian cancer by activating the Wnt/beta-catenin signaling pathway. *Oncotarget* **2015**, *6*, 28882–28894. [CrossRef]

50. MacLean, J.A., 2nd; King, M.L.; Okuda, H.; Hayashi, K. WNT7A Regulation by miR-15b in Ovarian Cancer. *PLoS ONE* **2016**, *11*, e0156109. [CrossRef]

51. Chen, M.W.; Yang, S.T.; Chien, M.H.; Hua, K.T.; Wu, C.J.; Hsiao, S.M.; Lin, H.; Hsiao, M.; Su, J.L.; Wei, L.H. The STAT3-miRNA-92-Wnt Signaling Pathway Regulates Spheroid Formation and Malignant Progression in Ovarian Cancer. *Cancer Res.* **2017**, *77*, 1955–1967. [CrossRef] [PubMed]

52. Wei, C.; Zhang, X.; He, S.; Liu, B.; Han, H.; Sun, X. MicroRNA-219-5p inhibits the proliferation, migration, and invasion of epithelial ovarian cancer cells by targeting the Twist/Wnt/beta-catenin signaling pathway. *Gene* **2017**, *637*, 25–32. [CrossRef] [PubMed]

53. Liu, Y.; Lin, J.; Zhai, S.; Sun, C.; Xu, C.; Zhou, H.; Liu, H. MicroRNA-214 Suppresses Ovarian Cancer by Targeting beta-Catenin. *Cell Physiol. Biochem.* **2018**, *45*, 1654–1662. [CrossRef] [PubMed]

54. Tu, J.; Cheung, H.H.; Lu, G.; Chen, Z.; Chan, W.Y. MicroRNA-10a promotes granulosa cells tumor development via PTEN-AKT/Wnt regulatory axis. *Cell Death Dis.* **2018**, *9*, 1076. [CrossRef] [PubMed]

55. Zhang, L.Y.; Chen, Y.; Jia, J.; Zhu, X.; He, Y.; Wu, L.M. MiR-27a promotes EMT in ovarian cancer through active Wnt/-catenin signalling by targeting FOXO1. *Cancer Biomark.* **2019**, *24*, 31–42. [CrossRef]

56. Li, N.; Yang, L.; Sun, Y.; Wu, X. MicroRNA-16 inhibits migration and invasion via regulation of the Wnt/beta-catenin signaling pathway in ovarian cancer. *Oncol. Lett.* **2019**, *17*, 2631–2638. [CrossRef]

57. Wang, Y.; Yang, X.; Yuan, M.; Xian, S.; Zhang, L.; Yang, D.; Cheng, Y. Promotion of ovarian cancer cell invasion, migration and colony formation by the miR21/Wnt/CD44v6 pathway. *Oncol. Rep.* **2019**, *42*, 91–102. [CrossRef]

58. Majem, B.; Parrilla, A.; Jimenez, C.; Suarez-Cabrera, L.; Barber, M.; Marin, A.; Castellvi, J.; Tamayo, G.; Moreno-Bueno, G.; Ponce, J.; et al. MicroRNA-654-5p suppresses ovarian cancer development impacting on MYC, WNT and AKT pathways. *Oncogene* **2019**, *38*, 6035–6050. [CrossRef]

59. Hu, J.; Zhao, W.; Huang, Y.; Wang, Z.; Jiang, T.; Wang, L. MiR-1180 from bone marrow MSCs promotes cell proliferation and glycolysis in ovarian cancer cells via SFRP1/Wnt pathway. *Cancer Cell Int.* **2019**, *19*, 66. [CrossRef]

60. Li, J.; Yang, S.; Su, N.; Wang, Y.; Yu, J.; Qiu, H.; He, X. Overexpression of long non-coding RNA HOTAIR leads to chemoresistance by activating the Wnt/beta-catenin pathway in human ovarian cancer. *Tumour Biol.* **2016**, *37*, 2057–2065. [CrossRef]

61. He, S.; Zhao, Y.; Wang, X.; Deng, Y.; Wan, Z.; Yao, S.; Shen, H. Up-regulation of long non-coding RNA SNHG20 promotes ovarian cancer progression via Wnt/beta-catenin signaling. *Biosci. Rep.* **2018**, *38*. [CrossRef] [PubMed]

62. Zhang, Y.; Dun, Y.; Zhou, S.; Huang, X.H. LncRNA HOXD-AS1 promotes epithelial ovarian cancer cells proliferation and invasion by targeting miR-133a-3p and activating Wnt/beta-catenin signaling pathway. *Biomed. Pharmacother.* **2017**, *96*, 1216–1221. [CrossRef] [PubMed]

63. Wang, B.; Liu, M.; Zhuang, R.; Jiang, J.; Gao, J.; Wang, H.; Chen, H.; Zhang, Z.; Kuang, Y.; Li, P. Long non-coding RNA CCAT2 promotes epithelial-mesenchymal transition involving Wnt/beta-catenin pathway in epithelial ovarian carcinoma cells. *Oncol. Lett.* **2018**, *15*, 3369–3375. [CrossRef]

64. Guo, C.; Wang, X.; Chen, L.P.; Li, M.; Li, M.; Hu, Y.H.; Ding, W.H.; Wang, X. Long non-coding RNA MALAT1 regulates ovarian cancer cell proliferation, migration and apoptosis through Wnt/beta-catenin signaling pathway. *Eur. Rev. Med. Pharmacol. Sci.* **2018**, *22*, 3703–3712. [CrossRef] [PubMed]

65. Yu, G.; Wang, W.; Deng, J.; Dong, S. LncRNA AWPPH promotes the proliferation, migration and invasion of ovarian carcinoma cells via activation of the Wnt/betacatenin signaling pathway. *Mol. Med. Rep.* **2019**, *19*, 3615–3621. [CrossRef] [PubMed]

66. Saegusa, M.; Okayasu, I. Frequent nuclear beta-catenin accumulation and associated mutations in endometrioid-type endometrial and ovarian carcinomas with squamous differentiation. *J. Pathol.* **2001**, *194*, 59–67. [CrossRef] [PubMed]

67. Moreno-Bueno, G.; Gamallo, C.; Perez-Gallego, L.; de Mora, J.C.; Suarez, A.; Palacios, J. beta-Catenin expression pattern, beta-catenin gene mutations, and microsatellite instability in endometrioid ovarian carcinomas and synchronous endometrial carcinomas. *Diagn. Mol. Pathol.* **2001**, *10*, 116–122. [CrossRef] [PubMed]

68. Wright, K.; Wilson, P.; Morland, S.; Campbell, I.; Walsh, M.; Hurst, T.; Ward, B.; Cummings, M.; Chenevix-Trench, G. beta-catenin mutation and expression analysis in ovarian cancer: Exon 3 mutations and nuclear translocation in 16% of endometrioid tumours. *Int. J. Cancer* **1999**, *82*, 625–629. [CrossRef]

69. Gamallo, C.; Palacios, J.; Moreno, G.; Calvo de Mora, J.; Suarez, A.; Armas, A. beta-catenin expression pattern in stage I and II ovarian carcinomas: Relationship with beta-catenin gene mutations, clinicopathological features, and clinical outcome. *Am. J. Pathol.* **1999**, *155*, 527–536. [CrossRef]

70. Palacios, J.; Gamallo, C. Mutations in the beta-catenin gene (CTNNB1) in endometrioid ovarian carcinomas. *Cancer Res.* **1998**, *58*, 1344–1347.

71. Zhuang, X.H.; Liu, Y.; Li, J.L. Overexpression of long noncoding RNA HOXB-AS3 indicates an unfavorable prognosis and promotes tumorigenesis in epithelial ovarian cancer via Wnt/beta-catenin signaling pathway. *Biosci. Rep.* **2019**, *39*. [CrossRef] [PubMed]

72. Ricken, A.; Lochhead, P.; Kontogiannea, M.; Farookhi, R. Wnt signaling in the ovary: Identification and compartmentalized expression of wnt-2, wnt-2b, and frizzled-4 mRNAs. *Endocrinology* **2002**, *143*, 2741–2749. [CrossRef] [PubMed]

73. Tothill, R.W.; Tinker, A.V.; George, J.; Brown, R.; Fox, S.B.; Lade, S.; Johnson, D.S.; Trivett, M.K.; Etemadmoghadam, D.; Locandro, B.; et al. Novel molecular subtypes of serous and endometrioid ovarian cancer linked to clinical outcome. *Clin. Cancer Res.* **2008**, *14*, 5198–5208. [CrossRef] [PubMed]

74. Badiglian Filho, L.; Oshima, C.T.; De Oliveira Lima, F.; De Oliveira Costa, H.; De Sousa Damiao, R.; Gomes, T.S.; Goncalves, W.J. Canonical and noncanonical Wnt pathway: A comparison among normal ovary, benign ovarian tumor and ovarian cancer. *Oncol. Rep.* **2009**, *21*, 313–320. [PubMed]

75. Popadiuk, C.M.; Xiong, J.; Wells, M.G.; Andrews, P.G.; Dankwa, K.; Hirasawa, K.; Lake, B.B.; Kao, K.R. Antisense suppression of pygopus2 results in growth arrest of epithelial ovarian cancer. *Clin. Cancer Res.* **2006**, *12*, 2216–2223. [CrossRef] [PubMed]

76. Duan, H.; Yan, Z.; Chen, W.; Wu, Y.; Han, J.; Guo, H.; Qiao, J. TET1 inhibits EMT of ovarian cancer cells through activating Wnt/beta-catenin signaling inhibitors DKK1 and SFRP2. *Gynecol. Oncol.* **2017**, *147*, 408–417. [CrossRef] [PubMed]

77. Gray, J.W.; Suzuki, S.; Kuo, W.L.; Polikoff, D.; Deavers, M.; Smith-McCune, K.; Berchuck, A.; Pinkel, D.; Albertson, D.; Mills, G.B. Specific keynote: Genome copy number abnormalities in ovarian cancer. *Gynecol. Oncol.* **2003**, *88*, S16–S21, discussion S22-14. [CrossRef]

78. Takada, T.; Yagi, Y.; Maekita, T.; Imura, M.; Nakagawa, S.; Tsao, S.W.; Miyamoto, K.; Yoshino, O.; Yasugi, T.; Taketani, Y.; et al. Methylation-associated silencing of the Wnt antagonist SFRP1 gene in human ovarian cancers. *Cancer Sci.* **2004**, *95*, 741–744. [CrossRef]

79. Matte, I.; Lane, D.; Laplante, C.; Rancourt, C.; Piche, A. Profiling of cytokines in human epithelial ovarian cancer ascites. *Am. J. Cancer Res.* **2012**, *2*, 566–580.

80. Thibault, B.; Castells, M.; Delord, J.P.; Couderc, B. Ovarian cancer microenvironment: Implications for cancer dissemination and chemoresistance acquisition. *Cancer Metastasis Rev.* **2014**, *33*, 17–39. [CrossRef]

81. Endo, H.; Hosono, K.; Uchiyama, T.; Sakai, E.; Sugiyama, M.; Takahashi, H.; Nakajima, N.; Wada, K.; Takeda, K.; Nakagama, H.; et al. Leptin acts as a growth factor for colorectal tumours at stages subsequent to tumour initiation in murine colon carcinogenesis. *Gut* **2011**, *60*, 1363–1371. [CrossRef] [PubMed]

82. Yan, D.; Avtanski, D.; Saxena, N.K.; Sharma, D. Leptin-induced epithelial-mesenchymal transition in breast cancer cells requires beta-catenin activation via Akt/GSK3- and MTA1/Wnt1 protein-dependent pathways. *J. Biol. Chem.* **2012**, *287*, 8598–8612. [CrossRef] [PubMed]

83. Asuthkar, S.; Gondi, C.S.; Nalla, A.K.; Velpula, K.K.; Gorantla, B.; Rao, J.S. Urokinase-type plasminogen activator receptor (uPAR)-mediated regulation of WNT/beta-catenin signaling is enhanced in irradiated medulloblastoma cells. *J. Biol. Chem.* **2012**, *287*, 20576–20589. [CrossRef] [PubMed]

84. Zhang, X.; Chen, L.; Wang, Y.; Ding, Y.; Peng, Z.; Duan, L.; Ju, G.; Ren, Y.; Wang, X. Macrophage migration inhibitory factor promotes proliferation and neuronal differentiation of neural stem/precursor cells through Wnt/beta-catenin signal pathway. *Int. J. Biol. Sci.* **2013**, *9*, 1108–1120. [CrossRef] [PubMed]

85. Glass, D.A., 2nd; Bialek, P.; Ahn, J.D.; Starbuck, M.; Patel, M.S.; Clevers, H.; Taketo, M.M.; Long, F.; McMahon, A.P.; Lang, R.A.; et al. Canonical Wnt signaling in differentiated osteoblasts controls osteoclast differentiation. *Dev. Cell* **2005**, *8*, 751–764. [CrossRef] [PubMed]

86. Masckauchan, T.N.; Shawber, C.J.; Funahashi, Y.; Li, C.M.; Kitajewski, J. Wnt/beta-catenin signaling induces proliferation, survival and interleukin-8 in human endothelial cells. *Angiogenesis* **2005**, *8*, 43–51. [CrossRef] [PubMed]

87. Burkhalter, R.J.; Symowicz, J.; Hudson, L.G.; Gottardi, C.J.; Stack, M.S. Integrin regulation of beta-catenin signaling in ovarian carcinoma. *J. Biol. Chem.* **2011**, *286*, 23467–23475. [CrossRef] [PubMed]

88. Gross, J.C.; Chaudhary, V.; Bartscherer, K.; Boutros, M. Active Wnt proteins are secreted on exosomes. *Nat. Cell Biol.* **2012**, *14*, 1036–1045. [CrossRef] [PubMed]

89. Beach, A.; Zhang, H.G.; Ratajczak, M.Z.; Kakar, S.S. Exosomes: An overview of biogenesis, composition and role in ovarian cancer. *J. Ovarian Res.* **2014**, *7*, 14. [CrossRef] [PubMed]

90. Latifi, A.; Abubaker, K.; Castrechini, N.; Ward, A.C.; Liongue, C.; Dobill, F.; Kumar, J.; Thompson, E.W.; Quinn, M.A.; Findlay, J.K.; et al. Cisplatin treatment of primary and metastatic epithelial ovarian carcinomas generates residual cells with mesenchymal stem cell-like profile. *J. Cell Biochem.* **2011**, *112*, 2850–2864. [CrossRef] [PubMed]

91. Bapat, S.A.; Mali, A.M.; Koppikar, C.B.; Kurrey, N.K. Stem and progenitor-like cells contribute to the aggressive behavior of human epithelial ovarian cancer. *Cancer Res.* **2005**, *65*, 3025–3029. [CrossRef] [PubMed]

92. Raghavan, S.; Mehta, P.; Xie, Y.; Lei, Y.L.; Mehta, G. Ovarian cancer stem cells and macrophages reciprocally interact through the WNT pathway to promote pro-tumoral and malignant phenotypes in 3D engineered microenvironments. *J. Immunother. Cancer* **2019**, *7*, 190. [CrossRef]

93. Ruan, X.; Liu, A.; Zhong, M.; Wei, J.; Zhang, W.; Rong, Y.; Liu, W.; Li, M.; Qing, X.; Chen, G.; et al. Silencing LGR6 Attenuates Stemness and Chemoresistance via Inhibiting Wnt/beta-Catenin Signaling in Ovarian Cancer. *Mol. Ther. Oncolytics* **2019**, *14*, 94–106. [CrossRef] [PubMed]

94. Ahmed, N.; Stenvers, K.L. Getting to know ovarian cancer ascites: Opportunities for targeted therapy-based translational research. *Front. Oncol.* **2013**, *3*, 256. [CrossRef] [PubMed]

95. Mills, G.B.; May, C.; Hill, M.; Campbell, S.; Shaw, P.; Marks, A. Ascitic fluid from human ovarian cancer patients contains growth factors necessary for intraperitoneal growth of human ovarian adenocarcinoma cells. *J. Clin. Investig.* **1990**, *86*, 851–855. [CrossRef] [PubMed]

96. Burleson, K.M.; Casey, R.C.; Skubitz, K.M.; Pambuccian, S.E.; Oegema, T.R., Jr.; Skubitz, A.P. Ovarian carcinoma ascites spheroids adhere to extracellular matrix components and mesothelial cell monolayers. *Gynecol. Oncol.* **2004**, *93*, 170–181. [CrossRef] [PubMed]

97. Shield, K.; Ackland, M.L.; Ahmed, N.; Rice, G.E. Multicellular spheroids in ovarian cancer metastases: Biology and pathology. *Gynecol. Oncol.* **2009**, *113*, 143–148. [CrossRef] [PubMed]

98. Naora, H.; Montell, D.J. Ovarian cancer metastasis: Integrating insights from disparate model organisms. *Nat. Rev. Cancer* **2005**, *5*, 355–366. [CrossRef] [PubMed]

99. Loret, N.; Denys, H.; Tummers, P.; Berx, G. The Role of Epithelial-to-Mesenchymal Plasticity in Ovarian Cancer Progression and Therapy Resistance. *Cancers (Basel)* **2019**, *11*. [CrossRef] [PubMed]

100. Pradeep, S.; Kim, S.W.; Wu, S.Y.; Nishimura, M.; Chaluvally-Raghavan, P.; Miyake, T.; Pecot, C.V.; Kim, S.J.; Choi, H.J.; Bischoff, F.Z.; et al. Hematogenous metastasis of ovarian cancer: Rethinking mode of spread. *Cancer Cell* **2014**, *26*, 77–91. [CrossRef] [PubMed]

101. Kuhlmann, J.D.; Wimberger, P.; Bankfalvi, A.; Keller, T.; Scholer, S.; Aktas, B.; Buderath, P.; Hauch, S.; Otterbach, F.; Kimmig, R.; et al. ERCC1-positive circulating tumor cells in the blood of ovarian cancer patients as a predictive biomarker for platinum resistance. *Clin. Chem.* **2014**, *60*, 1282–1289. [CrossRef] [PubMed]

102. Nieto, M.A.; Huang, R.Y.; Jackson, R.A.; Thiery, J.P. Emt: 2016. *Cell* **2016**, *166*, 21–45. [CrossRef] [PubMed]

103. Dongre, A.; Rashidian, M.; Reinhardt, F.; Bagnato, A.; Keckesova, Z.; Ploegh, H.L.; Weinberg, R.A. Epithelial-to-Mesenchymal Transition Contributes to Immunosuppression in Breast Carcinomas. *Cancer Res.* **2017**, *77*, 3982–3989. [CrossRef] [PubMed]

104. Mani, S.A.; Guo, W.; Liao, M.J.; Eaton, E.N.; Ayyanan, A.; Zhou, A.Y.; Brooks, M.; Reinhard, F.; Zhang, C.C.; Shipitsin, M.; et al. The epithelial-mesenchymal transition generates cells with properties of stem cells. *Cell* **2008**, *133*, 704–715. [CrossRef] [PubMed]

105. Smith, B.N.; Bhowmick, N.A. Role of EMT in Metastasis and Therapy Resistance. *J. Clin. Med.* **2016**, *5*. [CrossRef]

106. Brabletz, T.; Jung, A.; Spaderna, S.; Hlubek, F.; Kirchner, T. Opinion: Migrating cancer stem cells - an integrated concept of malignant tumour progression. *Nat. Rev. Cancer* **2005**, *5*, 744–749. [CrossRef]

107. Wu, Z.Q.; Li, X.Y.; Hu, C.Y.; Ford, M.; Kleer, C.G.; Weiss, S.J. Canonical Wnt signaling regulates Slug activity and links epithelial-mesenchymal transition with epigenetic Breast Cancer 1, Early Onset (BRCA1) repression. *Proc. Natl. Acad. Sci. USA* **2012**, *109*, 16654–16659. [CrossRef]

108. Yook, J.I.; Li, X.Y.; Ota, I.; Fearon, E.R.; Weiss, S.J. Wnt-dependent regulation of the E-cadherin repressor snail. *J. Biol. Chem.* **2005**, *280*, 11740–11748. [CrossRef]

109. Bernaudo, S.; Salem, M.; Qi, X.; Zhou, W.; Zhang, C.; Yang, W.; Rosman, D.; Deng, Z.; Ye, G.; Yang, B.B.; et al. Cyclin G2 inhibits epithelial-to-mesenchymal transition by disrupting Wnt/beta-catenin signaling. *Oncogene* **2016**, *35*, 4816–4827. [CrossRef]

110. Deng, Z.; Wang, L.; Hou, H.; Zhou, J.; Li, X. Epigenetic regulation of IQGAP2 promotes ovarian cancer progression via activating Wnt/beta-catenin signaling. *Int. J. Oncol.* **2016**, *48*, 153–160. [CrossRef]

111. Hojo, N.; Huisken, A.L.; Wang, H.; Chirshev, E.; Kim, N.S.; Nguyen, S.M.; Campos, H.; Glackin, C.A.; Ioffe, Y.J.; Unternaehrer, J.J. Snail knockdown reverses stemness and inhibits tumour growth in ovarian cancer. *Sci. Rep.* **2018**, *8*, 8704. [CrossRef] [PubMed]

112. Sun, J.; Yang, X.; Zhang, R.; Liu, S.; Gan, X.; Xi, X.; Zhang, Z.; Feng, Y.; Sun, Y. GOLPH3 induces epithelial-mesenchymal transition via Wnt/beta-catenin signaling pathway in epithelial ovarian cancer. *Cancer Med.* **2017**, *6*, 834–844. [CrossRef] [PubMed]

113. Aiello, N.M.; Maddipati, R.; Norgard, R.J.; Balli, D.; Li, J.; Yuan, S.; Yamazoe, T.; Black, T.; Sahmoud, A.; Furth, E.E.; et al. EMT Subtype Influences Epithelial Plasticity and Mode of Cell Migration. *Dev. Cell* **2018**, *45*, 681–695.e4. [CrossRef] [PubMed]

114. Pastushenko, I.; Brisebarre, A.; Sifrim, A.; Fioramonti, M.; Revenco, T.; Boumahdi, S.; Van Keymeulen, A.; Brown, D.; Moers, V.; Lemaire, S.; et al. Identification of the tumour transition states occurring during EMT. *Nature* **2018**, *556*, 463–468. [CrossRef] [PubMed]

115. Hudson, L.G.; Zeineldin, R.; Stack, M.S. Phenotypic plasticity of neoplastic ovarian epithelium: Unique cadherin profiles in tumor progression. *Clin. Exp. Metastasis* **2008**, *25*, 643–655. [CrossRef] [PubMed]

116. Huang, R.Y.; Wong, M.K.; Tan, T.Z.; Kuay, K.T.; Ng, A.H.; Chung, V.Y.; Chu, Y.S.; Matsumura, N.; Lai, H.C.; Lee, Y.F.; et al. An EMT spectrum defines an anoikis-resistant and spheroidogenic intermediate mesenchymal state that is sensitive to e-cadherin restoration by a src-kinase inhibitor, saracatinib (AZD0530). *Cell Death Dis.* **2013**, *4*, e915. [CrossRef]

117. Klymenko, Y.; Johnson, J.; Bos, B.; Lombard, R.; Campbell, L.; Loughran, E.; Stack, M.S. Heterogeneous Cadherin Expression and Multicellular Aggregate Dynamics in Ovarian Cancer Dissemination. *Neoplasia* **2017**, *19*, 549–563. [CrossRef]

118. Strauss, R.; Li, Z.Y.; Liu, Y.; Beyer, I.; Persson, J.; Sova, P.; Moller, T.; Pesonen, S.; Hemminki, A.; Hamerlik, P.; et al. Analysis of epithelial and mesenchymal markers in ovarian cancer reveals phenotypic heterogeneity and plasticity. *PLoS ONE* **2011**, *6*, e16186. [CrossRef]

119. Chikazawa, N.; Tanaka, H.; Tasaka, T.; Nakamura, M.; Tanaka, M.; Onishi, H.; Katano, M. Inhibition of Wnt signaling pathway decreases chemotherapy-resistant side-population colon cancer cells. *Anticancer Res.* **2010**, *30*, 2041–2048.

120. Zheng, H.C. The molecular mechanisms of chemoresistance in cancers. *Oncotarget* **2017**, *8*, 59950–59964. [CrossRef]

121. Eyre, R.; Harvey, I.; Stemke-Hale, K.; Lennard, T.W.; Tyson-Capper, A.; Meeson, A.P. Reversing paclitaxel resistance in ovarian cancer cells via inhibition of the ABCB1 expressing side population. *Tumour Biol.* **2014**, *35*, 9879–9892. [CrossRef] [PubMed]

122. Hu, L.; McArthur, C.; Jaffe, R.B. Ovarian cancer stem-like side-population cells are tumourigenic and chemoresistant. *Br. J. Cancer* **2010**, *102*, 1276–1283. [CrossRef] [PubMed]

123. Su, H.Y.; Lai, H.C.; Lin, Y.W.; Liu, C.Y.; Chen, C.K.; Chou, Y.C.; Lin, S.P.; Lin, W.C.; Lee, H.Y.; Yu, M.H. Epigenetic silencing of SFRP5 is related to malignant phenotype and chemoresistance of ovarian cancer through Wnt signaling pathway. *Int. J. Cancer* **2010**, *127*, 555–567. [CrossRef] [PubMed]

124. Zhang, B.; Chen, F.; Xu, Q.; Han, L.; Xu, J.; Gao, L.; Sun, X.; Li, Y.; Li, Y.; Qian, M.; et al. Revisiting ovarian cancer microenvironment: A friend or a foe? *Protein Cell* **2018**, *9*, 674–692. [CrossRef] [PubMed]

125. Shiga, K.; Hara, M.; Nagasaki, T.; Sato, T.; Takahashi, H.; Takeyama, H. Cancer-Associated Fibroblasts: Their Characteristics and Their Roles in Tumor Growth. *Cancers (Basel)* **2015**, *7*, 2443–2458. [CrossRef] [PubMed]

126. Ferrari, N.; Ranftl, R.; Chicherova, I.; Slaven, N.D.; Moeendarbary, E.; Farrugia, A.J.; Lam, M.; Semiannikova, M.; Westergaard, M.C.W.; Tchou, J.; et al. Dickkopf-3 links HSF1 and YAP/TAZ signalling to control aggressive behaviours in cancer-associated fibroblasts. *Nat. Commun.* **2019**, *10*, 130. [CrossRef] [PubMed]

127. Armstrong, D.K.; Bundy, B.; Wenzel, L.; Huang, H.Q.; Baergen, R.; Lele, S.; Copeland, L.J.; Walker, J.L.; Burger, R.A.; Gynecologic Oncology Group. Intraperitoneal cisplatin and paclitaxel in ovarian cancer. *N. Engl. J. Med.* **2006**, *354*, 34–43. [CrossRef] [PubMed]

128. Burger, R.A.; Brady, M.F.; Bookman, M.A.; Fleming, G.F.; Monk, B.J.; Huang, H.; Mannel, R.S.; Homesley, H.D.; Fowler, J.; Greer, B.E.; et al. Incorporation of bevacizumab in the primary treatment of ovarian cancer. *N. Engl. J. Med.* **2011**, *365*, 2473–2483. [CrossRef]

129. Katsumata, N.; Yasuda, M.; Takahashi, F.; Isonishi, S.; Jobo, T.; Aoki, D.; Tsuda, H.; Sugiyama, T.; Kodama, S.; Kimura, E.; et al. Dose-dense paclitaxel once a week in combination with carboplatin every 3 weeks for advanced ovarian cancer: A phase 3, open-label, randomised controlled trial. *Lancet* **2009**, *374*, 1331–1338. [CrossRef]

130. Saha, S.; Aranda, E.; Hayakawa, Y.; Bhanja, P.; Atay, S.; Brodin, N.P.; Li, J.; Asfaha, S.; Liu, L.; Tailor, Y.; et al. Macrophage-derived extracellular vesicle-packaged WNTs rescue intestinal stem cells and enhance survival after radiation injury. *Nat. Commun.* **2016**, *7*, 13096. [CrossRef]

131. Azmi, A.S.; Bao, B.; Sarkar, F.H. Exosomes in cancer development, metastasis, and drug resistance: A comprehensive review. *Cancer Metastasis Rev.* **2013**, *32*, 623–642. [CrossRef] [PubMed]

132. Keller, S.; Konig, A.K.; Marme, F.; Runz, S.; Wolterink, S.; Koensgen, D.; Mustea, A.; Sehouli, J.; Altevogt, P. Systemic presence and tumor-growth promoting effect of ovarian carcinoma released exosomes. *Cancer Lett.* **2009**, *278*, 73–81. [CrossRef] [PubMed]

133. Liang, B.; Peng, P.; Chen, S.; Li, L.; Zhang, M.; Cao, D.; Yang, J.; Li, H.; Gui, T.; Li, X.; et al. Characterization and proteomic analysis of ovarian cancer-derived exosomes. *J. Proteom.* **2013**, *80*, 171–182. [CrossRef] [PubMed]

134. Runz, S.; Keller, S.; Rupp, C.; Stoeck, A.; Issa, Y.; Koensgen, D.; Mustea, A.; Sehouli, J.; Kristiansen, G.; Altevogt, P. Malignant ascites-derived exosomes of ovarian carcinoma patients contain CD24 and EpCAM. *Gynecol. Oncol.* **2007**, *107*, 563–571. [CrossRef] [PubMed]

135. Taylor, D.D.; Gercel-Taylor, C. MicroRNA signatures of tumor-derived exosomes as diagnostic biomarkers of ovarian cancer. *Gynecol. Oncol.* **2008**, *110*, 13–21. [CrossRef] [PubMed]

136. Luga, V.; Zhang, L.; Viloria-Petit, A.M.; Ogunjimi, A.A.; Inanlou, M.R.; Chiu, E.; Buchanan, M.; Hosein, A.N.; Basik, M.; Wrana, J.L. Exosomes mediate stromal mobilization of autocrine Wnt-PCP signaling in breast cancer cell migration. *Cell* **2012**, *151*, 1542–1556. [CrossRef] [PubMed]

137. Hu, Y.B.; Yan, C.; Mu, L.; Mi, Y.L.; Zhao, H.; Hu, H.; Li, X.L.; Tao, D.D.; Wu, Y.Q.; Gong, J.P.; et al. Exosomal Wnt-induced dedifferentiation of colorectal cancer cells contributes to chemotherapy resistance. *Oncogene* **2019**, *38*, 1951–1965. [CrossRef]

138. Dovrat, S.; Caspi, M.; Zilberberg, A.; Lahav, L.; Firsow, A.; Gur, H.; Rosin-Arbesfeld, R. 14-3-3 and beta-catenin are secreted on extracellular vesicles to activate the oncogenic Wnt pathway. *Mol. Oncol.* **2014**, *8*, 894–911. [CrossRef]

139. Zhang, S.; Zhang, X.; Fu, X.; Li, W.; Xing, S.; Yang, Y. Identification of common differentially-expressed miRNAs in ovarian cancer cells and their exosomes compared with normal ovarian surface epithelial cell cells. *Oncol. Lett.* **2018**, *16*, 2391–2401. [CrossRef]

140. Alharbi, M.; Lai, A.; Guanzon, D.; Palma, C.; Zuniga, F.; Perrin, L.; He, Y.; Hooper, J.D.; Salomon, C. Ovarian cancer-derived exosomes promote tumour metastasis in vivo: An effect modulated by the invasiveness capacity of their originating cells. *Clin. Sci. (Lond)* **2019**, *133*, 1401–1419. [CrossRef]

141. Nakamura, K.; Sawada, K.; Kinose, Y.; Yoshimura, A.; Toda, A.; Nakatsuka, E.; Hashimoto, K.; Mabuchi, S.; Morishige, K.I.; Kurachi, H.; et al. Exosomes Promote Ovarian Cancer Cell Invasion through Transfer of CD44 to Peritoneal Mesothelial Cells. *Mol. Cancer Res.* **2017**, *15*, 78–92. [CrossRef] [PubMed]

142. Liu, Y.; Cao, X. Characteristics and Significance of the Pre-metastatic Niche. *Cancer Cell* **2016**, *30*, 668–681. [CrossRef] [PubMed]

143. Lee, W.; Ko, S.Y.; Mohamed, M.S.; Kenny, H.A.; Lengyel, E.; Naora, H. Neutrophils facilitate ovarian cancer premetastatic niche formation in the omentum. *J. Exp. Med.* **2019**, *216*, 176–194. [CrossRef] [PubMed]

144. Yokoi, A.; Yoshioka, Y.; Yamamoto, Y.; Ishikawa, M.; Ikeda, S.I.; Kato, T.; Kiyono, T.; Takeshita, F.; Kajiyama, H.; Kikkawa, F.; et al. Malignant extracellular vesicles carrying MMP1 mRNA facilitate peritoneal dissemination in ovarian cancer. *Nat. Commun.* **2017**, *8*, 14470. [CrossRef] [PubMed]

145. Wielenga, V.J.; Smits, R.; Korinek, V.; Smit, L.; Kielman, M.; Fodde, R.; Clevers, H.; Pals, S.T. Expression of CD44 in Apc and Tcf mutant mice implies regulation by the WNT pathway. *Am. J. Pathol.* **1999**, *154*, 515–523. [CrossRef]

146. Schmitt, M.; Metzger, M.; Gradl, D.; Davidson, G.; Orian-Rousseau, V. CD44 functions in Wnt signaling by regulating LRP6 localization and activation. *Cell Death Differ.* **2015**, *22*, 677–689. [CrossRef]

147. De la Fuente, A.; Alonso-Alconada, L.; Costa, C.; Cueva, J.; Garcia-Caballero, T.; Lopez-Lopez, R.; Abal, M. M-Trap: Exosome-Based Capture of Tumor Cells as a New Technology in Peritoneal Metastasis. *J. Natl. Cancer Inst.* **2015**, *107*. [CrossRef] [PubMed]

148. De Lau, W.B.; Snel, B.; Clevers, H.C. The R-spondin protein family. *Genome Biol.* **2012**, *13*, 242. [CrossRef] [PubMed]

149. Chartier, C.; Raval, J.; Axelrod, F.; Bond, C.; Cain, J.; Dee-Hoskins, C.; Ma, S.; Fischer, M.M.; Shah, J.; Wei, J.; et al. Therapeutic Targeting of Tumor-Derived R-Spondin Attenuates beta-Catenin Signaling and Tumorigenesis in Multiple Cancer Types. *Cancer Res.* **2016**, *76*, 713–723. [CrossRef] [PubMed]

150. Boone, J.D.; Arend, R.C.; Johnston, B.E.; Cooper, S.J.; Gilchrist, S.A.; Oelschlager, D.K.; Grizzle, W.E.; McGwin, G., Jr.; Gangrade, A.; Straughn, J.M., Jr.; et al. Targeting the Wnt/beta-catenin pathway in primary ovarian cancer with the porcupine inhibitor WNT974. *Lab. Investig.* **2016**, *96*, 249–259. [CrossRef] [PubMed]

151. Yo, Y.T.; Lin, Y.W.; Wang, Y.C.; Balch, C.; Huang, R.L.; Chan, M.W.; Sytwu, H.K.; Chen, C.K.; Chang, C.C.; Nephew, K.P.; et al. Growth inhibition of ovarian tumor-initiating cells by niclosamide. *Mol. Cancer Ther.* **2012**, *11*, 1703–1712. [CrossRef] [PubMed]

152. Li, Y.; Li, P.K.; Roberts, M.J.; Arend, R.C.; Samant, R.S.; Buchsbaum, D.J. Multi-targeted therapy of cancer by niclosamide: A new application for an old drug. *Cancer Lett.* **2014**, *349*, 8–14. [CrossRef] [PubMed]

153. Yang, K.; Wang, X.; Zhang, H.; Wang, Z.; Nan, G.; Li, Y.; Zhang, F.; Mohammed, M.K.; Haydon, R.C.; Luu, H.H.; et al. The evolving roles of canonical WNT signaling in stem cells and tumorigenesis: Implications in targeted cancer therapies. *Lab. Investig.* **2016**, *96*, 116–136. [CrossRef] [PubMed]
154. Moore, K.N.; Gunderson, C.C.; Sabbatini, P.; McMeekin, D.S.; Mantia-Smaldone, G.; Burger, R.A.; Morgan, M.A.; Kapoun, A.M.; Brachmann, R.K.; Stagg, R.; et al. A phase 1b dose escalation study of ipafricept (OMP54F28) in combination with paclitaxel and carboplatin in patients with recurrent platinum-sensitive ovarian cancer. *Gynecol. Oncol.* **2019**, *154*, 294–301. [CrossRef] [PubMed]
155. Harb, J.; Lin, P.J.; Hao, J. Recent Development of Wnt Signaling Pathway Inhibitors for Cancer Therapeutics. *Curr. Oncol. Rep.* **2019**, *21*, 12. [CrossRef]
156. Li, R.; Dong, T.; Hu, C.; Lu, J.; Dai, J.; Liu, P. Salinomycin repressed the epithelial-mesenchymal transition of epithelial ovarian cancer cells via downregulating Wnt/beta-catenin pathway. *Onco. Targets Ther.* **2017**, *10*, 1317–1325. [CrossRef]
157. Baryawno, N.; Sveinbjornsson, B.; Eksborg, S.; Chen, C.S.; Kogner, P.; Johnsen, J.I. Small-molecule inhibitors of phosphatidylinositol 3-kinase/Akt signaling inhibit Wnt/beta-catenin pathway cross-talk and suppress medulloblastoma growth. *Cancer Res.* **2010**, *70*, 266–276. [CrossRef]
158. Zhang, K.; Song, H.; Yang, P.; Dai, X.; Li, Y.; Wang, L.; Du, J.; Pan, K.; Zhang, T. Silencing dishevelled-1 sensitizes paclitaxel-resistant human ovarian cancer cells via AKT/GSK-3beta/beta-catenin signalling. *Cell Prolif.* **2015**, *48*, 249–258. [CrossRef]
159. Zhang, C.; Zhang, Z.; Zhang, S.; Wang, W.; Hu, P. Targeting of Wnt/beta-Catenin by Anthelmintic Drug Pyrvinium Enhances Sensitivity of Ovarian Cancer Cells to Chemotherapy. *Med. Sci. Monit.* **2017**, *23*, 266–275. [CrossRef]
160. Thorne, C.A.; Hanson, A.J.; Schneider, J.; Tahinci, E.; Orton, D.; Cselenyi, C.S.; Jernigan, K.K.; Meyers, K.C.; Hang, B.I.; Waterson, A.G.; et al. Small-molecule inhibition of Wnt signaling through activation of casein kinase 1alpha. *Nat. Chem. Biol.* **2010**, *6*, 829–836. [CrossRef]
161. Arend, R.C.; Londono-Joshi, A.I.; Gangrade, A.; Katre, A.A.; Kurpad, C.; Li, Y.; Samant, R.S.; Li, P.K.; Landen, C.N.; Yang, E.S.; et al. Niclosamide and its analogs are potent inhibitors of Wnt/beta-catenin, mTOR and STAT3 signaling in ovarian cancer. *Oncotarget* **2016**, *7*, 86803–86815. [CrossRef] [PubMed]
162. Arend, R.C.; Londono-Joshi, A.I.; Samant, R.S.; Li, Y.; Conner, M.; Hidalgo, B.; Alvarez, R.D.; Landen, C.N.; Straughn, J.M.; Buchsbaum, D.J. Inhibition of Wnt/beta-catenin pathway by niclosamide: A therapeutic target for ovarian cancer. *Gynecol. Oncol.* **2014**, *134*, 112–120. [CrossRef] [PubMed]
163. King, M.L.; Lindberg, M.E.; Stodden, G.R.; Okuda, H.; Ebers, S.D.; Johnson, A.; Montag, A.; Lengyel, E.; MacLean Ii, J.A.; Hayashi, K. WNT7A/beta-catenin signaling induces FGF1 and influences sensitivity to niclosamide in ovarian cancer. *Oncogene* **2015**, *34*, 3452–3462. [CrossRef] [PubMed]
164. Barnard, M.E.; Poole, E.M.; Curhan, G.C.; Eliassen, A.H.; Rosner, B.A.; Terry, K.L.; Tworoger, S.S. Association of Analgesic Use With Risk of Ovarian Cancer in the Nurses' Health Studies. *JAMA Oncol.* **2018**, *4*, 1675–1682. [CrossRef] [PubMed]

Journal of
Clinical Medicine

Review

Comparison of the Genomic Profile of Cancer Stem Cells and Their Non-Stem Counterpart: The Case of Ovarian Cancer

Elena Laura Mazzoldi [1,†,‡], Anna Pastò [1,†,§], Giorgia Pilotto [1], Sonia Minuzzo [2], Ilaria Piga [1,2], Pietro Palumbo [3], Massimo Carella [3], Simona Frezzini [4], Maria Ornella Nicoletto [4], Alberto Amadori [1,2] and Stefano Indraccolo [1,*]

[1]   Immunology and Molecular Oncology Unit, Veneto Institute of Oncology IOV - IRCCS, 35128 Padova, Italy; elena.mazzoldi@libero.it (E.L.M.); annapasto.phd@gmail.com (A.P.); giorgia1210@hotmail.it (G.P.); ilaria.piga1992@gmail.com (I.P.); albido@unipd.it (A.A.)
[2]   Department of Surgery, Oncology and Gastroenterology, University of Padova, 35128 Padova, Italy; soniaanna.minuzzo@unipd.it
[3]   Medical Genetics Unit, Fondazione IRCCS Casa Sollievo della Sofferenza, 71013 San Giovanni Rotondo, Italy; p.palumbo@operapadrepio.it (P.P.); m.carella@operapadrepio.it (M.C.)
[4]   Medical Oncology 2, Veneto Institute of Oncology IOV - IRCCS, 35128 Padova, Italy; simona.frezzini@iov.veneto.it (S.F.); ornella.nicoletto@iov.veneto.it (M.O.N.)
*   Correspondence: stefano.indraccolo@unipd.it; Tel.: +390498215875
†   These authors contributed equally to this work.
‡   Current address: Department of Molecular and Translational Medicine, University of Brescia, 25123 Brescia, Italy.
§   Current address: Department of Inflammation and Immunology, Humanitas Clinical and Research Center, Rozzano, 20089 Milan, Italy.

Received: 20 December 2019; Accepted: 26 January 2020; Published: 29 January 2020

Abstract: The classical cancer stem cell (CSC) model places CSCs at the apex of a hierarchical scale, suggesting different genetic alterations in non-CSCs compared to CSCs, since an ill-defined number of cell generations and time intervals separate CSCs from the more differentiated cancer cells that form the bulk of the tumor. Another model, however, poses that CSCs should be considered a functional state of tumor cells, hence sharing the same genetic alterations. Here, we review the existing literature on the genetic landscape of CSCs in various tumor types and as a case study investigate the genomic complexity of DNA obtained from matched CSCs and non-CSCs from five ovarian cancer patients, using a genome-wide single-nucleotide polymorphism (SNP) microarray.

Keywords: ovarian cancer; cancer stem cell; genetic heterogeneity; SNP array

---

1. Cancer Stem Cell (CSC) Theory

It has been established and it is well accepted that tumors are composed of heterogeneous cell populations. According to the classical hierarchical model, tumors arise from a small population of cells, called CSCs, derived from the malignant transformation of a normal stem cell. One of the main features of stem cells is their unlimited proliferative potential to sustain renewal and repair needs of normal tissues after injury. However, following exposure to environmental carcinogens or due to stochastic effects, normal stem cells can accumulate genetic mutations in cancer-associated genes (e.g., oncogenes and onco-suppressor genes), as well as defects in the DNA repair machinery. These mutated stem cells can become resistant to apoptosis and undergo malignant transformation in CSCs [1]. According to this model, CSCs inherit all the key features from their normal stem counterparts, including the unlimited proliferation rate and self-renewal capability. This implies that CSCs can accumulate a number of genetic or epigenetic changes, subsequently inherited by tumor cells derived from CSC asymmetric division and differentiation (Figure 1A, left panel) [2,3]. Due to the genetic heterogeneity of cancer, it is likely that the mutational burden of CSCs varies among different tumors, as occurs in non-CSCs. On the other hand, other studies indicate that CSCs develop mechanisms to lower radical oxygen species (ROS) accumulation and to extrude drugs, thus reducing the risk of damage to their genetic content [4]. These features let us speculate a reduced number of genetic alterations in CSCs compared to non-CSCs (Figure 1A, right panel). Moreover, non-CSCs could possibly accumulate private mutations—defined as mutations not shared by CSCs—during subsequent rounds of proliferation of trans-amplifying cells and bulk tumor cells (Figure 1A).

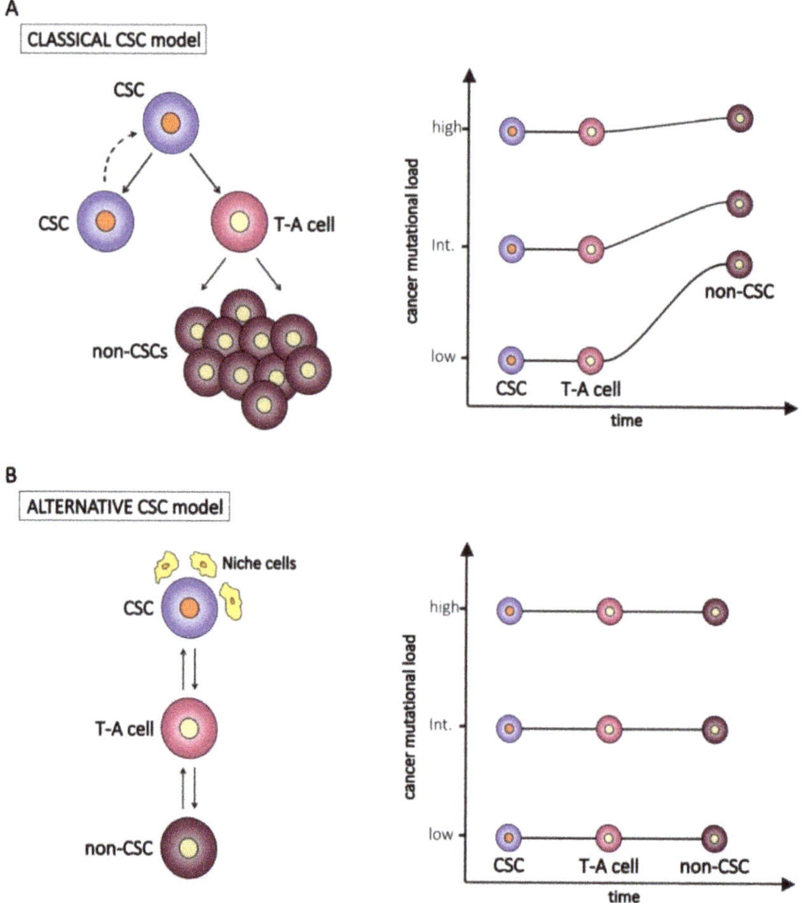

Figure 1. The classical (**A**) and the alternative (**B**) "Cancer Stem Cell (CSC)" models and hypothetical implications for the tumor mutational burden of transient-amplifying (T-A) and bulk tumor cells (non-CSCs). In the right panels, examples of CSC with different levels of genetic alterations (low, intermediate, high) are represented.

In contrast to the hierarchical model, an alternative theory poses CSC as a functional state of a tumor cell (reviewed in [5]). According to this alternative model (Figure 1B, left panel), stemness can not only be a cellular intrinsic property but also the result of extrinsic stimuli and of the cross-talk between CSCs and the complex network of cells, matrices, and vesicles that cooperate to maintain a tumor permissive microenvironment [6]. For example, in a study on breast cancer in which cell lines were used as models, only one subpopulation of cells out of three (stem cell-like, basal-like, and luminal-like cells) was tumorigenic *in vivo*, whereas in another setting, in which irradiated stromal cells were provided, all three considered subpopulations where equally tumorigenic [7]. Within the tumor microenvironment, CSCs, as their normal adult counterpart, reside in specialized niches, since they need to receive cues from the surrounding cells in order to activate and maintain their stemness program (self-renewal, proliferation and apoptosis resistance) and stemness regulating pathways (i.e., Notch, Wnt/β-catenin and Hedgehog). The niche consists of different cell types, including tumor-associated fibroblasts (TAFs), endothelial cells, pericytes, and immune cells, especially tumor-associated macrophages (TAMs) and myeloid-derived suppressor cells (MDSCs), as well as non-cellular elements, such as the extracellular matrix and the complex network of growth factors and cytokines [6,8]. All these components regulate stemness in different ways, i.e., by secreting cytokines which activate signaling pathways and stimulate self-renewal [6], or by promoting epithelial-to-mesenchymal transition [9].

According to this alternative model, the genetic profile of CSCs and non-CSCs is predicted to be quite similar (Figure 1B, right panel).

2. Genetic Profile of Cancer Stem Cells Versus the Bulk of Tumor Cells

In the last decade, CSCs have been identified in most tumor types (both solid and liquid) based on the expression of specific surface markers, lack of certain differentiation markers and high tumorigenic potential. In addition, CSCs have the ability to grow in vitro under serum-free conditions as rounded structures called spheroids and undergo asymmetric cell division, thus originating one daughter cell (maintaining stemness features) and one differentiated cell.

To date, only a few attempts have been made to describe the cytogenetic complexity of CSCs, mainly comparing their karyotype to that of non-CSCs (Table 1). Most studies analyzed either tumor cell lines or a small number of primary samples, due to the difficulty of obtaining enough genomic DNA from CSCs. Moreover, CSC isolation was often not performed according to stemness marker expression, but rather by using in vitro specific cell culture protocols. Along this line, Gasparini and colleagues performed a complete cytogenetic characterization of sphere-growing stem-like cells from six different cell lines of melanoma, breast, lung, and ovarian cancer [10].

Results indicated a more rearranged karyotype of CSCs, compared to the parental cell lines maintained under standard adhesion culture conditions. CSCs showed a higher number and complexity of chromosomal alterations in all cell lines evaluated. One bias of this study, acknowledged by the authors, was the use of immortalized cell lines and in vitro culture techniques for CSC isolation and expansion. Indeed, CSC-enriched spheroids are mainly composed of highly dividing progenitors that could affect results of genetic analysis. In any case, the main finding of this study was that growth culture conditions affect the genetic landscape of tumor cells.

In another study, Lee and colleagues demonstrated that glioblastoma (GBM) cells from primary tumor samples, cultured in serum-free spheroid-forming conditions, harbored extensive genotype similarity to parental tumor cells [13]. In contrast, when cultured in serum-supplemented media, glioblastoma cells underwent genomic rearrangements in terms of loss of heterozygosity (LOH), pseudo-tetraploidy and chromosomal deletions, suggesting a key role of in vitro culture conditions in the acquisition of genetic instability. Unfortunately, it is unknown whether these in vitro findings can be relevant to CSCs from patients.

Few other studies were performed in different tumor types, including GBM, breast cancer, head-and-neck squamous cell carcinoma (HNSCC), and bladder cancer, in order to compare either FACS-sorted CSCs or spheroid-forming cells with either their non-CSC counterpart or the tumor bulk (Table 1). Piccirillo et al. analyzed 12 primary GBM tumor sample-derived neurospheres by single nucleotide polymorphism (SNP) array and found that four out of 12 presented at least three copy number alterations (CNAs). Then, the authors analyzed one of these samples in detail, by using whole-exome sequencing (WES) and single-cell analysis. They concluded that the selected somatic mutations and CNAs highlighted in the primary tumor were also present in the derived neurospheres. Moreover, some genetic alterations were subclonal and correlated with different tumorigenic potential of single tumor cells in mice. Thus, conclusions suggest that CSCs and bulk tumor cells share the same genetic alterations but GBM presents substantial intra-tumor genetic heterogeneity [11].

Table 1. Studies on genetic alterations in CSCs.

Cancer Type	Materials	Methods	Main Conclusions	Reference
GBM	Primary tumor cells	SNP array, WES, single-cell analysis	CNAs found in primary tumor cells are also present in neurospheres in different subclones	Piccirillo S.M. et al., 2015 [11]
GBM	Primary tumor cells	aCGH	High similarity between GBM mass and spheroids	Pesenti C. et al., 2019 [12]
GBM	Primary tumor cells	SKY; SNP array	Cells cultured in serum-containing medium underwent genomic rearrangements, while spheroids did not	Lee J. et al., 2006 [13]
Breast	Primary tumor cells	WES; ultra-deep amplicon sequencing	Mutations are shared between tumor bulk and spheres	Klevebring D. et al., 2014 [14]
Breast	Tumor cells from pleural effusions	low-coverage WGS	Same alteration in sorted CSCs and bulk tumor	Tiran V. et al., 2017 [15]
Breast	MDA-MB-231 cell line	WGS; target deep sequencing	No differences in VAF between monolayer and spheres	Tong M. et al., 2018 [16]
HNSCC	Primary tumor cells	WES	From LOH analysis, it is hypothesized that CSCs may originate either from normal tissue or from tumor cell dedifferentiation	Salazar-Garcia L. et al., 2018 [17]
Bladder	One primary tumor and lymph node metastases	WES	SNPs are mainly shared by sorted CSCs and bulk tumor cells; a small number is enriched either in CSCs or in bulk cells	Prado K. et al., 2017 [18]
Various	Cell lines	SKY	More rearranged genotype of spheres compared to parental cell lines	Gasparini P. et al., 2010 [10]

Abbreviations: CSCs, cancer stem cell; GBM = glioblastoma multiforme; SNP = single nucleotide polymorphism; WES = whole exome sequencing; can = copy number alteration; aCGH = array comparative genomic hybridization; SKY = spectral karyotyping imaging; WGS = whole genome sequencing; VAF = variant allele frequency; HNSCC = head and neck squamous cell carcinoma; LOH = loss of heterozygosity.

In an independent study, Pesenti et al. obtained spheroids from three out of 10 primary GBM tumor samples and analyzed them by array comparative genomic hybridization (aCGH). In each case, they observed that tumor cells and their derived spheres shared the same genetic alterations [12], in line with previous findings [19]. Altogether, all these studies demonstrated that neurospheres share mutations and alterations with parental GBM cells.

Three other studies focused on breast cancer. Kleverbring and colleagues compared bulk tumor cells and spheroids in ten primary samples, sorted CD44$^+$/CD24$^-$ and CD44$^-$/CD24$^-$ cells, and sorted aldehyde dehydrogenase positive (ALDH$^+$) and negative (ALDH$^-$) cells in two additional samples, by WES, and validated their findings by ultra-deep amplicon sequencing [14]. In all the analyses, the Authors observed that CSCs and non-CSCs shared most mutations; thus, they concluded that the existence of CSCs and non-CSCs is the result of a continuous dynamic transition between a stem- and a non-stem functional state due to cell plasticity, rather than being distinct cell populations irreversibly characterized by different genomic landscapes.

Tiran et al. compared FACS-sorted CD44$^+$/CD24$^-$ and ALDH$^+$ cells with bulk tumor cells, obtained from pleural effusions, by low-coverage whole genome sequencing (WGS), thus demonstrating that FACS-sorted CSCs and the bulk tumor shared the same alterations [15].

Tong and colleagues compared, by WGS, MDA-MB-231 cells maintained either in adherent culture conditions or in spheroid-forming conditions and validated their findings by target deep DNA sequencing [16]. They demonstrated that the observed SNVs were characterized by a similar allelic frequency in both cell culture conditions, thus indicating that no variant was specifically associated with the stem status in their experimental model. Rather, spheroid-forming cells distinguished themselves from the adherent counterpart for a distinct gene expression profile.

Once again, putative breast CSCs, either sorted on the basis of surface markers or enzymatic activity or enriched by serum-free conditions, seem to be genetically similar to the bulk tumor cells, even though they differ for the functional properties or their transcriptomic profile.

Finally, sporadic studies reported the CSC genomic profile of other tumor types. Salazar-Garcia et al. tried to rebuild the CSC evolutionary history by WES analysis of ALDH$^+$ and ALDH$^-$ cells FACS-sorted from four HNSCC primary tumors, as well as normal cells. By evaluating LOH, they hypothesized that in some patients CSCs derived from the neoplastic transformation of normal tissue, whereas in other patients they derived from dedifferentiation of tumor cells. Indeed, if a variant was present in normal cells and in CSCs in heterozygosity, but heterozygosity was lost in the differentiated tumor cells, it is likely, according to the authors, that tumor cells derived from CSCs, which derived from normal tissue. On the contrary, if a variant was present in heterozygosity in both normal and tumor cells, but not in CSCs, such cells probably derived from tumor cell dedifferentiation. Thus, in HNSCC, the authors observed a certain extent of genetic difference between CSCs and non-CSCs.

In one case of invasive urothelial bladder carcinoma, Prado et al. FACS-isolated CD44$^+$/CD49f$^+$/EpCAM$^+$ CSCs from both primary tumor and lymph node metastases and analyzed them by WES, comparing results with the bulk tumor cells and normal lymph nodes [18]. They found 51 SNVs, of which the majority was shared by CSCs and tumor cells. Only a small number were uniquely found in the bulk tumor cells or in CSCs. The authors concluded that the SNVs unique in bulk were the result of clonal evolution and the SNVs unique in CSCs belonged to a quiescent subpopulation that had not yet generated a progeny big enough to be detectable among the other bulk cells.

3. The Case of Ovarian Cancer

Ovarian cancer is the fifth most frequent female cancer and the most common cause of death from gynecological tumors [20]. Epithelial ovarian cancer (EOC) comprises almost 90% of all cases [21]. Even if several models have been proposed to explain EOC pathogenesis [22], origin of EOC is still debated. Indeed, although serous tubal intraepithelial carcinoma (STIC)—a non-invasive tumor formed preferentially in the distal fallopian tube epithelium—has been considered by the scientific community as a precursor of high-grade serous carcinoma, in many cases it is unclear whether advanced stage disease results from progression from an early stage.

Known risk factors for ovarian cancer include the number of ovulations [23], inflammatory conditions [23,24], factors of hormonal nature [25], and genetic predisposition. Indeed, women with Lynch syndrome and with germline mutations in BRCA-1 and BRCA-2 have increased lifetime risk of several cancers, including EOC [19].

As for other tumors, CSCs in ovarian cancer has been identified according to the expression of specific surface markers. One of the most widely used markers is CD133 or Prominin, introduced by Curley and colleagues who demonstrated the higher tumorigenic potential of CD133$^+$ cells, compared to CD133$^-$, isolated from primary samples of EOCs and injected into immunocompromised mice [26]. However, more recently, the use of CD133 has been debated and novel markers were added for CSCs identification, such as ALDH, a detoxifying enzyme that enables CSCs to survive chemotherapeutic drugs. Recently, CD44 and CD117 (c-kit) have been proposed by Zhang and colleagues as markers for CSCs in EOCs [27]. These results were also confirmed by Pastò et al., who demonstrated that CD44$^+$CD117$^+$ double positive cells presented all canonical features of CSCs including ability to grow as spheroids, expression of stemness-associated markers (e.g., Nanog, Sox2 and Oct4), expression of multidrug-resistant pumps involved in drug extrusion, and high tumorigenic potential when injected into immunocompromised mice [28].

With regard to ovarian cancer, it is unknown whether CSCs disclose a genetic fingerprint similar to non-CSCs. We consider this question important, as its answer can help to discriminate whether epithelial ovarian CSCs fit the standard CSC hierarchical model or whether they are to be considered a functional state of tumor cells, as has been advanced for other tumor types. This lack of knowledge motivated our choice to investigate in a pilot experiment the genomic complexity of DNA obtained from matched CSCs and non-CSCs from ovarian cancer patients using a genome-wide SNP microarray. We performed karyotype profile analysis on CSCs isolated from human primary cultures of EOC established from high grade serous ovarian cancer ascitic fluid samples. The study was approved by the local ethical committee. CSCs were FACS-sorted as CD44$^+$CD117$^+$ from the ascitic effusions of eight patients and the extracted DNA was analyzed by high-density SNP arrays (CytoScan® HD Array, Affymetrix, Santa Clara, California, USA) and compared to CD44$^+$CD117$^-$ cells (non-CSCs). The essential clinical features of these patients are summarized in Table 2.

Table 2. Clinical features of EOC patients involved in the study.

Sample	Histotype	Stage	Grade	Chemotherapy	CD117 enrichment *
49 III	Serous-papillary	3C	3	Yes	7.08
49 V	Serous-papillary	3C	3	Yes	2.41
84 IV	Serous-papillary	3C	3	Yes	4.02
98	Serous	3B	1	No	21.53
101	Serous	3C	3	Yes	4.37
106	Bilateral serous-papillary	4	3	No	5.99

* CD117 mRNA expression (fold change) between CD44$^+$CD117$^+$ and CD44$^+$CD117$^-$ FACS-sorted populations. This parameter is used to check by an orthogonal technique CSC enrichment after sorting, EOC, Epithelial ovarian cancer.

In three out of eight pairs of CSCs and non-CSCs analyzed quality of the DNA samples was too low and precluded comparison of the genetic fingerprints of the two subpopulations. In the remaining five pairs (49, 84, 98, 101, and 106), SNP analysis was successful and did not disclose any genetic difference between CSC and non-CSC DNA samples, with the exception of a mosaic rearrangement on chromosome 2 (arr[GRCh37] 2p21p11.2(45744391_84671244)× 2-3) detected in sample 84_CSC which was not detected in the matched 84_non-CSC sample. Additional genetic studies, such as mutation profiling of matched CSC and non-CSC samples could not be performed, due to limited amount of tumor gDNA. Finally, in the case of patient #49, it was possible to obtain two samples of ascitic fluid at a 12 month-interval (#49_III and #49_V) and SNP array analysis was performed in both cases confirming an identical genotype of the CSC and non-CSC subpopulations. Notably, although within each pair no or only marginal genetic differences were found, each tumor sample presented multiple genetic alterations in terms of LOH, deletions or amplifications, compared with normal genomes. The representative profile of one of these samples is shown in Figure 2.

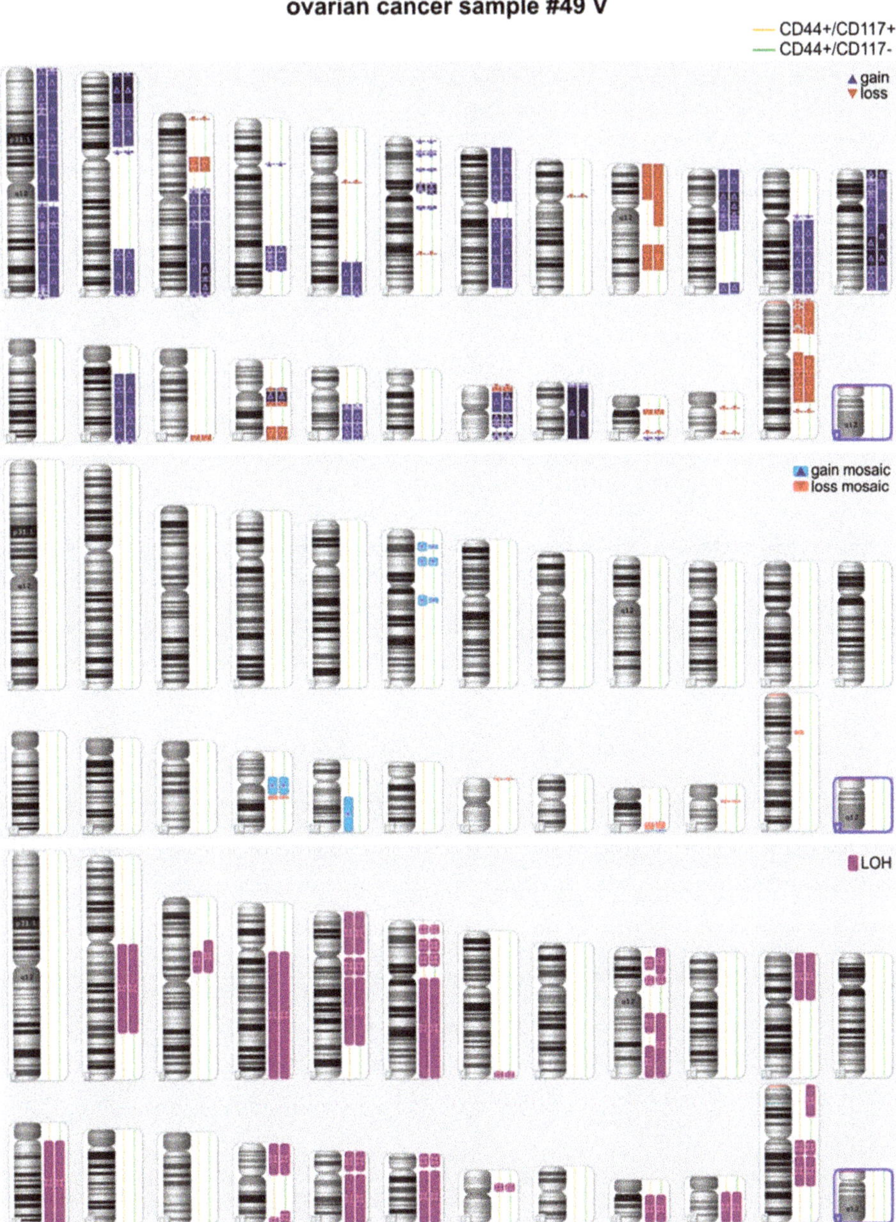

Figure 2. Comparison of chromosomal alterations in CSCs versus non-CSCs by Cytoscan SNP-array in one representative ovarian cancer patient (#49 V).

In summary, although the number of pairs analyzed was low, we found that patient-derived ovarian CSCs present very similar genetic features compared with non-CSCs. Therefore, it is likely that epigenetic differences account for the marked functional differences between these two sub-populations described elsewhere [28].

4. Conclusions

The classical CSC model places CSCs at the apex of a hierarchical scale, implying different genetic alterations in non-CSCs compared to CSCs, since a number of cell generations and time intervals separate CSCs from the more differentiated cancer cells that form the bulk of the tumor (Figure 1). In addition, CSCs seem to be endowed with more efficient DNA repair mechanisms, which partially shield their genome from genotoxic events. In this study, we tested this prediction in the case of ovarian cancer. Our results suggest that CSCs are genetically very similar to more differentiated cancer cells. Altogether, our findings agree with the majority of previous studies in other tumor types (Table 1) and support the alternative theory which poses CSC as a functional state of a tumor cell rather than a specific cell type.

How is this functional state of CSCs induced? Although a full answer to this question is currently not possible and is beyond the scope of this article, several intriguing hypotheses deserve to be mentioned. In a recent work, Canova described three stemness-associated genes: Nanog, Sox2, and Oct4 [29]. Each one controls the differentiation into a specific cell lineage, repressing the alternative; when co-expressed, they actually block differentiation in all lineages, thus promoting a stemness phenotype. We and others have found that ovarian CSCs co-expressed Nanog, Sox2, and Oct4, whereas expression of these genes is reduced or absent in tumor non-CSCs. Thus, stemness seems to be maintained because the differentiation pathways are blocked, rather than because the stemness pathways are activated. In addition, it has been demonstrated that some cells (i.e., hepatocytes and pancreatic islet cells) can de-differentiate under specific stimuli [30]. These results imply that tumor cells (non-CSCs) could potentially be forced to stemness by pressure from surrounding cells (i.e., within the tumor microenvironment) or stress conditions (i.e., anti-tumor drug treatment, metabolic substrate, or oxygen restriction). In order to survive, these cells block differentiation pathways and de-differentiate into CSCs, thus acquiring the ability to enter quiescence, overexpress multi-drug resistant pumps to extrude toxic compounds, or activate altered metabolic pathways, all canonical features of CSCs [28,31].

In conclusion, our results, albeit limited to a small number of cases, support the alternative CSC model shown in Figure 1 and suggest stemness as a dynamic state, a state of plasticity between tumor cells and CSCs. The genetic similarity of CSCs and non-CSCs should be taken into account in the development of successful new therapeutic approaches for ovarian cancer.

Author Contributions: Conceptualization, validation, writing—original draft preparation, and writing—review and editing: E.L.M. and A.P.; data curation: G.P.; methodology and writing—review and editing: S.M.; methodology and validation: I.P.; data curation and formal analysis: P.P. and M.C.; resources: S.F. and M.O.N.; writing—review and editing: A.A.; writing—original draft preparation, writing—review and editing, project administration, and funding acquisition: S.I. All authors have read and agreed to the published version of the manuscript.

Funding: This work was funded by AIRC (Grant n. IG18803 to S.I.).

Acknowledgments: Authors are grateful to Christina Drace for the English revision of this manuscript.

Conflicts of Interest: The authors declare no conflict of interest.

References

1. Vaish, M. Mismatch repair deficiencies transforming stem cells into cancer stem cells and therapeutic implications. *Mol. Cancer* **2007**, *6*, 26. [CrossRef]
2. Lagana, A.S.; Colonese, F.; Colonese, E.; Sofo, V.; Salmeri, F.M.; Granese, R.; Chiofalo, B.; Ciancimino, L.; Triolo, O. Cytogenetic analysis of epithelial ovarian cancer's stem cells: An overview on new diagnostic and therapeutic perspectives. *Eur. J. Gynaecol. Oncol.* **2015**, *36*, 495–505. [PubMed]
3. Tomao, F.; Papa, A.; Rossi, L.; Strudel, M.; Vici, P.; Lo Russo, G.; Tomao, S. Emerging role of cancer stem cells in the biology and treatment of ovarian cancer: Basic knowledge and therapeutic possibilities for an innovative approach. *J. Exp. Clin. Cancer Res.* **2013**, *32*, 48. [CrossRef] [PubMed]
4. Wang, Q.E. DNA damage responses in cancer stem cells: Implications for cancer therapeutic strategies. *World J. Biol. Chem.* **2015**, *6*, 57–64. [CrossRef] [PubMed]

5. Batlle, E.; Clevers, H. Cancer stem cells revisited. *Nat. Med.* **2017**, *23*, 1124–1134. [CrossRef] [PubMed]

6. Pattabiraman, D.R.; Weinberg, R.A. Tackling the cancer stem cells—What challenges do they pose? *Nat. Rev. Drug Discov.* **2014**, *13*, 497–512. [CrossRef]

7. Gupta, P.B.; Fillmore, C.M.; Jiang, G.; Shapira, S.D.; Tao, K.; Kuperwasser, C.; Lander, E.S. Stochastic state transitions give rise to phenotypic equilibrium in populations of cancer cells. *Cell* **2011**, *146*, 633–644. [CrossRef]

8. Sica, A.; Porta, C.; Amadori, A.; Pasto, A. Tumor-associated myeloid cells as guiding forces of cancer cell stemness. *Cancer Immunol. Immunother.* **2017**, *66*, 1025–1036. [CrossRef]

9. Nassar, D.; Blanpain, C. Cancer Stem Cells: Basic Concepts and Therapeutic Implications. *Annu. Rev. Pathol.* **2016**, *11*, 47–76. [CrossRef]

10. Gasparini, P.; Bertolini, G.; Binda, M.; Magnifico, A.; Albano, L.; Tortoreto, M.; Pratesi, G.; Facchinetti, F.; Abolafio, G.; Roz, L.; et al. Molecular cytogenetic characterization of stem-like cancer cells isolated from established cell lines. *Cancer Lett.* **2010**, *296*, 206–215. [CrossRef]

11. Piccirillo, S.G.M.; Colman, S.; Potter, N.E.; van Delft, F.W.; Lillis, S.; Carnicer, M.J.; Kearney, L.; Watts, C.; Greaves, M. Genetic and functional diversity of propagating cells in glioblastoma. *Stem Cell Rep.* **2015**, *4*, 7–15. [CrossRef] [PubMed]

12. Pesenti, C.; Navone, S.E.; Guarnaccia, L.; Terrasi, A.; Costanza, J.; Silipigni, R.; Guarneri, S.; Fusco, N.; Fontana, L.; Locatelli, M.; et al. The Genetic Landscape of Human Glioblastoma and Matched Primary Cancer Stem Cells Reveals Intratumour Similarity and Intertumour Heterogeneity. *Stem Cells Int.* **2019**, *2019*, 2617030. [CrossRef] [PubMed]

13. Lee, J.; Kotliarova, S.; Kotliarov, Y.; Li, A.; Su, Q.; Donin, N.M.; Pastorino, S.; Purow, B.W.; Christopher, N.; Zhang, W.; et al. Tumor stem cells derived from glioblastomas cultured in bFGF and EGF more closely mirror the phenotype and genotype of primary tumors than do serum-cultured cell lines. *Cancer Cell* **2006**, *9*, 391–403. [CrossRef] [PubMed]

14. Klevebring, D.; Rosin, G.; Ma, R.; Lindberg, J.; Czene, K.; Kere, J.; Fredriksson, I.; Bergh, J.; Hartman, J. Sequencing of breast cancer stem cell populations indicates a dynamic conversion between differentiation states in vivo. *Breast Cancer Res.* **2014**, *16*, R72. [CrossRef] [PubMed]

15. Tiran, V.; Stanzer, S.; Heitzer, E.; Meilinger, M.; Rossmann, C.; Lax, S.; Tsybrovskyy, O.; Dandachi, N.; Balic, M. Genetic profiling of putative breast cancer stem cells from malignant pleural effusions. *PLoS ONE* **2017**, *12*, e0175223. [CrossRef]

16. Tong, M.; Deng, Z.; Yang, M.; Xu, C.; Zhang, X.; Zhang, Q.; Liao, Y.; Deng, X.; Lv, D.; Zhang, Y.; et al. Transcriptomic but not genomic variability confers phenotype of breast cancer stem cells. *Cancer Commun. (Lond)* **2018**, *38*, 56. [CrossRef]

17. Salazar-Garcia, L.; Perez-Sayans, M.; Garcia-Garcia, A.; Carracedo, A.; Cruz, R.; Lozano, A.; Sobrino, B.; Barros, F. Whole exome sequencing approach to analysis of the origin of cancer stem cells in patients with head and neck squamous cell carcinoma. *J. Oral Pathol. Med.* **2018**, *47*, 938–944. [CrossRef]

18. Prado, K.; Zhang, K.X.; Pellegrini, M.; Chin, A.I. Sequencing of cancer cell subpopulations identifies micrometastases in a bladder cancer patient. *Oncotarget* **2017**, *8*, 45619–45625. [CrossRef]

19. Bartosch, C.; Clarke, B.; Bosse, T. Gynaecological neoplasms in common familial syndromes (Lynch and HBOC). *Pathology* **2018**, *50*, 222–237. [CrossRef] [PubMed]

20. Siegel, R.L.; Miller, K.D.; Jemal, A. Cancer statistics, 2018. *CA Cancer J. Clin.* **2018**, *68*, 7–30. [CrossRef]

21. Garces, A.H.; Dias, M.S.; Paulino, E.; Ferreira, C.G.; de Melo, A.C. Treatment of ovarian cancer beyond chemotherapy: Are we hitting the target? *Cancer Chemother Pharmacol.* **2015**, *75*, 221–234. [CrossRef]

22. Kuhn, E.; Kurman, R.J.; Shih, I.M. Ovarian Cancer Is an Imported Disease: Fact or Fiction? *Curr. Obstet. Gynecol. Rep.* **2012**, *1*, 1–9. [CrossRef] [PubMed]

23. Ozols, R.F.; Bookman, M.A.; Connolly, D.C.; Daly, M.B.; Godwin, A.K.; Schilder, R.J.; Xu, X.; Hamilton, T.C. Focus on epithelial ovarian cancer. *Cancer Cell* **2004**, *5*, 19–24. [CrossRef]

24. Ness, R.B.; Cottreau, C. Possible role of ovarian epithelial inflammation in ovarian cancer. *J. Natl. Cancer Inst.* **1999**, *91*, 1459–1467. [CrossRef] [PubMed]

25. Pearce, C.L.; Chung, K.; Pike, M.C.; Wu, A.H. Increased ovarian cancer risk associated with menopausal estrogen therapy is reduced by adding a progestin. *Cancer* **2009**, *115*, 531–539. [CrossRef] [PubMed]

26. Curley, M.D.; Therrien, V.A.; Cummings, C.L.; Sergent, P.A.; Koulouris, C.R.; Friel, A.M.; Roberts, D.J.; Seiden, M.V.; Scadden, D.T.; Rueda, B.R.; et al. CD133 expression defines a tumor initiating cell population in primary human ovarian cancer. *Stem Cells* **2009**, *27*, 2875–2883. [CrossRef] [PubMed]
27. Kryczek, I.; Liu, S.; Roh, M.; Vatan, L.; Szeliga, W.; Wei, S.; Banerjee, M.; Mao, Y.; Kotarski, J.; Wicha, M.S.; et al. Expression of aldehyde dehydrogenase and CD133 defines ovarian cancer stem cells. *Int. J. Cancer* **2012**, *130*, 29–39. [CrossRef]
28. Pasto, A.; Bellio, C.; Pilotto, G.; Ciminale, V.; Silic-Benussi, M.; Guzzo, G.; Rasola, A.; Frasson, C.; Nardo, G.; Zulato, E.; et al. Cancer stem cells from epithelial ovarian cancer patients privilege oxidative phosphorylation, and resist glucose deprivation. *Oncotarget* **2014**, *5*, 4305–4319. [CrossRef]
29. Casanova, J. Stemness as a cell default state. *EMBO Rep.* **2012**, *13*, 396–397. [CrossRef]
30. Zipori, D. The stem state: Mesenchymal plasticity as a paradigm. *Curr. Stem Cell Res. Ther.* **2006**, *1*, 95–102. [CrossRef]
31. Pagotto, A.; Pilotto, G.; Mazzoldi, E.L.; Nicoletto, M.O.; Frezzini, S.; Pasto, A.; Amadori, A. Autophagy inhibition reduces chemoresistance and tumorigenic potential of human ovarian cancer stem cells. *Cell Death Dis.* **2017**, *8*, e2943. [CrossRef] [PubMed]

Review

Implication for Cancer Stem Cells in Solid Cancer Chemo-Resistance: Promising Therapeutic Strategies Based on the Use of HDAC Inhibitors

Maria Serena Roca, Elena Di Gennaro and Alfredo Budillon *

ExperimentalPharmacology Unit, Istituto Nazionale Tumori–IRCCS–Fondazione G. Pascale, 80131 Naples, Italy
* Correspondence: a.budillon@istitutotumori.na.it; Tel.: +39-081-5903292

Received: 17 May 2019; Accepted: 20 June 2019; Published: 26 June 2019

Abstract: Resistance to therapy in patients with solid cancers represents a daunting challenge that must be addressed. Indeed, current strategies are still not effective in the majority of patients; which has resulted in the need for novel therapeutic approaches. Cancer stem cells (CSCs), a subset of tumor cells that possess self-renewal and multilineage differentiation potential, are known to be intrinsically resistant to anticancer treatments. In this review, we analyzed the implications for CSCs in drug resistance and described that multiple alterations in morphogenetic pathways (i.e., Hippo, Wnt, JAK/STAT, TGF-β, Notch, Hedgehog pathways) were suggested to be critical for CSC plasticity. By interrogating The Cancer Genome Atlas (TCGA) datasets, we first analyzed the prevalence of morphogenetic pathways alterations in solid tumors with associated outcomes. Then, by highlighting epigenetic relevance in CSC development and maintenance, we selected histone deacetylase inhibitors (HDACi) as potential agents of interest to target this subpopulation based on the pleiotropic effects exerted specifically on altered morphogenetic pathways. In detail, we highlighted the role of HDACi in solid cancers and, specifically, in the CSC subpopulation and we pointed out some mechanisms by which HDACi are able to overcome drug resistance and to modulate stemness. Although, further clinical and preclinical investigations should be conducted to disclose the unclear mechanisms by which HDACi modulate several signaling pathways in different tumors. To date, several lines of evidence support the testing of novel combinatorial therapeutic strategies based on the combination of drugs commonly used in clinical practice and HDACi to improve therapeutic efficacy in solid cancer patients.

Keywords: cancer stem cells; solid cancer; chemo-resistance; HDAC inhibitors

---

1. Introduction

Drug resistance is a well-known phenomenon that arises when a disease becomes tolerant to treatment. This concept was described first in bacteria when they became resistant to antibiotics, but since then, this phenomenon has been observed in other diseases, including cancer. Although many types of cancers are initially susceptible to antitumor approaches including chemotherapy and target-therapy and immunotherapy, over time, cancers can develop resistance through several mechanisms, such as DNA mutations or metabolic changes that promote drug inhibition and degradation [1]. Solid tumors are biologically complex structures with strong intratumor heterogeneity that arises among cancer cells within the same tumor as a consequence of genetic changes, environmental differences, and epigenetic and reversible changes in cell features [2]. Two main conceptual frameworks have been elaborated to conceptualize the link between intratumor heterogeneity and therapy resistance [3]. The first and most supported idea is clonal evolution, where a single mutated cell creates a tumor and over time acquires additional mutations, resulting in several subpopulations with evolutionary advantages [4]. An example of this comes from the analysis of circulating tumor DNA in the blood of colorectal cancer (CRC) patients with primary or acquired resistance to epidermal growth factor receptor (EGFR) blockade [5]. Siravegna and colleagues exploited the circulating tumor DNA to genotype colorectal tumors and tracked clonal evolution during treatment with EGFR-specific antibodies, discovering that the percentage of mutated KRAS clones declines in blood when EGFR-specific antibodies are withdrawn. This result suggests that resistant cell populations are highly dynamic and that specific resistant clones arise in specific therapeutic conditions [5].

The second concept is the cancer stem cell (CSC) model. In the last decade, several lines of evidence have suggested the presence of CSCs within the plethora of heterogeneous cells in solid cancers. The CSC paradigm implies that the tumor is organized into a hierarchy of subpopulations of tumorigenic CSCs and their non-tumorigenic progeny [6]. Among other described CSC features, self-renewal potential and the capability to generate progenitor/daughter cells with various degrees of differentiation make CSCsresponsible for tumor heterogeneity, drug resistance and tumor relapse (Figure 1) [7]. However, although several papers have identified CSCs as being responsible for drug resistance, the evidences are based on the identification and characterization of CSCs made with numerous non-homogeneous experiments that are difficult to compare. Thus, to confirm this theory, different approaches (*in vitro*, *in vivo* and *in silico*) must be implemented. Moreover, understanding the features and the complex signaling mechanisms that underlie the CSC state is a key point in highlighting new possible therapeutic strategies to target CSCs and to overcome resistance (Figure 1).

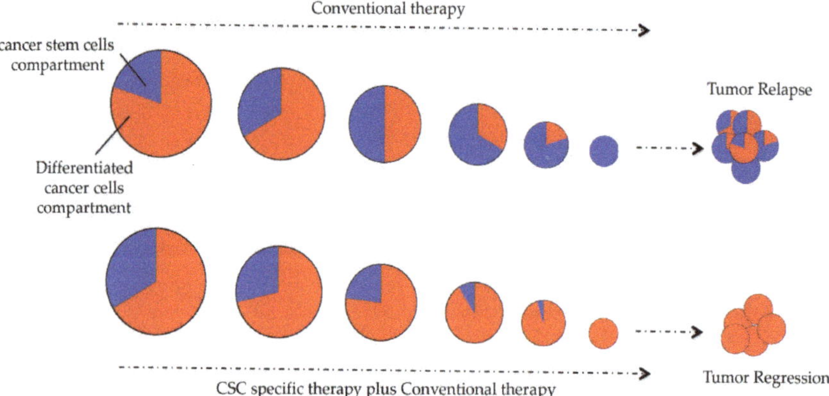

Figure 1. Cancer stem cells (CSCs) model of drug resistance:CSCs are responsible for tumor heterogeneity, drug resistance and tumor relapse. Indeed, they may survive chemotherapy lead to tumor relapse. Only by taking advantages of a CSCs specific targeted therapy, the outcome could result in tumor regression and patients' complete survival.

To date, it is widely recognized that alterations of the "epigenome" are therapeutically relevant as well as DNA mutations. Indeed, in contrast to DNA mutations, "epimutations" must be actively maintained through DNA replication for their dynamic nature; thus, their functional effects are reversible and, consequently, targetable [8,9].

Increasing evidence supports the significance of the epigenetic regulation of CSCs' features [10]. DNA methylation as well as histone acetylation are two epigenetic modifications that participate in the modulation of expression of many genes, regulating important cellular activities such as proliferation, differentiation and migration. While the role of DNA methylation in CSCs is relatively well established, the role of histone and non-histone protein acetylation is still not completely clear.

It has been suggested that, globally, hypoacetylated chromatin is associated with cancer [11]. In detail, as reviewed by Liu et al. [12], the dysregulation of two classes of enzymes, histone acetyltransferases (HAT) and histone deacetylases (HDAC), implicated in the regulation of acetylation levels is related to carcinogenesis and the regulation of the stemness properties of both normal and cancer cells. Thus, acetylation and deacetylation of histones and non-histone proteins regulate important signaling pathways involved in the maintenance of cancer stem-like cell traits such as self-renewal or differentiation.

In invasive breast cancer patients, Sulaiman et al. recently observed a direct link between HDAC expression and CSCs. By analyzing 887 samples, they found that high Wnt and HDAC activity was associated with estrogen receptor 1 (ESR1) and progesterone receptor (PGR) repression, poor survival and increased relapse. Interestingly, clinically achievable doses of Wnt, HDAC, and ESR1 inhibitors were able to inhibit both bulk and CSC subpopulations, inducing the differentiation of CSCs in non-CSCs without affecting normal mammary epithelial cells (MCF-10A) [13]. The overexpression of individual HDACs has been correlated to predict poor patients' prognosis independently of tumor type and disease stage and a role of specific HDACs has been also reported in CSCs. The genetic knockdown of individual isoforms such as HDAC1, 2, 3 and 6 have been shown to induce cell cycle arrest and apoptosis in several tumor types including breast, lung, colon cancer and leukemia as review by West et al. [14]. Witt et al. [15] taking advantage of two independent pairs of genetically matched immortalized breast cancer cell lines, derived one from normal human breast precursor epithelial cells with a mixed luminal-myoepithelial phenotype resembling CSC characteristics, and the one from normal human mammary epithelial cells that exhibit a more differentiated myoepithelial phenotype, demonstrating that HDAC1 and HDAC7 play an essential role in the stem-like phenotype, maintaining the CSC population in breast and ovarian cancer models. Another work reported that the levels of HDAC 1, 7 and 8 were overexpressed in pancreatic ductal adenocarcinoma (PDAC) compared to those of adjacent non-tumor tissues and that patients with high levels of HDAC 1, 7 or 8 had significantly worse overall survival compared to those with low expression levels [16]. A specific role of HDAC8 was also demonstrated in breast cancer, where a selective HDAC8 inhibitor suppressed Notch 1 expression [17]. Moreover, An et al. reported that, in triple negative breast cancer, HDAC8 induced cell migration by Hippo–Yap signaling. In detail, HDAC8 suppressed the phosphorylation of YAPSer127 to promote its nuclear localization [18].

In contrast, Zimberlin et al. taking advantage of a mouse model in which HDAC1 and HDAC2 were simultaneously deleted in the intestine of adult mice, showed a rapid loss of intestinal homeostasis and a decrease in stem-like features when both HDACs were deleted [19]. This latter observation suggests a rationale for the intestinal side effects often observed during treatment with HDAC inhibitors (HDACi). Similarly, Jamaladdin et al. by conditional knock-down in embryonic stem cellsdemonstrated the essential role of HDAC1 and 2 in cellular proliferation and stem cell self-renewal throughthe regulation of key pluripotent transcription factors [20]. Indeed, severe phenotypes were observed following specific deletion of both HDAC1 and HDAC2 in the heart, brain, smooth muscle and neuronal endocrine cells, suggesting a fundamental role for these HDACs in organ maintenance [21]. However, the hypothesis of positive or negative effect exerted by HDACi on normal stem cells is controversial. In this regard, Yusoff and colleagues published a systematic review about preclinical studies evaluating the organ protection effects of HDACi. They conclude that HDACi reduce mortality in experimental models by conferring multi-organ protection, often following a single treatment administered in some cases post injury, suggesting the design of early phase clinical trials in order to confirm the protective role exerted by these agents [22].

From this perspective, the increased dependency of tumor cells on HDACs and their upregulation during transformation could help identifying the therapeutic window's width of HDACi. At the same time, we should be aware of the potential toxicity induced by this class of agent.

Below, we provide an overview of promising therapeutic strategies based on HDACi, to sensitize CSCs to antitumor approaches including chemotherapy and target and immunotherapy. We first outlined the multiple alterations within morphogenetic pathways that show a critical role in CSC plasticity and are potentially targeted by HDACi. Then, considering all intrinsic and extrinsic features of CSCs responsible for chemo-toxicity escape, we discuss the capability of HDACi, alone or in combination, to overcome chemo-resistance.

2. Morphogenetic Pathways Are Dysregulated in Solid Cancers

As extensively demonstrated, stem cell proliferation and cell fate are under the control of several morphogenetic pathways. Among them, we focused on those that have been extensively characterized in cancer: Wnt/β-catenin, Notch, Hedgehog, TGF-β/BMP, JAK-STAT and Hippo pathways (Figure 2). These pathways can be altered in solid cancer and participate in phenomena such as drug resistance. However, these pathways undertake a different role depending on which solid cancer is considered. To identify specific morphogenetic pathways that are more relevant than others in specific solid tumors, we performed a bioinformatics analysis, taking advantage of 16 public solid tumor expression datasets (Table 1 reports The Cancer Genome Atlas (TCGA) Id, the sample size and the patient status for each dataset) by using the online tool R2 Genomics Analysis and Visualization Platform (https://hgserver1.amc.nl/cgi-bin/r2/main.cgi). R2 calculates, for all the genes in the differentiation-related KEGG pathways (718), whether they are differentially expressed considering overall survival, which is the best endpoint to consider to compare multiple cancer types with the goal of identifying common themes that transcend the tissue of origin and may inform precision oncology, as described in a recent publication by Liu et al. [23].

We are aware that our pan-cancer systematic analysis is performed across all TCGA tumor types without considering patients clinical features. In details, we did not considered age of diagnosis and grade, when available in TCGA datasets. Thus we anticipated that ouranalysis hassome limitations.

In a subsequent calculation, the overrepresentation of these genes in the individual pathways was determined. A valuewas reported depending on how many related genes were identified withineach pathway as compared with all 718 differentiation-KEGG pathways analyzed. In detail, we tested the significance of the 6 chosen morphogenetic pathways described above. In Table 1, the significance values are reported for the 6 pathways for each solid cancer, and the differences among the 6 pathways analyzed are represented by a color scale. From the resulting list, it is obvious that any specific pathway can be indicated as most relevant in the poor prognosis of patients for all cancer types. Notably, the Hippo pathway had a strong over-representation in patients with a poor prognosis among solid tumors (first in 7/16 tumors), playing a critical role in pancreatic, prostate, stomach adenocarcinomas, lung squamous and cervical cell carcinomas, sarcoma and skin cutaneous melanoma. Similarly, the Wnt/β-catenin pathway seems to play a very important role, representing the most important pathway in 4/16 cases, such as mixed colon adenocarcinoma, liver hepatocellular, head and neck and bladder urothelial carcinomas and third most-important pathway in 8/16 cases (Table 1). Jak/ STAT resulted as key pathway in rectum adenocarcinoma and glioblastoma, while TGFβ in lung adenocarcinoma, Notch in kidney renal clear cell carcinoma and finally Hedgehog in breast invasive cancers.

Table 1. Prevalence of CSCs-activated signaling pathways in solid tumors accordingly with outcome: Results obtained by R2 on line tool. Clustering the patients in dead and alive for each solid cancer, the values are indicative of how enriched in dead patients are the Hippo, Wnt, Jak/STAT, TGFb, Notch, Hedgehog pathways, based on gene expression data. The values are between 0 and 1 where 0 is the best over-representation. The color scale (from violet to white) groups the 6 CSCs pathways from the highest to the lowest enriched, among each TCGA (The Cancer Genome Atlas) dataset.

			Patient Status			CSCs-Activated Signaling Pathways					
Solid Cancer	TCGA_Id	Whole Sample	Alive	Dead	Hippo	Wnt	Jak/STAT	TGF	Notch	Hedgehog
Mixed Colon Adenocarcinoma	COAD	174	140	15	0.880	0.310	0.470	0.640	0.950	0.910
Rectum adenocarcinoma	READ	95	87	7	0.360	0.800	0.00041	0.150	1.000	0.360
Pancreatic adenocarcinoma	PAAD	178	119	59	0.270	0.330	0.430	0.550	0.390	0.520
Lung Adenocarcinoma	LUAD	515	389	126	0.900	0.300	0.400	0.210	0.800	0.420
Lung Squamous Cell Carcinoma	LUSC	81	61	19	0.390	0.530	0.610	0.440	0.580	0.620
Prostate Adenocarcinoma	PRAD	497	489	8	0.016	0.300	0.670	0.210	0.360	0.360
Stomach adenocarcinoma	STAD	415	336	79	0.260	0.410	0.500	0.310	0.460	0.520
Liver Hepatocellular Carcinoma	LIHC	371	282	89	0.840	0.510	0.580	0.680	0.550	0.680
Kidney Renal Clear Cell Carcinoma	KIRC	533	363	160	0.780	0.620	0.810	0.120	0.080	0.950
Head Neck Squamous Cell Carcinoma	HNSC	520	353	167	0.140	0.006	0.030	0.930	0.440	0.500
Cervical Squamous Cell Carcinoma	CESC	305	244	60	0.400	0.610	0.520	0.630	0.480	0.530
Bladder Urothelial Carcinoma	BLCA	408	300	108	0.260	0.100	0.500	0.310	0.440	0.520
Sarcoma	SARC	259	184	75	0.580	0.680	0.740	0.620	0.720	0.750
Breast Invasive Carcinoma	BRCA	1097	992	104	0.100	0.300	0.400	0.670	0.800	0.000008
Glioblastoma	TARGET_NBL	153	52	99	0.840	0.400	0.002	0.410	0.550	0.160
Skin Cutaneous Melanoma	SKCM	470	313	156	0.360	0.500	0.580	0.410	0.550	0.600

2.1. Wnt Signaling

Wnt signaling is one of the main pathways regulating development and stemness, and it has also been closely related to cancer. The canonical Wnt signaling pathway, which through β–catenin modulates the expression of specific target genes, is an important regulator of stem cells and CSCs and is aberrantly activated during the development of various human cancers as reviewed by Fodde et al. [24]. Gain-of-function mutations of the *CTNNB1* gene (encoding β–catenin) and loss-of function mutations of *APC* and *AXIN* genes were identified as the main mechanisms associated with Wnt signaling dysfunction in cancers [25]. The role of this pathway in carcinogenesis has been most prominently described for colorectal cancer (CRC). In the gut, the Wnt pathway is essential for sustaining cell proliferation within the crypt base and shows a gradient in expression along the crypt axis. Manipulation of Wnt signaling can drastically alter crypt integrity, while the stimulation of Wnt activity by, for instance, R-spondin leads to crypt proliferation [6], and its inhibition leads to the loss of crypt formation and progenitor offspring [7]. Moreover, the presence of Wnt in conjunction with other morphogenetic pathways is critical for maintaining normal and cancer cycling stem cells and Paneth cells, as reviewed by Reya and Clevers [26].

2.2. Hippo Pathway

In mammals, cell–cell junctions and apicobasal polarity are involved in upstream activation of the Hippo cascade. The core of the Hippo pathway is a kinase cascade: the nuclear result is that Lats1/2, nuclear dbf2-related family kinases, phosphorylate two major downstream effectors, Yes-associated protein (YAP) and transcriptional coactivator with PDZ-binding motif (TAZ), resulting in their ubiquitination and proteolysis [27]. Nguyen et al. showed that YAP1 induced transcriptional changes and resulted in age-related prostate tumors in mice [28]. Similarly, Zhang et al. showed a significant upregulation and hyperactivation of YAP in castration-resistant prostate tumors compared to their levels in hormone-responsive prostate tumors. They claimed that the enhanced expression of YAP was able to transform immortalized prostate epithelial cells and promote migration and invasion in both immortalized and cancerous prostate cells [29].

2.3. Stats

Stats are a family of transcription factors with additional functions in the cytoplasm, the mitochondria and the nucleus. They participate in chromatin conformation and epigenetic marking in the nucleus. In addition, they affect oxidative metabolism in the mitochondria and in the cytoplasm, and they interact with microtubule components to regulate cellular motility [30]. However, in our bioinformatics analysis, the JAK/STAT pathway was highly represented in glioblastoma and rectum adenocarcinoma (Table 1). Signal transducer and activator of transcription 3 (STAT3) is aberrantly activated in glioblastoma and has been identified as a relevant therapeutic target in this disease and many other human cancers [31]. Moreover, Gangulyand colleagues demonstrated high levels of STAT 3-phosphorylation in Tyrosine 705 and Serine 727 in glioma-initiating cells compared to their differentiating counterparts, suggesting a clear involvement of STAT pathway activation in gliomagenesis [32]. In addition, a strong link between IL-6-STAT3 signaling and O6-methylguanine DNA methyl transferase expression and methylation and to temozolomide sensitivity in glioblastoma was found, suggesting a possible combination therapeutic approach based on IL-6/STAT3 inhibitors [33,34]. Nevertheless, STAT3 is over-activated in many breast cancers, while STAT5 promotes both survival and differentiation of mammary epithelium. Moreover, it is known that, in the context of breast cancer, STAT3 activity can be modulated through STAT5 activity, and their combined functions can have an impact on breast cancer progression [35].

2.4. TGF-β Signaling

TGF-β signaling controls the cell cycle, differentiation and the microenvironment in epithelial cells through both the SMAD protein family and non-SMAD signaling pathways. In colon cancer, Yusra et al. showed that the expression levels of TGF-β receptor genes, TGFBR1 and TGFBR3, were higher in CD133 positive (+) colon CSCs, suggesting that CD133+ colon CSCs have increased sensitivity to TGF-β [36]. They showed high expression of the *TGFBR1* gene in the tumor budding structures derived from the mouse xenograft with TGF-β transfected cells. Importantly, Yu and colleagues showed that *TGFBR2* was responsible for triggering the TGF-β signaling pathway through recruitment and phosphorylation of Type I receptor, which has been shown to act as a tumor suppressor by assisting in the regulation of stemness through downregulated signaling effects of the Wnt/β-catenin signaling pathway [37].

2.5. Notch Pathway

Regarding the Notch pathway, the literature provides many data about its fundamental role in development and maintenance of the hematopoietic system. Less is available on its role in solid cancers. In hematopoiesis, Notch plays a central role in cell fate proliferation and differentiation of stem cells. There is also evidence demonstrating that the Notch pathway can leads hematopoietic stem cells and common lymphoid precursors to undergo T- or B-cell differentiation [38]. Moreover, Notch alteration has been associated with tumorigenesis, since it can be considered, depending on cell model, as an oncogene or a tumor suppressor [39].

2.6. Hedgehog Pathway

Finally, the Hedgehog (Hh) pathway is mainly involved in tissue repair, embryonic development, and epithelial-to-mesenchymal transition in cells. In the signaling cascade, Hh ligands, such as Sonic Hedgehog (Shh), Indian Hedgehog (Ihh), and Desert Hedgehog (Dhh), undergo cleavage and produce a signaling protein with dual lipid modifications. Subsequently, the signaling cascade initiated by smoothened (SMO) leads to activation and nuclear localization of GLI transcription factors, which drive the expression of Hh target genes, mostly involved in proliferation, survival, and angiogenesis [40]. This signaling transduction pathway has been shown to be required for self-renewal and proliferation of cerebellar, retinal, and pancreatic CSCs [41]. Recently, Duan et al. have shown that Hh signaling is upregulated in breast cancer cells through nuclear factor-κB (NF-κB) activation and promoter hypomethylation. In addition, overexpression of Shh enhances the self-renewal capacity and migration ability of breast cancer cells and is a poor prognostic indicator in breast cancer [42].

Because of a clear deregulation is described in multiple cancer types, various clinical approaches have been aimed to target the aforementioned pathways. Although several types of inhibitors of the Wnt-pathway are under development as single-agent anticancer therapies, they have not led to exciting results thus far. Indeed, even if PRI-724 (CBP/β-catenin antagonist), LGK-974 (Porcupine inhibitor), vantictumab (anti-Frizzled-1/2/5/7/8 antibody), OMP-54F28 (Frizzled-8–Fc decoy fusion protein) and TSA101 (a radiolabeled anti-Frizzled-10 antibody) are all currently in early clinical development, the results obtained in three concluded studies in monotherapy with PRI-724 (NCT01302405), vantictumab (NCT01345201) and OMP-54F28 (NCT01608867) are not very promising [42].

In solid tumors, several clinical trials are ongoing to test the efficacy of two major classes of Notch inhibitors such as γ-secretase inhibitors, which obstruct Notch receptor cleavage, and monoclonal antibodies that interfere with the Notch ligand-receptor interaction, as reviewed by Venkatesh et al. [38]. Thus far, the best results have been obtained in a clinical trial that ended in 2012, where a γ-secretase inhibitor (MK-0752) was used in adult patients with advanced solid tumors, mostly gliomas. Only one objective complete response and stable disease longer than 4 months in 10 patients were observed among 113 patients enrolled [43].

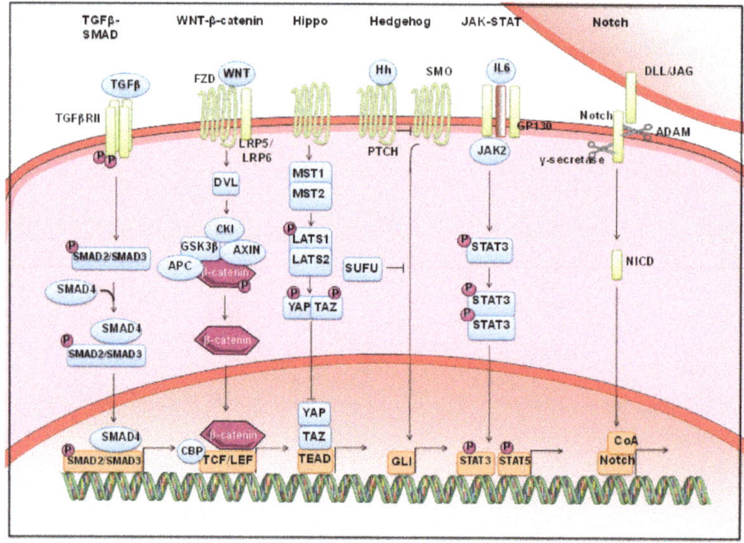

Figure 2. CSCs-activated signaling pathways: Schematic representation of the signaling pathway responsible for induction and maintenance of CSC state. The transcription factors, co-factors, main nodes reported are targetable by drugs. (Adapted from Pattabiraman and Weinberg, Nat Rev Drug Discovery, 2014 [44]).

Limited results have been obtained with vismodegib, an Hh pathway inhibitor, which was approved by the European Medicines Agency (EMA) in 2013 for the treatment of metastatic basal cell carcinoma (BCC) or locally advanced BCC in patients who are not candidates for surgery or radiotherapy. Indeed, the gain in overall survival in these patients compared to standard therapy, is only 0.8 years [42].

Currently, several other attempts are ongoing, targeting CSCs to deplete the tumor; however, to date, no specific CSC targeted therapy is yet available. In contrast, it may be more effective to use an approach based on pushing CSCs to differentiate and on sensitizing them to chemotherapy by modulating their epigenetic profile. Indeed, by highlighting the significance of the epigenetic profile in CSC maintenance, in the next sections we reviewed the role of HDACi to target CSCs, both directly and through the modulation of the morphogenetic pathways described above. Actually, by demonstrating that morphogenetic pathway alterations correlating with poor prognosis often coexist in a solid cancer type, we anticipated the limited efficacy of targeted agents inhibiting a single pathway as reported above. This could shed a new light on therapeutic agents with pleiotropic effect such as HDACi. Therefore, we next discussed how HDACi could play a role in overcoming resistance to some of the most commonly used chemotherapeutics and their ability to enhance specific cancer therapeutic agents by targeting CSCs.

3. Histone Deacetylase (HDAC) Inhibitors Target Cancer Stem Cells (CSCs) and Morphogenetic Pathways

It is well known that the morphogenetic pathways described above are frequently dysregulated in cancer by epigenetic mechanisms as reviewed by Munoz et al. [45]. Since trichostatin A (TSA), a natural product isolated from *Streptomyces hygroscopicus*, was described as an HDACi [46], several agents of this class of epigenetic drugs have been synthetized and tested as anticancer agents, both in preclinical and clinical models. HDACi are truly pleiotropic agents, which act through a wide variety of disparate and mutually interactive mechanisms. Indeed, by influencing the chromatin structure, HDACi regulate gene expression. Moreover, by acting on non-histone proteins deacetylation, HDACi can also modulate cellular functions independent of gene expression. In this way, HDACi regulate different altered pathways in cancer, such as apoptosis, DNA repair, growth arrest, terminal differentiation, senescence, apoptosis, anti-angiogenesis, anti-metastasis and immune responses, all of which represent hallmarks of cancer and critical characteristics of CSCs [11]. Finally, HDACi have a strong effect on CSCs by pushing them from a stem-like and resistant phenotype to a more differentiated and sensible phenotype (Figure 3).

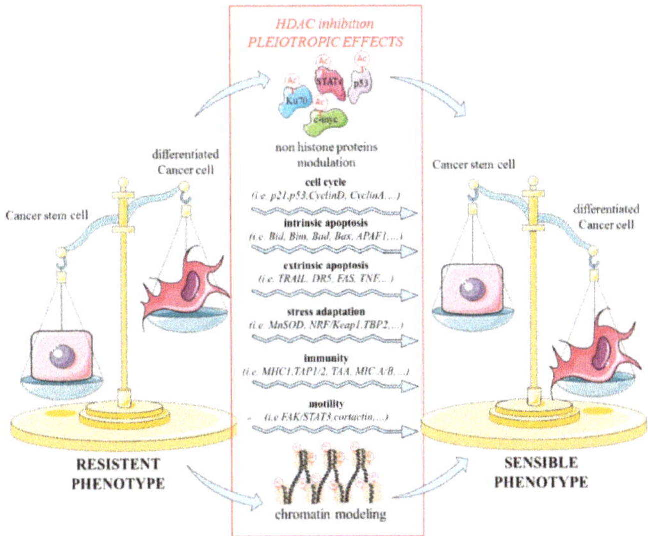

Figure 3. Histone deacetylase inhibitors (HDACi) pleiotropic effects toward CSCs: HDAC are described to modulate several pathways involved in cell cycle, metabolism, stress adaptation, intrinsic and extrinsic apoptosis, motility and immunity through chromatin modeling and non histone proteins modulation. The plasticity of epigenetic regulation makes HDAC inhibition a good strategy to target CSCs chemo-resistant subpopulation in solid cancer pushing the CSCs to differentiate and so gain a chemo-sensible phenotype.

In recent years, we demonstrated that HDACi, such as vorinostat and valproic acid (VPA), were able to induce selective death of the CSC subpopulation in non-small-cell lung cancer (NSCLC) and CRC models, as evaluated by colony and sphere formation assays [47,48].

Similar results were obtained by Salvador et al. who demonstrated that the HDACi abexinostat induced cytotoxicity in 16 breast cancer cell lines (BCLs), as evaluated by Aldefluor and tumorsphere formation assays. They found that abexinostat induced CSC differentiation, identifying the long non-coding RNA Xist (Xinactive specific transcript) as a biomarker that can predict the BCL response to HDACi [49]. In line with these findings, Cai et al. demonstrated that TSA or vorinostat, downregulating HDACs 1, 7 and 8, repressed epithelial to mesenchymal transition (EMT) in PDAC and targeted well-known CSC phenotypes, including resistance to therapy and metastasis. Notably, they also suggested that HDACi such as mocetinostat, which specifically inhibits isoforms 1, 2, 3 and 11 HDACs, were not good candidates for PDAC treatment [16]. Moreover, we obtained preliminary data showing that HDACi sensitized cancer cells to chemotherapeutics by pushing them in a differentiation state in colorectal and prostate cancer cell models by modulating morphogenetic pathways [50].

In addition to the preclinical studies that focused on HDACi and their functional effects on modulating CSC subpopulations, several attempts are aiming to explain the molecular mechanisms underlying HDACi effects of HDACi on them. We most likely believe that HDACi strongly modulate the CSC phenotype through their ability to modulate one or more morphogenetic pathway main components, as described above.

Similarly, several data support the close connection between HDAC6 and multiple components of the Wnt, Hippo, TGFβ and Hh pathways. Concerning the Wnt pathway, two papers showed that HDAC6 inhibition led to a decrease in β-catenin nuclear localization, resulting in a strong inhibition of cell proliferation [51,52]. In breast cancer, the involvement of HDAC6 in the Hippo pathway-regulating network was demonstrated by downregulation of YAP protein levels [53] and by a strong acetylation and destabilization of the tumor suppressor MST1 [54]. Some evidence suggests that HDAC6 plays a role in theTGFβ–SMAD activation pathway, and thus its inhibition establishes a decrease in EMT induction [55,56]. Finally, most recently, it has been shown that HDAC6 inhibition leads to inactivation of Gli1, resulting in glioma cancer cell radio-sensitization [57]. In the same manner, HDAC1 has been shownto be essential for SMAD2-3 and Gli1 transcriptional activation, consequently also quenching related pathways [58,59]. As expected, the molecular mechanism becomes complicated when we consider the effect of HDACi, which target more than one HDAC isoform and, simultaneously, more than one morphogenetic pathway.

For example, valproic acid (VPA), a class I- II HDAC inhibitor, is clearly able to inhibit TGF-β, Yap and Notch signaling by dual suppression of SMAD4 and SMAD3 [60], by depletion of nuclear YAP [61] and by downregulation of the transcription factor Notch1 and its target gene HES-1 [62]. Moreover, it has been reported that the pan-HDAC inhibitor trichostatin A (TSA) and the selective HDAC inhibitor entinostat (isoforms 1, 2 and 3) are able to reduce Wnt pathway activation by induction of DKK1, a negative regulator of β–catenin [63] and by inhibition of HDAC1-2 [19].

Finally, as VPA and TSA, vorinostat modulates SMAD4 localization and Nocth3 expression [64,65], but Fan and colleagues also showed that a small molecule SMO-HDAC antagonist was able to retain inhibitory activity for Gli transcription induced in SMO-dependent and SMO-independent ways [66]. Furthermore, TSA and domatinostat, two specific class I and II HDAC inhibitors, switched off Gli signaling by downregulating a transcriptional factor of FoxM1 [67] in the first case and by inhibition of HDAC1/2/3 in the second case [68]. However, it should be noted that some papers have reported that VPA treatment is responsible for Wnt pathway activation by β–catenin nuclear stabilization [13,69].

However, the exact function and interactions governing HDACi activity remain elusive. Thus, further investigationsare necessary to understand the mechanism by whichHDACi target CSCs. Only a clear view of their actions will lead to the identification of a biologically efficacious dose. Thus, it should always be considered that the effects of HDACi depend not only on the cancer type but also on the context, dosing and schedule of treatment. For example, the association of entinostat with the aromatase inhibitor exemestane has been designated as a breakthrough therapy for the treatment of recurrent/metastatic estrogen receptor-positive breast cancer based on the results of a phase II randomized trial [70]. In contrast, the clinical activity of mocetinostat, tested in phase I/II in association with gemcitabine in patients with solid tumors including pancreatic cancer, has been considered insufficient to merit further testing in this setting [71], as well as due to the significant toxicities in the phase II cohort [71].

4. CSC Chemo-Toxicity Escape Mechanisms

Cancer drug-resistance can be due to intrinsic or acquired factors. The intratumor heterogeneity of tumor cells and the tumor microenvironment and the presence of cancer stem cells are all intrinsic characteristics involved in cancer drug resistance. But latter, has also been observed in most drug-sensitive tumor types for most classes of drugs as an acquired mechanism.

Several mechanisms of cancer drug resistance are well described in both tumor cells and CSCs, and the most important and targetable ones, such as multidrug resistance [72], resistance to apoptosis program, increased repair of drug-induced DNA damage and quiescence phenotype induction [73], are presented schematically in Figure 4. In addition to the common and well-known mechanisms by which normal stem cells and CSCs overcome drug toxicity, some others have recently been described, such as metabolism adaptation or activation of the immune escape program, while others remain to be uncovered.

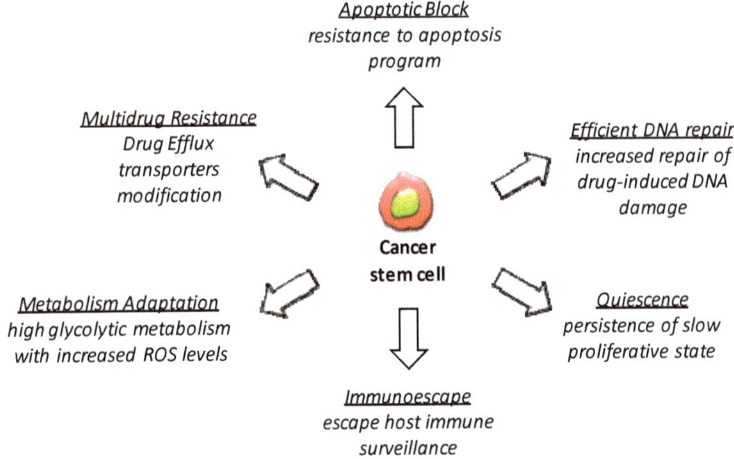

Figure 4. CSCs model of drug resistance: CSCs features mostly responsible for drug resistance.

The CSC-resistant phenotype is likely to be the result of a complex but specific mixture of molecular circuitries, and it is the complex nature of this phenomenon that explains the difficulties encountered in trying to overcome drug resistance by targeting this specific cell population.

4.1. Multidrug Resistance

Over 40 years ago, multidrug resistance, the high expression of drug efflux pumps, such as ATP-binding cassette (ABC) transporter family proteins, was the first described innate resistance mechanism. Overexpression of ABCB1 confers cancer cell resistance to multiple drugs (i.e., DNA-toxic antitumor agents, reactive oxygen species –ROS- inducers), including colchicine, doxorubicin, etoposide, vinblastine and paclitaxel [74]. The critical role of ABC transporters in CSC drug resistance was reported by Chau et al. in c-Kit+ ovarian CSCs. The authors demonstrated that Wnt/β-catenin regulates ABCG2 expression and resistance to cisplatin/paclitaxel. LEF/TCF binding sites are within the *ABCG2* gene promoter, indicating that ABCG2 is a transcriptional target of β-catenin [75].

Moreover, in a prospective clinical study of 142 CRC patients, Guo et al. found that ABCB5 mRNA transcripts are significantly enriched in patient peripheral blood specimens compared with those in non-CRC controls and correlate with CRC disease progression. Notably, ABCB5 regulates CRC invasiveness, at least in part by enhancing AXL receptor tyrosine kinase signaling [76]. Additionally, in glioblastoma and melanoma, high expression of drug efflux pumps, such as ABCG2 and ABCB5, is reported in CSCs [77,78].

4.2. Apoptosis

In addition to other mechanisms, CSC resistance to drug cytotoxicity is commonly associated with intrinsic or acquired defects and/or inefficient signaling in either the extrinsic or intrinsic pathway of apoptosis. Regarding the extrinsic pathway, lung, pancreatic, breast and glioma CSC models exhibit resistance to TRAIL-induced apoptosis by genetic and epigenetic silencing of pro-apoptotic factors, such as caspase 8 or c-FLIP, or by upregulation of TRAIL-R2 receptor, as reviewed by Fulda [79]. Regarding the intrinsic pathway, the role of Bcl-2 family members in tumorigenesis and cancer cell survival has been known for a long time, and its effects on CSC biology are well described. Bcl-2 is highly expressed in CD44$^+$/CD24$^{-/low}$ breast CSCs [80] and in quiescent leukemic CD34$^+$ progenitor cells [81]. Similarly, in colon CSCs, high levels of Bcl-xL activity were found in a BH3 mimetic screen [82]. Reductions of Bcl-xL expression resulted in increased sensitivity to oxaliplatin and 5-fluorouracil (5-FU) [83]. Other members of the Bcl-2 family can be involved in CSC apoptotic escape, such as MCL-1, that was found to be downregulated by MiR-519d in cisplatin-resistance breast CSCs [84]. However, it is clear that the plethora of intrinsic and extrinsic apoptosis factors involved in the fine-tuning of the apoptotic process could be affected. Indeed, Rouhrazi et al., performing a quantitative real-time polymerase chain reaction (qRT-PCR) screen for 84 key apoptosis-related genes in zoledronic acid-resistant CD133$^+$/CD44$^+$ prostate CSCs, showed significant over/underexpression of a cluster of anti-apoptotic and pro-apoptotic genes [85]. Our group confirmed the increase in expression of antiapoptotic proteins in another preclinical model of zoledronic acid resistant prostate cancer cells in which the acquisition of CSCs features has been described. Interestingly, activation of the p38-MAPK in these cells was found to be a crucial mechanism in the regulation of several biological processes, such as antiapoptotic, prosurvival, proinflammatory and proangiogenic events, as well as EMT and invasion [86,87]. Similarly, a proteomic analysis of colonosphere cultures derived from resection specimens of liver metastases in patients with colon cancer highlighted that 20 of 32 proteins upregulated two-fold in CSCs were classified as regulating "Cell Death" [88]. From this perspective, in contrast to the observed CSC high apoptotic threshold, only combination therapies and drugs with pleiotropic effects could be effective at targeting the multiple resistance mechanisms.

4.3. Alteration of DNA Damage Repair System

Since radiotherapy and the majority of chemotherapeutic agents induce DNA damage, perturbation of the DNA damage repair system is another well-described mechanism of CSCs to escape genotoxic effects. As reviewed by Vitale et al. various molecular alterations lead to a robust DNA damage response (DDR) in CSCs compared with the relatively more differentiated malignant cells in glioblastoma, breast, lung and CRC [73]. However, each CSC could establish a specific alteration that increases the basal dependence on specific DDR components for proliferation and survival. Moreover, each DDR protein is under the control of several factors and in turn modulates several phenomena such as ATM (ataxia-telangiectasia mutated). Indeed, Bao et al. highlight that the CSC population exhibits an upregulation of phosphorylated ATM, Chk2, RAD51 and RAD17, either at baseline or in response to radiation [89]. Moreover, Carruthers et al. indicated that CSCs possessed ATM-independent mechanisms for activation and maintenance of the G2/M checkpoint, whereas differentiated tumor cell populations appeared to be more reliant on ATM function for G2/M checkpoint integrity [90]. In addition, Zhang et al. showed that ZEB1 participated in an ATM-dependent mechanism in the DDR response in breast CSCs by stabilizing CHK1 [91].

4.4. Quiescence State

In addition to a robust DDR, CSCs display a persistent quiescence state. The notion that standard chemotherapeutic resistance results from the persistence of quiescent CSCs has emerged recently by genetic-fate mapping in several solid tumor types. A slow proliferative state is essential for the survival of cells (resistance) in the presence of oxaliplatin or temozolomide treatments in CRC and in glioblastoma, respectively [92,93]. Knowledge concerning activated or silenced mechanisms relying on this state could be useful to employ combinatorial therapeutic strategies to manipulate and to sensitize CSCs to chemotherapeutics. For example, activated TGF-β signaling (among others, responsible for triggering cytostatic signals) drives the dormancy of CSCs in mouse squamous cell carcinoma, leading to cisplatin resistance [94]. Similarly, a subpopulation of CSCs undergoing EMT is associated with a slow proliferative state that confers resistance to anti-proliferative drugs in a model of breast and skin cancer [95]. Moreover, Soeda et al. reported that inhibition of the p38 MAPK pathway led to an increase in EGFR expression but reduced proliferation and cell death induction and promoted maintenance of an undifferentiated state [96]. Recently, it has been demonstrated that leukemia stem cells express the highest levels of enhancer of zeste homolog 1 (EZH1) and 2 (EZH2), two histone-lysine N-methyltransferases that mediate methylation of histone H3 at lysine 27 (H3K27), to maintain the quiescent state. The key role of these two enzymes was confirmed, showing that inactivation of EZH1/2 eradicated quiescent leukemia stem cells, inducing cell cycle progression and differentiation [97]. Notably, EZH2 was also associated with radioresistance of glioma stem cells [98] as well as $CD24^{-/low}CD44^{+}$ population amplification in breast cancer upon treatment [99].

4.5. Metabolism Adaptation

In recent years, conventional wisdom indicates that CSC metabolism adaptation could play a main role in chemoresistance and radioresistance. Metabolism based on oxidative phosphorylation is crucial for the generation of energy needed to support the maintenance of tumors; however, this process also produces reactive oxygen species (ROS), which have the potential to cause stem cell dysfunction. For this reason, CSCs usually grow in a hypoxic niche and employ a glycolytic metabolism [100,101]. Additionally, CSCs are characterized by increased ROS levels, reduced oxidative damage and, thus, longer survival, potentially due to a combination of mechanisms that arise in the tumor, including the modulation of multiple antioxidative enzyme systems or redox-sensitive signaling pathways, such as NRF2, NF-κB, c-Jun, and HIFs, leading to increased expression of antioxidant molecules, as recently reviewed in detail by our group [102].

Another altered metabolic feature described in the CSC subpopulation is the high glycolytic metabolism recently demonstrated to be related to the high mitochondrial mass in MCF-7 breast cancer cells [103]. In detail, an unbiased proteomic approach has allowed the establishment that mitochondrial proteins are among the most strongly upregulated in cells overexpressing WNT1 and FGF3, which are responsible for the stemness phenotype. Interestingly, the mito-high MCF7 cells are also resistant to paclitaxel, resulting in little or no DNA damage [103]. Moreover, an enrichment in mitochondrial content has been associated with a higher DNA repair capacity in human breast cancer stem cells, suggesting that an increased mitochondrial mass may enable CSCs to cope efficiently with the action of certain anticancer drugs [104]. In line with these findings, a metabolic profile of $CD34^{+}$ and $CD34^{-}$ chronic myeloid leukemia (CML) cells, derived from fourindividuals by recording steady-state levels of 70 metabolites through liquid chromatography–mass spectrometry (LC–MS), highlighted a selective increase in glucose oxidation and anaplerosis, the process of replenishment of depleted metabolic pathway intermediates, in CML cells resistant to imatinib treatment [105].

4.6. Immune Evasion

Finally, a recently described CSC-resistance feature is immune evasion. CSCs have evolved sophisticated strategies to escape host immune surveillance that are targets of current therapeutic efforts (as reviewed by Fiori and Maccalli) [106,107]. A bioinformatic approach using The Cancer Genome Atlas (TCGA) revealed an enrichment of tumor-intrinsic WNT/β-catenin signaling in non-T cell inflamed tumors, providing a strong rationale for the development of pharmacologic inhibitors of this pathway with the aim of restoring immune cell infiltration and augmenting the immunotherapeutic response [108]. Molecularly, Agudo et al. showed that cycling epithelial stem cells, including $Lgr5^{+}$ intestinal stem cells, as well as ovary and mammary stem cells, were eliminated by activated T cells, but quiescent stem cells in the hair follicle and muscle were resistant to T cell killing. Mechanistically, the authors highlighted that the quiescent stem cells downregulated the antigen presentation machinery, including MHC class I and TAP proteins, through the trans-activator NLRC5 [109]. Similarly, the slow-cycling CSC subpopulation in CRC, with the loss of the major histocompatibility complex by overexpression of costimulation molecules and CSC-specific antigens, were resistant to the cytotoxic effect of dendritic and cytokine-induced killer cells [110]. Interestingly, Peng et al. found that myeloid-derived suppressor cells (MDSC) promoted tumor formation by enhancing breast CSC-like properties as well as by suppressing T cell activation, claiming a cross-talk between MDSC and breast CSCs. Specifically, MDSC inducing IL6-dependent phosphorylation of STAT3 and activation of NOTCH through nitric oxide lead to prolonged STAT3 activation in CSCs [111]. However, our knowledge about specific immunological properties of distinct CSC populations is still limited and requires further study to implement new targetingtherapeutic strategies.

Overall, these observations suggest that CSCs have several ways to escape anti-tumor approaches and that it is critical to identify CSC-specific alterations for targeting within a single tumor or rather to use drugs with pleiotropic effects to successfully target this difficult to kill subpopulation.

5. HDAC Inhibitors Are Able to Overcome Chemo-Resistance

As pleiotropic agents, HDACi can modulate a wide variety of molecules and target several related molecular pathways. Thus, it is obvious that this class of drug could affect many escape chemotoxicity strategies implemented by tumor cells and CSCs, which are described above.

Indeed, it has been demonstrated that entinostat, as a type of HDACi, reverses cisplatin resistance, among various mechanisms, by the induction of apoptosis with an increase in cleaved PARP and a decrease in MDR1 in esophageal squamous cell carcinoma [112]. Similarly, Zhao et al. observed that specific inhibition of HDAC8 mediates the upregulation of miR-137 and inhibition of MDR1 to sensitize neuroblastoma cells to doxorubicin (Dox) [113]. In a lung cancer cell model, To et al. demonstrated that belinostat reverts cisplatin resistance by the inhibition of both ABCC2 and the DNA repair gene ERCC1 [114].

Concerning P-glycoprotein (P-gp), another crucial drug efflux transporter, although it has been reported that HDACi upregulates P-gp in colorectal cancer cells through STAT3 induction and ABCB1 posttranscriptional stabilization [115], Tomono et al. demonstrated that HDACi inhibited Snail-induced activation of P-gp in lung cancer Snail-overexpressing cells [116]. Moreover, in squamous carcinoma cell lines, Chikamatsu et al. reported that the expression of CSC markers, such as CD44 and ABCG2, decreased upon vorinostat and TSA treatment and that the combination of these HDACi with cisplatin or docetaxel had a synergistic cytotoxic effect [117].

As pointed out above, HDACi usually work as sensitizers and modulators of the entire gene pattern, synergizing with several chemotherapeutics and molecular targeted agents. Thus, the doses used in *in vitro* and *in vivo* experiments are not expected to provide clear apoptosis induction. Nevertheless, Aztopal et al. reported that relatively low doses (2.5 and 5 mM) of VPA prevented mammosphere formation, inducing apoptosis [118]. Similarly, Di Pompo et al. reported that new compounds with selective HDACi activity affect CSCs generated from three different histotypes of human sarcomas, inducing, among others, apoptosis [119]. The author demonstrated that concurrent class I and IIb HDAC inhibition is crucial to obtain anticancer effects. In pancreatic CSCs, vorinostat epigenetically restores the expression of miR-34a, leading to apoptosis through caspase-3/7 activation [120]. Moreover, HDAC1 inhibition contributes to NANOG-mediated TRIM17 and *NOXA* gene expression, leading to a downregulation of antiapoptotic MCL-1, conferring immuno- and chemo-sensitization [121].

Since a clear characterization of CSCs has not yet been defined completely, we believe that the majority of findings regarding overcoming chemoresistance were not described in a specific cell subpopulation; therefore, we have also reviewed the literature on HDACi and their ability to sensitize resistance in subpopulations in many cancer types regardless of stem-like features.

Fluoropyrimidine-based therapy still represents a classic therapeutic strategy in several solid tumors, such as colorectal, breast and pancreatic cancers; however, chemo-resistance remains a big open question to resolve. Resistance to 5-FU or to its pro-drugs may result from deficient drug uptake, alterations of targets, activation of DNA repair pathways, resistance to apoptosis andalterations of the tumor microenvironment.

We and others have previously demonstrated synergistic antitumor effects of different HDACi in combination with fluoropyrimidines in different tumors, such as breast, colorectal [48,122–125] and head and neck squamous cell carcinomas (HNSCC) [126]. Our results demonstrated that the synergistic interaction between HDACi and 5-FU is dependent on both the downregulation of thymidylate synthase (TS), the key enzyme in the mechanism of action of 5-FU, and on the upregulation of thymidine phosphorylase (TP), the key enzyme converting capecitabine to 5-FU. Interestingly, HDAC3 is the

HDAC isoform principally involved in TP upregulation. These observations could be clinically relevant since HDAC3 hasrecently emerged as a critical anticancer target [127–129], and more selective HDAC3 inhibitors may have a more favorable side-effect profile compared with class-I or non-selective HDACi. Intrinsic or acquired resistance to 5-FU is often related to TS protein overexpression. Indeed, high levels of TS expression have been correlated with poorer overall patient survival in several tumors [130]. Interestingly, we have previously demonstrated a synergistic antitumor effect of vorinostat with 5-FU in CRC cells selected for resistance to 5-FU (HT29FU cells) and in cells carrying an amplification of the TS gene (H630-R10 cells), suggesting a potential mechanism by which vorinostat may overcome the resistance to 5-FU [122]. Similarly, we recently showed that VPA/capecitabine combination treatment synergizes with radiotherapy (RT), confirming the modulation of both TS and TP protein levels by VPA in CRC models, even in the presence of RT [48]. Based on these data, a phase 1/2 study is currently ongoing exploring VPA at an antiepileptic dosage, in combination with capecitabine, during preoperative radiotherapy in locally advanced rectal cancers patients (V-shoRT-R3 trial) [131]. Notably, data from the completed phase 1 of the study support the feasibility of VPA in combination with chemoradiotherapy [132].

In the same context, Huang et al. demonstrated that TSA was able to decrease colon CSC properties and, in combination with 5-FU, suppress colon cancer viability *in vitro* and colon tumorigenesis *in vivo* [133].

Another challenging class of chemotherapeutics is represented by the "platinum complexes", which includes anticancer drugs such as oxaliplatin, cisplatin and carboplatin. Cisplatin is currently employed for several tumors, such as testicular, ovarian, bladder, head and neck, esophageal, small and non-small-cell lung, cervical, stomach and others, regiments containing oxaliplatin, with or without a biologic agents in combination, are the optimal choice for metastatic CRC treatment, leading to sufficient disease reduction and allowing patients to become eligible for resection of metastatic diseases.

However, cisplatin or oxaliplatin treatments are associated with drug resistance and several adverse side effects. For these reasons, a combinational strategy aimed to revert resistance, improve outcomes and reduce side effects, could be of great benefit. Unpublished results from our group demonstrated that VPA induces cellular differentiation and sensitization of colorectal CSCs to oxaliplatin. In details, by using CRC primary spheroid cultures, transduced with a TOP-GFP Wnt reporter, we monitored apoptosis and cell proliferation in both differentiated cells and CSCs within the same population, treated with VPA alone and/or standard chemotherapy [50]. Based on these preliminary results, a randomized phase II trial has been designed by our group to evaluate whether the combination of VPA with bevacizumab and oxaliplatin/fluoropyrimidine regimens (mFOLFOX6/mOXXEL) prolongs progression-free survival (PFS) compared with bevacizumab and oxaliplatin/fluoropyrimidine regimens alone, as the first-line treatment in patients with metastatic CRC with mutation of RAS (Revolution Trial–Randomized phase II study of VPA in combination with bevacizumab and Oxaliplatin/fLUoropyrimidine regimens in patients with ras-mutated metastaTIc cOlorectal cancer). Correlative studies will be performed on patient materials to study the impact of the treatment on the CSCs population.

Cisplatin resistance has been reported in HNSCC cells to be related to enhanced stem cell properties, tumor metastasis, and increased HDACs expression [134]. Wang et al. demonstrated that cisplatin treatment, but not paclitaxel and doxorubicin treatment, result in the enrichment of CSCs, conferring multidrug resistance in NSCLC cell lines by the induction of TRIB1 and HDACs [135]. It has been demonstrated that TRIB1 enhances histone deacetylase 1 (HDAC1)-mediated p53 deacetylation and decreases DNA binding of p53, decreasing its tumor suppressor activity [136]. Moreover, high levels of TRIB1 show a significantly poorer prognosis in CRC patients, in NSCLC cisplatin-treated patients and in a Chinese Han population with pancreatic cancer [135,137,138]. In metastatic NSCLC, pancreatic and bladder cancer cisplatin treatment is associated with gemcitabine. In this regard, it has been described that the pan-HDACi CG200745, decreasing the transcript for multidrug resistance protein (MRP) 4, a member of the MRP/ABCC subfamily of the ATP-binding cassette, controls drug efflux and sensitivity in gemcitabine-resistant pancreatic cancer cells [139], for which gemcitabine-based regimens are still the major treatment. Moreover, the inhibition of HDACs 1, 7 and 8 by TSA or vorinostat results in an upregulation of e-cadherin in PDAC cells and downregulation of Oct-4, Sox-2, and Nanog, as well as inhibition of PDAC tumor sphere formation, resulting in a strong potentiation of gemcitabine therapeutic activity [16]. Interestingly, overexpression of HDAC7 has been demonstrated in pancreatic cancer and suggested as clinical biomarker for pancreatic cancer diagnosis and prognosis [140].

Additionally, CG200745, by inducing miR-509-3p expression, selectively targets Hippo signaling in cholangiocarcinoma cells and synergistically interacts with conventional chemotherapeutic drugs, including gemcitabine, cisplatin, oxaliplatin and 5-FU, even enhancing the sensitivity of gemcitabine-resistant cholangiocarcinoma cells to these drugs [141]. However, although co-treatment of biliary tract cancer cells with HDACi, including TSA, VPA or vorinostat, and gemcitabine suppresses EMT with tolerable cytotoxicity [62,142], it has also been reported that HDACi increase the expression of both e-cadherin and vimentin in different cholangiocarcinoma cell lines, suggesting that further analyses are needed before using these drugs in the clinic for the treatment of this tumor [142]. The preclinical synergistic interaction between HDACi and gemcitabine has also been described in leiomyosarcoma, for which Lopez et al. demonstrated that the selective class I/IV HDACi mocetinostat combined with gemcitabine exhibits synergistic effects *in vitro* and *in vivo* [143].

Interestingly, in a phase I study in which vorinostat was combined with carboplatin and gemcitabine in women with recurrent, platinum-sensitive epithelial ovarian, fallopian tube, or peritoneal cancer, despite no maximum tolerated dose determined due to toxicities, six of the seven patients evaluable for RECIST (Response Evaluation Criteria in Solid Tumours) assessment had partial responses (PR) [144]. These results indicate that the use of other HDACi with a better safety profile, such as VPA, could be further investigated in combination with gemcitabine and/or platinum compounds.

It has been already described in the literature that HDACi modulate the EGFR family in several ways, as also reported by our group [145,146]. This evidence explains how HDACi improve the EGFR-based target therapy and how they overcome the drugresistance that is commonly reported after the first anti-EGFR target therapy cycles of treatment. Indeed, Wang et al. showed that HDAC6 overexpression confers resistance to gefitinib via the stabilization of EGFR. Moreover, inhibition of HDAC6 by CAY10603, a selective inhibitor, represses the proliferation and synergizes with gefitinib to induce apoptosis in lung adenocarcinoma cell lines, via the destabilization of EGFR [147]. In the same model of NSCLC, it has been reported that the knockdown of HDAC1 sensitize resistant cells to paclitaxel *in vitro* and that SNOH-3, a selective HDAC1 inhibitor, induces apoptosis and suppresses angiogenesis in preclinical models [148]. In addition, we found that vorinostat or VPA sensitize primary NSCLC cell lines to anti-ErbB3 monoclonal antibody by modulating the EGFR family also in 2D and in 3D culture models, enriched in CSCs [47]. Interestingly, a phase I study of CUDC-101, a multitarget inhibitor of HDACs, EGFR, and HER2, in combination with chemoradiation in patients with HNSCC, showed an increase of 1.5 years in median duration of response and 9/15 patients free with increased PFS [149]. Other examples of combined treatment of HDACi with targeted drugs have also been reported. For example, Gruber et al. showed that 4SC-202, a class I HDACi, abrogates GLI activation and Hh target gene expression in both SMO-inhibitor-sensitive and -resistant cells. Significantly, it has been reported that treatment with SMO inhibitors leads to rapid and frequent development of drug resistance in basal cell carcinoma and medulloblastoma [68].

Finally, it is important to highlight that, despite a good and durable clinical benefit obtained by immune checkpoint blockers (ICB) in several tumors, particularly "hot" or "immunogenic tumors", in other solid tumors, the responses with ICB are quite modest, representing a critical therapeutic challenge [150]. A possible solution to address such challenges could be to combine ICB with specific drugs aimed to prime the immune response and increase the tumor immune profile. Published data suggest that HDACi enhance the immunogenicity of cancer cells. Indeed, HDACi are involved in the regulation of NK cell-activating ligands, MHC class I and II molecules, elevation of NK and CD8+ cytotoxicity and proinflammatory cytokines, and modulation of Treg and Treg Foxp3 expression [151].

Terranova-Barberio et al. showed that HDACi treatment leads to a significant decrease in tumor growth through epigenetic priming of the immune system, with increased tumor antigen presentation and immune cell activation. Interestingly, both pan-HDACi and class-selective HDACi promote an upregulation of PD-L1 and HLA-DR in triple negative breast cancer cell models when co-cultured with peripheral blood mononuclear cells, associated with a downregulation of CD4$^+$ Foxp3$^+$ Treg *in vitro* and *in vivo* [152]. In line with this finding, Miyashita et al. found that the combination of low-dose VPA and gemcitabine enhances the susceptibility of the PANC-1 cell line to $\gamma\delta$T cell-mediated tumor cell lysis through the upregulation of major histocompatibility complex class 1-related chain molecules [153]. Interestingly, several phase I and II clinical trials, based on HDACi treatment in combination with immunotherapy, are recruiting with the aim to identify optimal biological dosing strategies [151]. The first results are not yet available, and additional preclinical studies need to be performed to disclose the mechanisms underlying the controversial HDACi-dependent effects.

6. Conclusions and Future Perspectives

In solid cancers, multiple factors contribute to the failure of commonly used therapies, leading to relapse. Here, we emphasized the aspects related to CSCs and their involvement in this phenomenon, reviewing what is known in the literature. By highlighting the pathways and mechanisms responsible for the resistance to the commonly used chemotherapeutics, we suggest that only the addition of pleiotropic molecules will target the CSCs population efficiently. From this perspective, HDACi could be among the best candidate drugs. We have reviewed the role of HDACi in solid cancers, specifically in the CSC subpopulation, and have pointed out some mechanisms by which HDACi are able to overcome drug resistance. In clinicaltrials.gov, 377 registered studies in solid tumors with HDACi were retrieved. Most of them are phase-I and phase-II trials and several are currently recruiting as a single agent or in combination therapies (Table 2). Only seven phase-III studies are reported and only one have published results (Table 3). In the latter study, the vorinostat effect was evaluated in monotherapy, in a double-blind, randomized placebo-controlled trial, involving 90 international centers and enrolling 661 patients with measurable advanced malignant pleural mesothelioma and disease progression after one or two previous systemic regimens. The study was negative, with no statistical significant difference (median overall survival of 30.7 weeks; 95% CI 26.7–36.1) versus the placebo arm (27.1 weeks; 95% 23.1–31.9) [154] confirming negative results obtained by this class of agent in monotherapy in several solid tumor types [155]. Another phase -III trial (ClinicalTrial.gov identifier: NCT00473889) exploring the combination of vorinostat plus carboplatin and paclitaxel in NSCLC patients was terminated due to negative results and increased toxicity, although a phase II trial demonstrated that vorinostat was able to improve the efficacy of chemotherapy [156]. Even as single agent, drug-induced side-effects of HDACi were observed, associated with several toxicities including cardiotoxicities, hematological and gastrointestinal toxicities [157]. However, despite being able to affect a multitude of physiological cellular process, thus being potentially very toxic, several HDACi have gained FDA approval for use in hematological malignancies [155].

Table 2. Phase and status of HDACi clinical trials: clinical trials registered in https://clinicaltrials.gov/ in solid tumors with HDACi.

	Recruiting	Active, Not Recruiting	Not Yet Recruiting	Completed	Terminated	Suspended	Withdrawn	Unknown Status	Sum
Early Phase 1	2	0	0	0	0	0	1	1	4
Phase 1	24	34	8	109	45	2	5	3	230
Phase 2	15	16	3	55	37	0	1	3	130
Phase 3	1	2	0	1	1	0	0	2	7
Not Applicable	0	1	0	4	0	0	1	0	6

Table 3. Phase-III HDACi clinical trials: characteristics of Seven phase-III clinical trials registered in https://clinicaltrials.gov/ that involve HDACi.

Phase	Title	Status	Completion Date	Description	Condition	Url
Phase 3	Hydralazine Valproate for Ovarian Cancer	Unknown status	December 2009	Randomized, double-blind phase III trial. A total of 211 patients (alpha 0.5, power 0.8) with cisplatin-resistant recurrent or persistent cancer will be randomized to topotecan + placebo or topotecan + hydralazine + valproate for 6 courses every 4 weeks.	Ovarian Cancer	https://ClinicalTrials.gov/show/NCT00533299
Phase 3	Hydralazine Valproate for Cervical Cancer	Unknown status	December 2010	Randomized, double-blind phase III trial. A total of 143 patients (alpha 0.5, power 0.8) with metastatic, persistent or recurrent cervical cancer without previous systemic treatment will be randomized to cisplatin topotecan + placebo or cisplatin topotecan hydralazine valproate for 6 courses every 3 weeks.	Metastatic Cervical Cancer	https://ClinicalTrials.gov/show/NCT00532818
Phase 3	Anticancer Activity of Nicotinamide on Lung Cancer	Active, not recruiting	June 2020	Randomized Double-blinded Comparative Trial to study the Add-on Activity of Combination Treatment of Nicotinamide on Progression Free Survival for EGFR Mutated Lung Cancer Terminal Stage Patients Being Treated With Gefitinib or Erlotinib.	Non-Small-Cell Lung Carcinoma	https://ClinicalTrials.gov/show/NCT02416739
Phase 3	Exemestane With or Without Entinostat in Treating Patients With Recurrent Hormone Receptor-Positive Breast Cancer That is Locally Advanced or Metastatic	Active, not recruiting	-	Randomized phase III trial studies exemestane and entinostat to see how well they work compared to exemestane alone in treating patients with hormone receptor-positive breast cancer that has spread to nearby tissue or lymph nodes or another place in the body.	Breast Adenocarcinoma	https://ClinicalTrials.gov/show/NCT02115282
Phase 3	Exemestane With or Without Entinostat in Chinese Patients With Hormone Receptor-Positive, Locally Advanced or Metastatic Breast Cancer	Recruiting	August 2021	A Randomized Phase III Clinical Study of Entinostat/Placebo in Combination With Exemestane in Chinese Patients With Hormone Receptor-positive Advanced Breast Cancer.	Advanced Breast Carcinoma	https://ClinicalTrials.gov/show/NCT03538171
Phase 3	A Clinical Trial of Vorinostat (MK0683, SAHA) in Combination With FDA Approved Cancer Drugs in Patients With Advanced Non-Small Cell Lung Cancer (NSCLC)(0683-056)	Terminated	December 2008	A Phase II/III Randomized, Double-Blind Study of Paclitaxel Plus Carboplatin in Combination With Vorinostat or Placebo in Patients With Stage IIIB (With Pleural Effusion) or Stage IV Non-Small-Cell Lung Cancer (NSCLC).	Stage IIIB or IV Non-Small Cell Lung Cancer	https://ClinicalTrials.gov/show/NCT00473889
Phase 3	Suberoylanilide Hydroxamic Acid (Vorinostat, MK-0683) Versus Placebo in Advanced Malignant Pleural Mesothelioma (0683-014 AM5, EXT1)	Completed	November 2011	A Phase III, Randomized, Double-Blind, Placebo-Controlled Trial of Oral Suberoylanilide Hydroxamic Acid (Vorinostat, MK-0683) in Patients With Advanced Malignant Pleural Mesothelioma Previously Treated With Systemic Chemotherapy.	Mesothelioma/Lung Cancer	https://ClinicalTrials.gov/show/NCT00128102

J. Clin. Med. **2019**, *8*, 912

Moreover, we strongly believe that HDACi should be use in clinical practice as biological modifiers and not as cytotoxic drugs. Consequently, by contrast with clinical trials conducted so far, low doses of these drugs should be used to avoid direct cytotoxic effects. In addition, we believe that the therapeutic window of this class of drug is only in combination with standard chemotherapeutics.

On these bases, further clinical and preclinical investigations should be conducted to better disclose the mechanisms by which HDACi modulate several signaling pathways in different tumors. In summary, as highlighted in this review, the promising data obtained until now could represent the foundation to test novel combinatorial therapeutic strategies, where HDACi would be combined with commonly used drugs to improve therapeutic efficacy in solid cancer tumors.

Author Contributions: Conceptualization, M.S.R., E.D.G. and A.B.; software, M.S.R.; writing—original draft preparation, M.S.R., E.D.G. and A.B.; writing—review and editing M.S.R., E.D.G. and A.B.; visualization, M.S.R.; funding acquisition, M.S.R. and A.B.

Funding: This research was funded by grants to Alfredo Budillon by the Italian Ministry of Health (RF-2011-02346914). Maria Serena Roca is supported by a triennial AIRC Fellowship (AIRC id. 21113).

Conflicts of Interest: The authors declare no conflict of interest.

References

1. Burrell, R.A.; Swanton, C. Tumour heterogeneity and the evolution of polyclonal drug resistance. *Mol. Oncol.* **2014**, *8*, 1095–1111. [CrossRef] [PubMed]
2. Meacham, C.E.; Morrison, S.J. Tumour heterogeneity and cancer cell plasticity. *Nature* **2013**, *501*, 328–337. [CrossRef] [PubMed]
3. O'Connor, J.P.; Rose, C.J.; Waterton, J.C.; Carano, R.A.; Parker, G.J.; Jackson, A. Imaging intratumor heterogeneity: Role in therapy response, resistance, and clinical outcome. *Clin. Cancer Res.* **2015**, *21*, 249–257. [CrossRef] [PubMed]
4. Greaves, M.; Maley, C.C. Clonal evolution in cancer. *Nature* **2012**, *481*, 306–313. [CrossRef] [PubMed]
5. Siravegna, G.; Mussolin, B.; Buscarino, M.; Corti, G.; Cassingena, A.; Crisafulli, G.; Ponzetti, A.; Cremolini, C.; Amatu, A.; Lauricella, C.; et al. Clonal evolution and resistance to EGFR blockade in the blood of colorectal cancer patients. *Nat. Med.* **2015**, *21*, 795–801. [CrossRef] [PubMed]
6. Clevers, H. The cancer stem cell: Premises, promises and challenges. *Nat. Med.* **2011**, *17*, 313–319. [CrossRef]
7. Medemas, J.P. Cancer stem cells: The challenges ahead. *Nat. Cell Biol.* **2013**, *15*, 338–344. [CrossRef]
8. De Smedt, E.; Lui, H.; Maes, K.; De Veirman, K.; Menu, E.; Vanderkerken, K.; De Bruyne, E. The epigenome in multiple myeloma: Impact on tumor cell plasticity and drug response. *Front. Oncol.* **2018**, *8*, 566. [CrossRef]
9. Jones, P.A.; Issa, J.P.; Baylin, S. Targeting the cancer epigenome for therapy. *Nat. Rev. Genet.* **2016**, *17*, 630–641. [CrossRef]
10. Wainwright, E.N.; Scaffidi, P. Epigenetics and cancer stem cells: Unleashing, hijacking, and restricting cellular plasticity. *Trends Cancer* **2017**, *3*, 372–386. [CrossRef]
11. Budillon, A.; Di Gennaro, E.; Bruzzese, F.; Rocco, M.; Manzo, G.; Caraglia, M. Histone deacetylase inhibitors: A new wave of molecular targeted anticancer agents. *Recent Pat. Anticancer Drug Discov.* **2007**, *2*, 119–134. [CrossRef] [PubMed]
12. Liu, N.; Li, S.; Wu, N.; Cho, K.S. Acetylation and deacetylation in cancer stem-like cells. *Oncotarget* **2017**, *8*, 89315–89325. [CrossRef] [PubMed]
13. Sulaiman, A.; Sulaiman, B.; Khouri, L.; McGarry, S.; Nessim, C.; Arnaout, A.; Li, X.; Addison, C.; Dimitroulakos, J.; Wang, L. Both bulk and cancer stem cell subpopulations in triple-negative breast cancer are susceptible to Wnt, HDAC, and ERα coinhibition. *FEBS Lett.* **2016**, *590*, 4606–4616. [CrossRef] [PubMed]
14. West, A.C.; Johnstone, R.W. New and emerging HDAC inhibitors for cancer treatment. *J. Clin. Investig.* **2014**, *124*, 30–39. [CrossRef] [PubMed]
15. Witt, A.E.; Lee, C.W.; Lee, T.I.; Azzam, D.J.; Wang, B.; Caslini, C.; Petrocca, F.; Grosso, J.; Jones, M.; Cohick, E.B.; et al. Identification of a cancer stem cell-specific function for the histone deacetylases, HDAC1 and HDAC7, in breast and ovarian cancer. *Oncogene* **2017**, *36*, 1707–1720. [CrossRef] [PubMed]

16. Cai, M.H.; Xu, X.G.; Yan, S.L.; Sun, Z.; Ying, Y.; Wang, B.K.; Tu, Y.X. Depletion of HDAC1, 7 and 8 by histone deacetylase inhibition confers elimination of pancreatic cancer stem cells in combination with gemcitabine. *Sci. Rep.* **2018**, *8*, 1621. [CrossRef]

17. Chao, M.W.; Chu, P.C.; Chuang, H.C.; Shen, F.H.; Chou, C.C.; Hsu, E.C.; Himmel, L.E.; Huang, H.L.; Tu, H.J.; Kulp, S.K.; et al. Non-epigenetic function of HDAC8 in regulating breast cancer stem cells by maintaining Notch1 protein stability. *Oncotarget* **2016**, *7*, 1796–1807. [CrossRef]

18. An, P.; Li, J.; Lu, L.; Wu, Y.; Ling, Y.; Du, J.; Chen, Z.; Wang, H. Histone deacetylase 8 triggers the migration of triple negative breast cancer cells via regulation of YAP signals. *Eur. J. Pharmacol.* **2019**, *845*, 16–23. [CrossRef]

19. Zimberlin, C.D.; Lancini, C.; Sno, R.; Rosekrans, S.L.; McLean, C.M.; Vlaming, H.; van den Brink, G.R.; Bots, M.; Medema, J.P.; Dannenberg, J.H. HDAC1 and HDAC2 collectively regulate intestinal stem cell homeostasis. *FASEB J.* **2015**, *29*, 2070–2080. [CrossRef]

20. Jamaladdin, S.; Kelly, R.D.; O'Regan, L.; Dovey, O.M.; Hodson, G.E.; Millard, C.J.; Portolano, N.; Fry, A.M.; Schwabe, J.W.; Cowley, S.M. Histone deacetylase (HDAC) 1 and 2 are essential for accurate cell division and the pluripotency of embryonic stem cells. *Proc. Natl. Acad Sci. USA* **2014**, *111*, 9840–9845. [CrossRef]

21. Montgomery, R.L.; Davis, C.A.; Potthoff, M.J.; Haberland, M.; Fielitz, J.; Qi, X.; Hill, J.A.; Richardson, J.A.; Olson, E.N. Histone deacetylases 1 and 2 redundantly regulate cardiac morphogenesis, growth, and contractility. *Genes Dev.* **2007**, *21*, 1790–1802. [CrossRef] [PubMed]

22. Yusoff, S.I.; Roman, M.; Lai, F.Y.; Eagle-Hemming, B.; Murphy, G.J.; Kumar, T.; Wozniak, M. Systematic review and meta-analysis of experimental studies evaluating the organ protective effects of histone deacetylase inhibitors. *Transl. Res.* **2019**, *205*, 1–16. [CrossRef] [PubMed]

23. Liu, J.; Lichtenberg, T.; Hoadley, K.A.; Poisson, L.M.; Lazar, A.J.; Cherniack, A.D.; Kovatich, A.J.; Benz, C.C.; Levine, D.A.; Lee, A.V.; et al. An integrated TCGA pan-cancer clinical data resource to drive high-quality survival outcome analytics. *Cell* **2018**, *173*, 400–416. [CrossRef] [PubMed]

24. Fodde, R.; Brabletz, T. Wnt/beta-catenin signaling in cancer stemness and malignant behavior. *Curr. Opin. Cell Biol.* **2007**, *19*, 150–158. [CrossRef] [PubMed]

25. Barker, N.; Clevers, H. Mining the Wnt pathway for cancer therapeutics. *Nat. Rev. Drug Discov.* **2006**, *5*, 997–1014. [CrossRef] [PubMed]

26. Reya, T.; Clevers, H. Wnt signalling in stem cells and cancer. *Nature* **2005**, *434*, 843–850. [CrossRef]

27. Taha, Z.; Janse van Rensburg, H.J.; Yang, X. The Hippo pathway: Immunity and cancer. *Cancers* **2018**, *10*, 94. [CrossRef]

28. Nguyen, L.T.; Tretiakova, M.S.; Silvis, M.R.; Lucas, J.; Klezovitch, O.; Coleman, I.; Bolouri, H.; Kutyavin, V.I.; Morrissey, C.; True, L.D.; et al. ERG activates the YAP1 transcriptional program and induces the development of age-related prostate tumors. *Cancer Cell* **2015**, *27*, 797–808. [CrossRef]

29. Zhang, L.; Yang, S.; Chen, X.; Stauffer, S.; Yu, F.; Lele, S.M.; Fu, K.; Datta, K.; Palermo, N.; Chen, Y.; et al. The hippo pathway effector YAP regulates motility, invasion, and castration-resistant growth of prostate cancer cells. *Mol. Cell Biol.* **2015**, *35*, 1350–1362. [CrossRef]

30. Bourgeais, J.; Gouilleux-Gruart, V.; Gouilleux, F. Oxidative metabolism in cancer: A STAT affair? *Jak-Stat* **2013**, *2*, e25764. [CrossRef]

31. Ouedraogo, Z.G.; Biau, J.; Kemeny, J.L.; Morel, L.; Verrelle, P.; Chautard, E. Role of STAT3 in genesis and progression of human malignant gliomas. *Mol. Neurobiol.* **2017**, *54*, 5780–5797. [CrossRef] [PubMed]

32. Ganguly, D.; Fan, M.; Yang, C.H.; Zbytek, B.; Finkelstein, D.; Roussel, M.F.; Pfeffer, L.M. The critical role that STAT3 plays in glioma-initiating cells: STAT3 addiction in glioma. *Oncotarget* **2018**, *9*, 22095–22112. [CrossRef] [PubMed]

33. West, A.J.; Tsui, V.; Stylli, S.S.; Nguyen, H.P.T.; Morokoff, A.P.; Kaye, A.H.; Luwor, R.B. The role of interleukin-6-STAT3 signalling in glioblastoma. *Oncol. Lett.* **2018**, *16*, 4095–4104. [CrossRef]

34. Kohsaka, S.; Wang, L.; Yachi, K.; Mahabir, R.; Narita, T.; Itoh, T.; Tanino, M.; Kimura, T.; Nishihara, H.; Tanaka, S. STAT3 inhibition overcomes temozolomide resistance in glioblastoma by downregulating MGMT expression. *Mol. Cancer Ther.* **2012**, *11*, 1289–1299. [CrossRef] [PubMed]

35. Walker, S.R.; Xiang, M.; Frank, D.A. STAT3 activity and function in cancer: Modulation by STAT5 and miR-146b. *Cancers* **2014**, *6*, 958–968. [CrossRef] [PubMed]

36. Yusra; Semba, S.; Yokozaki, H. Biological significance of tumor budding at the invasive front of human colorectal carcinoma cells. *Int. J. Oncol.* **2012**, *41*, 201–210.

37. Yu, Y.; Kanwar, S.S.; Patel, B.B.; Oh, P.S.; Nautiyal, J.; Sarkar, F.H.; Majumdar, A.P. MicroRNA-21 induces stemness by downregulating transforming growth factor beta receptor 2 (TGFbetaR2) in colon cancer cells. *Carcinogenesis* **2012**, *33*, 68–76. [CrossRef] [PubMed]

38. Venkatesh, V.; Nataraj, R.; Thangaraj, G.S.; Karthikeyan, M.; Gnanasekaran, A.; Kaginelli, S.B.; Kuppanna, G.; Kallappa, C.G.; Basalingappa, K.M. Targeting Notch signalling pathway of cancer stem cells. *Stem Cell Investig.* **2018**, *5*, 5. [CrossRef]

39. Katoh, M. Networking of WNT, FGF, Notch, BMP, and Hedgehog signaling pathways during carcinogenesis. *Stem Cell Rev.* **2007**, *3*, 30–38. [CrossRef]

40. Jiang, J.; Hui, C.C. Hedgehog signaling in development and cancer. *Dev. Cell* **2008**, *15*, 801–812. [CrossRef]

41. Takebe, N.; Miele, L.; Harris, P.J.; Jeong, W.; Bando, H.; Kahn, M.; Yang, S.X.; Ivy, S.P. Targeting Notch, Hedgehog, and Wnt pathways in cancer stem cells: Clinical update. *Nat. Rev. Clin. Oncol.* **2015**, *12*, 445–464. [CrossRef]

42. Duan, Z.H.; Wang, H.C.; Zhao, D.M.; Ji, X.X.; Song, M.; Yang, X.J.; Cui, W. Cooperatively transcriptional and epigenetic regulation of sonic hedgehog overexpression drives malignant potential of breast cancer. *Cancer Sci.* **2015**, *106*, 1084–1091. [CrossRef] [PubMed]

43. Krop, I.; Demuth, T.; Guthrie, T.; Wen, P.Y.; Mason, W.P.; Chinnaiyan, P.; Butowski, N.; Groves, M.D.; Kesari, S.; Freedman, S.J.; et al. Phase I pharmacologic and pharmacodynamic study of the gamma secretase (Notch) inhibitor MK-0752 in adult patients with advanced solid tumors. *J. Clin. Oncol.* **2012**, *30*, 2307–2313. [CrossRef] [PubMed]

44. Pattabiraman, D.R.; Weinberg, R.A. Tackling the cancer stem cells—What challenges do they pose? *Nat. Rev. Drug Discov.* **2014**, *13*, 497–512. [CrossRef] [PubMed]

45. Munoz, P.; Iliou, M.S.; Esteller, M. Epigenetic alterations involved in cancer stem cell reprogramming. *Mol. Oncol.* **2012**, *6*, 620–636. [CrossRef] [PubMed]

46. Yoshida, M.; Kijima, M.; Akita, M.; Beppu, T. Potent and specific inhibition of mammalian histone deacetylase both in vivo and in vitro by trichostatin A. *J. Biol. Chem.* **1990**, *265*, 17174–17179.

47. Ciardiello, C.; Roca, M.S.; Noto, A.; Bruzzese, F.; Moccia, T.; Vitagliano, C.; Di Gennaro, E.; Ciliberto, G.; Roscilli, G.; Aurisicchio, L.; et al. Synergistic antitumor activity of histone deacetylase inhibitors and anti-ErbB3 antibody in NSCLC primary cultures via modulation of ErbB receptors expression. *Oncotarget* **2016**, *7*, 19559–19574. [CrossRef]

48. Terranova-Barberio, M.; Pecori, B.; Roca, M.S.; Imbimbo, S.; Bruzzese, F.; Leone, A.; Muto, P.; Delrio, P.; Avallone, A.; Budillon, A.; et al. Synergistic antitumor interaction between valproic acid, capecitabine and radiotherapy in colorectal cancer: Critical role of p53. *J. Exp. Clin. Cancer Res.* **2017**, *36*, 177. [CrossRef]

49. Salvador, M.A.; Wicinski, J.; Cabaud, O.; Toiron, Y.; Finetti, P.; Josselin, E.; Lelievre, H.; Kraus-Berthier, L.; Depil, S.; Bertucci, F.; et al. The histone deacetylase inhibitor abexinostat induces cancer stem cells differentiation in breast cancer with low Xist expression. *Clin. Cancer Res.* **2013**, *19*, 6520–6531. [CrossRef]

50. Roca, M.S.; Experimental Pharmacology Unit, Istituto Nazionale Tumori–IRCCS–Fondazione G. Pascale, Naples, Italy. Unpublished work. 2019.

51. Wang, Y.; Liu, M.; Jin, Y.; Jiang, S.; Pan, J. In vitro and in vivo anti-uveal melanoma activity of JSL-1, a novel HDAC inhibitor. *Cancer Lett.* **2017**, *400*, 47–60. [CrossRef]

52. Li, Y.; Zhang, X.; Polakiewicz, R.D.; Yao, T.P.; Comb, M.J. HDAC6 is required for epidermal growth factor-induced beta-catenin nuclear localization. *J. Biol. Chem.* **2008**, *283*, 12686–12690. [CrossRef]

53. Yu, S.; Cai, X.; Wu, C.; Liu, Y.; Zhang, J.; Gong, X.; Wang, X.; Wu, X.; Zhu, T.; Mo, L.; et al. Targeting HSP90-HDAC6 regulating network implicates precision treatment of breast cancer. *Int. J. Biol. Sci.* **2017**, *13*, 505–517. [CrossRef] [PubMed]

54. Li, L.; Fang, R.; Liu, B.; Shi, H.; Wang, Y.; Zhang, W.; Zhang, X.; Ye, L. Deacetylation of tumor-suppressor MST1 in Hippo pathway induces its degradation through HBXIP-elevated HDAC6 in promotion of breast cancer growth. *Oncogene* **2016**, *35*, 4048–4057. [CrossRef] [PubMed]

55. Shan, B.; Yao, T.P.; Nguyen, H.T.; Zhuo, Y.; Levy, D.R.; Klingsberg, R.C.; Tao, H.; Palmer, M.L.; Holder, K.N.; Lasky, J.A. Requirement of HDAC6 for transforming growth factor- beta1-induced epithelial-mesenchymal transition. *J. Biol. Chem.* **2008**, *283*, 21065–21073. [CrossRef] [PubMed]

56. Sferra, R.; Pompili, S.; Festuccia, C.; Marampon, F.; Gravina, G.L.; Ventura, L.; Di Cesare, E.; Cicchinelli, S.; Gaudio, E.; Vetuschi, A. The possible prognostic role of histone deacetylase and transforming growth factor beta/Smad signaling in high grade gliomas treated by radio-chemotherapy: A preliminary immunohistochemical study. *Eur. J. Histochem.* **2017**, *61*, 2732. [CrossRef] [PubMed]

57. Yang, W.; Liu, Y.; Gao, R.; Yu, H.; Sun, T. HDAC6 inhibition induces glioma stem cells differentiation and enhances cellular radiation sensitivity through the SHH/Gli1 signaling pathway. *Cancer Lett.* **2018**, *415*, 164–176. [CrossRef] [PubMed]

58. Tang, Y.N.; Ding, W.Q.; Guo, X.J.; Yuan, X.W.; Wang, D.M.; Song, J.G. Epigenetic regulation of Smad2 and Smad3 by profilin-2 promotes lung cancer growth and metastasis. *Nat. Commun.* **2015**, *6*, 8230. [CrossRef] [PubMed]

59. Canettieri, G.; Di Marcotullio, L.; Greco, A.; Coni, S.; Antonucci, L.; Infante, P.; Pietrosanti, L.; De Smaele, E.; Ferretti, E.; Miele, E.; et al. Histone deacetylase and Cullin3-REN(KCTD11) ubiquitin ligase interplay regulates Hedgehog signalling through Gli acetylation. *Nat. Cell Biol.* **2010**, *12*, 132–142. [CrossRef] [PubMed]

60. Wei, M.; Mao, S.; Lu, G.; Li, L.; Lan, X.; Huang, Z.; Chen, Y.; Zhao, M.; Zhao, Y.; Xia, Q. Valproic acid sensitizes metformin-resistant human renal cell carcinoma cells by upregulating H3 acetylation and EMT reversal. *BMC Cancer* **2018**, *18*, 434. [CrossRef] [PubMed]

61. Iannelli, F.; Experimental Pharmacology Unit, Istituto Nazionale Tumori–IRCCS–Fondazione G. Pascale, Naples, Italy. Unpublished work. 2019.

62. Sun, G.; Mackey, L.V.; Coy, D.H.; Yu, C.Y.; Sun, L. The histone deacetylase inhibitor vaproic acid induces cell growth arrest in hepatocellular carcinoma cells via suppressing notch signaling. *J. Cancer* **2015**, *6*, 996–1004. [CrossRef] [PubMed]

63. Cui, Y.; Ma, W.; Lei, F.; Li, Q.; Su, Y.; Lin, X.; Lin, C.; Zhang, X.; Ye, L.; Wu, S.; et al. Prostate tumour overexpressed-1 promotes tumourigenicity in human breast cancer via activation of Wnt/beta-catenin signalling. *J. Pathol.* **2016**, *239*, 297–308. [CrossRef] [PubMed]

64. Sakamoto, T.; Kobayashi, S.; Yamada, D.; Nagano, H.; Tomokuni, A.; Tomimaru, Y.; Noda, T.; Gotoh, K.; Asaoka, T.; Wada, H.; et al. A histone deacetylase inhibitor suppresses epithelial-mesenchymal transition and attenuates chemoresistance in biliary tract cancer. *PLoS ONE* **2016**, *11*, e0145985. [CrossRef] [PubMed]

65. Zhang, H.; Liu, L.; Liu, C.; Pan, J.; Lu, G.; Zhou, Z.; Chen, Z.; Qian, C. Notch3 overexpression enhances progression and chemoresistance of urothelial carcinoma. *Oncotarget* **2017**, *8*, 34362–34373. [CrossRef] [PubMed]

66. Fan, C.W.; Yarravarapu, N.; Shi, H.; Kulak, O.; Kim, J.; Chen, C.; Lum, L. A synthetic combinatorial approach to disabling deviant Hedgehog signaling. *Sci. Rep.* **2018**, *8*, 1133. [CrossRef] [PubMed]

67. Li, L.; Fan, B.; Zhang, L.H.; Xing, X.F.; Cheng, X.J.; Wang, X.H.; Guo, T.; Du, H.; Wen, X.Z.; Ji, J.F. Trichostatin A potentiates TRAIL-induced antitumor effects via inhibition of ERK/FOXM1 pathway in gastric cancer. *Tumour Biol.* **2016**, *37*, 10269–10278. [CrossRef] [PubMed]

68. Gruber, W.; Peer, E.; Elmer, D.P.; Sternberg, C.; Tesanovic, S.; Del Burgo, P.; Coni, S.; Canettieri, G.; Neureiter, D.; Bartz, R.; et al. Targeting class I histone deacetylases by the novel small molecule inhibitor 4SC-202 blocks oncogenic hedgehog-GLI signaling and overcomes smoothened inhibitor resistance. *Int. J. Cancer* **2018**, *142*, 968–975. [CrossRef] [PubMed]

69. Debeb, B.G.; Lacerda, L.; Xu, W.; Larson, R.; Solley, T.; Atkinson, R.; Sulman, E.P.; Ueno, N.T.; Krishnamurthy, S.; Reuben, J.M.; et al. Histone deacetylase inhibitors stimulate dedifferentiation of human breast cancer cells through WNT/beta-catenin signaling. *Stem Cells* **2012**, *30*, 2366–2377. [CrossRef] [PubMed]

70. Yardley, D.A.; Ismail-Khan, R.R.; Melichar, B.; Lichinitser, M.; Munster, P.N.; Klein, P.M.; Cruickshank, S.; Miller, K.D.; Lee, M.J.; Trepel, J.B. Randomized phase II, double-blind, placebo-controlled study of exemestane with or without entinostat in postmenopausal women with locally recurrent or metastatic estrogen receptor-positive breast cancer progressing on treatment with a nonsteroidal aromatase inhibitor. *J. Clin. Oncol.* **2013**, *31*, 2128–2135. [PubMed]

71. Chan, E.; Chiorean, E.G.; O'Dwyer, P.J.; Gabrail, N.Y.; Alcindor, T.; Potvin, D.; Chao, R.; Hurwitz, H. Phase I/II study of mocetinostat in combination with gemcitabine for patients with advanced pancreatic cancer and other advanced solid tumors. *Cancer Chemother. Pharmacol.* **2018**, *81*, 355–364. [CrossRef] [PubMed]

72. Szakacs, G.; Paterson, J.K.; Ludwig, J.A.; Booth-Genthe, C.; Gottesman, M.M. Targeting multidrug resistance in cancer. *Nat. Rev. Drug Discov.* **2006**, *5*, 219–234. [CrossRef] [PubMed]

73. Vitale, I.; Manic, G.; De Maria, R.; Kroemer, G.; Galluzzi, L. DNA damage in stem cells. *Mol. Cell* **2017**, *66*, 306–319. [CrossRef]

74. Robey, R.W.; Pluchino, K.M.; Hall, M.D.; Fojo, A.T.; Bates, S.E.; Gottesman, M.M. Revisiting the role of ABC transporters in multidrug-resistant cancer. *Nat. Rev. Cancer* **2018**, *18*, 452–464. [CrossRef] [PubMed]

75. Chau, W.K.; Ip, C.K.; Mak, A.S.; Lai, H.C.; Wong, A.S. c-Kit mediates chemoresistance and tumor-initiating capacity of ovarian cancer cells through activation of Wnt/beta-catenin-ATP-binding cassette G2 signaling. *Oncogene* **2013**, *32*, 2767–2781. [CrossRef] [PubMed]

76. Guo, Q.; Grimmig, T.; Gonzalez, G.; Giobbie-Hurder, A.; Berg, G.; Carr, N.; Wilson, B.J.; Banerjee, P.; Ma, J.; Gold, J.S.; et al. ATP-binding cassette member B5 (ABCB5) promotes tumor cell invasiveness in human colorectal cancer. *J. Biol. Chem.* **2018**, *293*, 11166–11178. [CrossRef] [PubMed]

77. Shervington, A.; Lu, C. Expression of multidrug resistance genes in normal and cancer stem cells. *Cancer Investig.* **2008**, *26*, 535–542. [CrossRef] [PubMed]

78. Schatton, T.; Murphy, G.F.; Frank, N.Y.; Yamaura, K.; Waaga-Gasser, A.M.; Gasser, M.; Zhan, Q.; Jordan, S.; Duncan, L.M.; Weishaupt, C.; et al. Identification of cells initiating human melanomas. *Nature* **2008**, *451*, 345–349. [CrossRef]

79. Fulda, S. Regulation of apoptosis pathways in cancer stem cells. *Cancer Lett.* **2013**, *338*, 168–173. [CrossRef] [PubMed]

80. Madjd, Z.; Mehrjerdi, A.Z.; Sharifi, A.M.; Molanaei, S.; Shahzadi, S.Z.; Asadi-Lari, M. CD44+ cancer cells express higher levels of the anti-apoptotic protein Bcl-2 in breast tumours. *Cancer Immun.* **2009**, *9*, 4.

81. Konopleva, M.; Zhao, S.; Hu, W.; Jiang, S.; Snell, V.; Weidner, D.; Jackson, C.E.; Zhang, X.; Champlin, R.; Estey, E.; et al. The anti-apoptotic genes Bcl-X(L) and Bcl-2 are over-expressed and contribute to chemoresistance of non-proliferating leukaemic CD34+ cells. *Br. J. Haematol.* **2002**, *118*, 521–534. [CrossRef]

82. Colak, S.; Zimberlin, C.D.; Fessler, E.; Hogdal, L.; Prasetyanti, P.R.; Grandela, C.M.; Letai, A.; Medema, J.P. Decreased mitochondrial priming determines chemoresistance of colon cancer stem cells. *Cell Death Differ.* **2014**, *21*, 1170–1177. [CrossRef]

83. Todaro, M.; Alea, M.P.; Di Stefano, A.B.; Cammareri, P.; Vermeulen, L.; Iovino, F.; Tripodo, C.; Russo, A.; Gulotta, G.; Medema, J.P.; et al. Colon cancer stem cells dictate tumor growth and resist cell death by production of interleukin-4. *Cell Stem Cell* **2007**, *1*, 389–402. [CrossRef]

84. Xie, Q.; Wang, S.; Zhao, Y.; Zhang, Z.; Qin, C.; Yang, X. MiR-519d impedes cisplatin-resistance in breast cancer stem cells by down-regulating the expression of MCL-1. *Oncotarget* **2017**, *8*, 22003–22013. [CrossRef] [PubMed]

85. Rouhrazi, H.; Turgan, N.; Oktem, G. Zoledronic acid overcomes chemoresistance by sensitizing cancer stem cells to apoptosis. *Biotech. Histochem.* **2018**, *93*, 77–88. [CrossRef] [PubMed]

86. Milone, M.R.; Pucci, B.; Bifulco, K.; Iannelli, F.; Lombardi, R.; Ciardiello, C.; Bruzzese, F.; Carriero, M.V.; Budillon, A. Proteomic analysis of zoledronic-acid resistant prostate cancer cells unveils novel pathways characterizing an invasive phenotype. *Oncotarget* **2015**, *6*, 5324–5341. [PubMed]

87. Milone, M.R.; Pucci, B.; Bruzzese, F.; Carbone, C.; Piro, G.; Costantini, S.; Capone, F.; Leone, A.; Di Gennaro, E.; Caraglia, M.; et al. Acquired resistance to zoledronic acid and the parallel acquisition of an aggressive phenotype are mediated by p38-MAP kinase activation in prostate cancer cells. *Cell Death Dis.* **2013**, *4*, e641. [CrossRef] [PubMed]

88. Van Houdt, W.J.; Emmink, B.L.; Pham, T.V.; Piersma, S.R.; Verheem, A.; Vries, R.G.; Fratantoni, S.A.; Pronk, A.; Clevers, H.; Borel Rinkes, I.H.; et al. Comparative proteomics of colon cancer stem cells and differentiated tumor cells identifies BIRC6 as a potential therapeutic target. *Mol. Cell Proteomics* **2011**, *10*, M111.011353. [CrossRef] [PubMed]

89. Bao, S.; Wu, Q.; McLendon, R.E.; Hao, Y.; Shi, Q.; Hjelmeland, A.B.; Dewhirst, M.W.; Bigner, D.D.; Rich, J.N. Glioma stem cells promote radioresistance by preferential activation of the DNA damage response. *Nature* **2006**, *444*, 756–760. [CrossRef] [PubMed]

90. Carruthers, R.; Ahmed, S.U.; Strathdee, K.; Gomez-Roman, N.; Amoah-Buahin, E.; Watts, C.; Chalmers, A.J. Abrogation of radioresistance in glioblastoma stem-like cells by inhibition of ATM kinase. *Mol. Oncol.* **2015**, *9*, 192–203. [CrossRef] [PubMed]

91. Zhang, P.; Wei, Y.; Wang, L.; Debeb, B.G.; Yuan, Y.; Zhang, J.; Yuan, J.; Wang, M.; Chen, D.; Sun, Y.; et al. ATM-mediated stabilization of ZEB1 promotes DNA damage response and radioresistance through CHK1. *Nat. Cell Biol.* **2014**, *16*, 864–875. [CrossRef]

92. Kreso, A.; O'Brien, C.A.; van Galen, P.; Gan, O.I.; Notta, F.; Brown, A.M.; Ng, K.; Ma, J.; Wienholds, E.; Dunant, C.; et al. Variable clonal repopulation dynamics influence chemotherapy response in colorectal cancer. *Science* **2013**, *339*, 543–548. [CrossRef]

93. Chen, J.; Li, Y.; Yu, T.S.; McKay, R.M.; Burns, D.K.; Kernie, S.G.; Parada, L.F. A restricted cell population propagates glioblastoma growth after chemotherapy. *Nature* **2012**, *488*, 522–526. [CrossRef]

94. Oshimori, N.; Oristian, D.; Fuchs, E. TGF-β promotes heterogeneity and drug resistance in squamous cell carcinoma. *Cell* **2015**, *160*, 963–976. [CrossRef] [PubMed]

95. Creighton, C.J.; Li, X.; Landis, M.; Dixon, J.M.; Neumeister, V.M.; Sjolund, A.; Rimm, D.L.; Wong, H.; Rodriguez, A.; Herschkowitz, J.I.; et al. Residual breast cancers after conventional therapy display mesenchymal as well as tumor-initiating features. *Proc. Natl. Acad Sci. USA* **2009**, *106*, 13820–13825. [CrossRef] [PubMed]

96. Soeda, A.; Lathia, J.; Williams, B.J.; Wu, Q.; Gallagher, J.; Androutsellis-Theotokis, A.; Giles, A.J.; Yang, C.; Zhuang, Z.; Gilbert, M.R.; et al. The p38 signaling pathway mediates quiescence of glioma stem cells by regulating epidermal growth factor receptor trafficking. *Oncotarget* **2017**, *8*, 33316–33328. [CrossRef] [PubMed]

97. Fujita, S.; Honma, D.; Adachi, N.; Araki, K.; Takamatsu, E.; Katsumoto, T.; Yamagata, K.; Akashi, K.; Aoyama, K.; Iwama, A.; et al. Dual inhibition of EZH1/2 breaks the quiescence of leukemia stem cells in acute myeloid leukemia. *Leukemia* **2018**, *32*, 855–864. [CrossRef] [PubMed]

98. Kim, S.H.; Joshi, K.; Ezhilarasan, R.; Myers, T.R.; Siu, J.; Gu, C.; Nakano-Okuno, M.; Taylor, D.; Minata, M.; Sulman, E.P.; et al. EZH2 protects glioma stem cells from radiation-induced cell death in a MELK/FOXM1-dependent manner. *Stem Cell Rep.* **2015**, *4*, 226–238. [CrossRef] [PubMed]

99. Chang, C.J.; Yang, J.Y.; Xia, W.; Chen, C.T.; Xie, X.; Chao, C.H.; Woodward, W.A.; Hsu, J.M.; Hortobagyi, G.N.; Hung, M.C. EZH2 promotes expansion of breast tumor initiating cells through activation of RAF1-β-catenin signaling. *Cancer Cell* **2011**, *19*, 86–100. [CrossRef] [PubMed]

100. Suda, T.; Takubo, K.; Semenza, G.L. Metabolic regulation of hematopoietic stem cells in the hypoxic niche. *Cell Stem Cell* **2011**, *9*, 298–310. [CrossRef]

101. Simsek, T.; Kocabas, F.; Zheng, J.; Deberardinis, R.J.; Mahmoud, A.I.; Olson, E.N.; Schneider, J.W.; Zhang, C.C.; Sadek, H.A. The distinct metabolic profile of hematopoietic stem cells reflects their location in a hypoxic niche. *Cell Stem Cell* **2010**, *7*, 380–390. [CrossRef]

102. Leone, A.; Roca, M.S.; Ciardiello, C.; Costantini, S.; Budillon, A. Oxidative stress gene expression profile correlates with cancer patient poor prognosis: Identification of crucial pathways might select novel therapeutic approaches. *Oxid. Med. Cell Longev.* **2017**, *2017*, 2597581. [CrossRef]

103. Farnie, G.; Sotgia, F.; Lisanti, M.P. High mitochondrial mass identifies a sub-population of stem-like cancer cells that are chemo-resistant. *Oncotarget* **2015**, *6*, 30472–30486. [CrossRef]

104. Prieto, J.; Leon, M.; Ponsoda, X.; Sendra, R.; Bort, R.; Ferrer-Lorente, R.; Raya, A.; Lopez-Garcia, C.; Torres, J. Early ERK1/2 activation promotes DRP1-dependent mitochondrial fission necessary for cell reprogramming. *Nat. Commun.* **2016**, *7*, 11124. [CrossRef] [PubMed]

105. Kuntz, E.M.; Baquero, P.; Michie, A.M.; Dunn, K.; Tardito, S.; Holyoake, T.L.; Helgason, G.V.; Gottlieb, E. Targeting mitochondrial oxidative phosphorylation eradicates therapy-resistant chronic myeloid leukemia stem cells. *Nat. Med.* **2017**, *23*, 1234–1240. [CrossRef] [PubMed]

106. Fiori, M.E.; Villanova, L.; De Maria, R. Cancer stem cells: At the forefront of personalized medicine and immunotherapy. *Curr. Opin. Pharmacol.* **2017**, *35*, 1–11. [CrossRef] [PubMed]

107. Maccalli, C.; Volonte, A.; Cimminiello, C.; Parmiani, G. Immunology of cancer stem cells in solid tumours. A review. *Eur. J. Cancer* **2014**, *50*, 649–655. [CrossRef] [PubMed]

108. Luke, J.J.; Bao, R.; Sweis, R.F.; Spranger, S.; Gajewski, T.F. WNT/beta-catenin pathway activation correlates with immune exclusion across human cancers. *Clin. Cancer Res.* **2019**, *25*, 3074–3083. [CrossRef]

109. Agudo, J.; Park, E.S.; Rose, S.A.; Alibo, E.; Sweeney, R.; Dhainaut, M.; Kobayashi, K.S.; Sachidanandam, R.; Baccarini, A.; Merad, M.; et al. Quiescent tissue stem cells evade immune surveillance. *Immunity* **2018**, *48*, 271–285. [CrossRef] [PubMed]

110. Wu, F.H.; Mu, L.; Li, X.L.; Hu, Y.B.; Liu, H.; Han, L.T.; Gong, J.P. Characterization and functional analysis of a slow-cycling subpopulation in colorectal cancer enriched by cell cycle inducer combined chemotherapy. *Oncotarget* **2017**, *8*, 78466–78479. [CrossRef] [PubMed]

111. Peng, D.; Tanikawa, T.; Li, W.; Zhao, L.; Vatan, L.; Szeliga, W.; Wan, S.; Wei, S.; Wang, Y.; Liu, Y.; et al. Myeloid-derived suppressor cells endow stem-like qualities to breast cancer cells through IL6/STAT3 and NO/NOTCH cross-talk signaling. *Cancer Res.* **2016**, *76*, 3156–3165. [CrossRef]

112. Huang, X.P.; Li, X.; Situ, M.Y.; Huang, L.Y.; Wang, J.Y.; He, T.C.; Yan, Q.H.; Xie, X.Y.; Zhang, Y.J.; Gao, Y.H.; et al. Entinostat reverses cisplatin resistance in esophageal squamous cell carcinoma via down-regulation of multidrug resistance gene 1. *Cancer Lett.* **2018**, *414*, 294–300. [CrossRef]

113. Zhao, G.; Wang, G.; Bai, H.; Li, T.; Gong, F.; Yang, H.; Wen, J.; Wang, W. Targeted inhibition of HDAC8 increases the doxorubicin sensitivity of neuroblastoma cells via up regulation of miR-137. *Eur. J. Pharmacol.* **2017**, *802*, 20–26. [CrossRef]

114. To, K.K.; Tong, W.S.; Fu, L.W. Reversal of platinum drug resistance by the histone deacetylase inhibitor belinostat. *Lung Cancer* **2017**, *103*, 58–65. [CrossRef] [PubMed]

115. Wang, H.; Huang, C.; Zhao, L.; Zhang, H.; Yang, J.M.; Luo, P.; Zhan, B.X.; Pan, Q.; Li, J.; Wang, B.L. Histone deacetylase inhibitors regulate P-gp expression in colorectal cancer via transcriptional activation and mRNA stabilization. *Oncotarget* **2016**, *7*, 49848–49858. [CrossRef]

116. Tomono, T.; Machida, T.; Kamioka, H.; Shibasaki, Y.; Yano, K.; Ogihara, T. Entinostat reverses P-glycoprotein activation in snail-overexpressing adenocarcinoma HCC827 cells. *PLoS ONE* **2018**, *13*, e0200015. [CrossRef] [PubMed]

117. Chikamatsu, K.; Ishii, H.; Murata, T.; Sakakura, K.; Shino, M.; Toyoda, M.; Takahashi, K.; Masuyama, K. Alteration of cancer stem cell-like phenotype by histone deacetylase inhibitors in squamous cell carcinoma of the head and neck. *Cancer Sci.* **2013**, *104*, 1468–1475. [CrossRef] [PubMed]

118. Aztopal, N.; Erkisa, M.; Erturk, E.; Ulukaya, E.; Tokullugil, A.H.; Ari, F. Valproic acid, a histone deacetylase inhibitor, induces apoptosis in breast cancer stem cells. *Chem. Biol. Interact.* **2018**, *280*, 51–58. [CrossRef] [PubMed]

119. Di Pompo, G.; Salerno, M.; Rotili, D.; Valente, S.; Zwergel, C.; Avnet, S.; Lattanzi, G.; Baldini, N.; Mai, A. Novel histone deacetylase inhibitors induce growth arrest, apoptosis, and differentiation in sarcoma cancer stem cells. *J. Med. Chem.* **2015**, *58*, 4073–4079. [CrossRef] [PubMed]

120. Nalls, D.; Tang, S.N.; Rodova, M.; Srivastava, R.K.; Shankar, S. Targeting epigenetic regulation of miR-34a for treatment of pancreatic cancer by inhibition of pancreatic cancer stem cells. *PLoS ONE* **2011**, *6*, e24099. [CrossRef]

121. Song, K.H.; Choi, C.H.; Lee, H.J.; Oh, S.J.; Woo, S.R.; Hong, S.O.; Noh, K.H.; Cho, H.; Chung, E.J.; Kim, J.H.; et al. HDAC1 upregulation by NANOG promotes multidrug resistance and a stem-like phenotype in immune edited tumor cells. *Cancer Res.* **2017**, *77*, 5039–5053. [CrossRef]

122. Di Gennaro, E.; Bruzzese, F.; Pepe, S.; Leone, A.; Delrio, P.; Subbarayan, P.R.; Avallone, A.; Budillon, A. Modulation of thymidilate synthase and p53 expression by HDAC inhibitor vorinostat resulted in synergistic antitumor effect in combination with 5FU or raltitrexed. *Cancer Biol. Ther.* **2009**, *8*, 782–791. [CrossRef]

123. Di Gennaro, E.; Piro, G.; Chianese, M.I.; Franco, R.; Di Cintio, A.; Moccia, T.; Luciano, A.; de Ruggiero, I.; Bruzzese, F.; Avallone, A.; et al. Vorinostat synergises with capecitabine through upregulation of thymidine phosphorylase. *Br. J. Cancer* **2010**, *103*, 1680–1691. [CrossRef]

124. Terranova-Barberio, M.; Roca, M.S.; Zotti, A.I.; Leone, A.; Bruzzese, F.; Vitagliano, C.; Scogliamiglio, G.; Russo, D.; D'Angelo, G.; Franco, R.; et al. Valproic acid potentiates the anticancer activity of capecitabine in vitro and in vivo in breast cancer models via induction of thymidine phosphorylase expression. *Oncotarget* **2016**, *7*, 7715–7731. [CrossRef] [PubMed]

125. Fazzone, W.; Wilson, P.M.; Labonte, M.J.; Lenz, H.J.; Ladner, R.D. Histone deacetylase inhibitors suppress thymidylate synthase gene expression and synergize with the fluoropyrimidines in colon cancer cells. *Int. J. Cancer* **2009**, *125*, 463–473. [CrossRef] [PubMed]

126. Piro, G.; Roca, M.S.; Bruzzese, F.; Carbone, C.; Iannelli, F.; Leone, A.; Volpe, M.G.; Budillon, A.; Di Gennaro, E. Vorinostat potentiates cisplatin-5-fluorouracil combination by inhibiting chemotherapy-induced EGFR nuclear translocation and increasing cisplatin uptake. *Mol. Cancer Ther.* **2019**, in press.

127. Muller, B.M.; Jana, L.; Kasajima, A.; Lehmann, A.; Prinzler, J.; Budczies, J.; Winzer, K.J.; Dietel, M.; Weichert, W.; Denkert, C. Differential expression of histone deacetylases HDAC1, 2 and 3 in human breast cancer—Overexpression of HDAC2 and HDAC3 is associated with clinicopathological indicators of disease progression. *BMC Cancer* **2013**, *13*, 215. [CrossRef] [PubMed]

128. Spurling, C.C.; Godman, C.A.; Noonan, E.J.; Rasmussen, T.P.; Rosenberg, D.W.; Giardina, C. HDAC3 overexpression and colon cancer cell proliferation and differentiation. *Mol. Carcinog.* **2008**, *47*, 137–147. [CrossRef] [PubMed]

129. Minami, J.; Suzuki, R.; Mazitschek, R.; Gorgun, G.; Ghosh, B.; Cirstea, D.; Hu, Y.; Mimura, N.; Ohguchi, H.; Cottini, F.; et al. Histone deacetylase 3 as a novel therapeutic target in multiple myeloma. *Leukemia* **2014**, *28*, 680–689. [CrossRef] [PubMed]

130. Ackland, S.P.; Clarke, S.J.; Beale, P.; Peters, G.J. Thymidylate synthase inhibitors. *Cancer Chemother. Biol. Response Modif.* **2002**, *20*, 1–36. [PubMed]

131. Avallone, A.; Piccirillo, M.C.; Delrio, P.; Pecori, B.; Di Gennaro, E.; Aloj, L.; Tatangelo, F.; D'Angelo, V.; Granata, C.; Cavalcanti, E.; et al. Phase 1/2 study of valproic acid and short-course radiotherapy plus capecitabine as preoperative treatment in low-moderate risk rectal cancer-V-shoRT-R3 (Valproic acid—Short Radiotherapy—Rectum 3rd trial). *BMC Cancer* **2014**, *14*, 875. [CrossRef] [PubMed]

132. Avallone, A.; Istituto Nazionale Tumori–IRCCS–Fondazione G. Pascale, 80131, Naples, Italy. Unpublished work. 2019.

133. Huang, T.H.; Wu, S.Y.; Huang, Y.J.; Wei, P.L.; Wu, A.T.; Chao, T.Y. The identification and validation of Trichosstatin A as a potential inhibitor of colon tumorigenesis and colon cancer stem-like cells. *Am. J. Cancer Res.* **2017**, *7*, 1227–1237.

134. Kumar, B.; Yadav, A.; Lang, J.C.; Teknos, T.N.; Kumar, P. Suberoylanilide hydroxamic acid (SAHA) reverses chemoresistance in head and neck cancer cells by targeting cancer stem cells via the downregulation of nanog. *Genes Cancer* **2015**, *6*, 169–181.

135. Wang, L.; Liu, X.; Ren, Y.; Zhang, J.; Chen, J.; Zhou, W.; Guo, W.; Wang, X.; Chen, H.; Li, M.; et al. Cisplatin-enriching cancer stem cells confer multidrug resistance in non-small cell lung cancer via enhancing TRIB1/HDAC activity. *Cell Death Dis.* **2017**, *8*, e2746. [CrossRef]

136. Miyajima, C.; Inoue, Y.; Hayashi, H. Pseudokinase tribbles 1 (TRB1) negatively regulates tumor-suppressor activity of p53 through p53 deacetylation. *Biol. Pharm. Bull.* **2015**, *38*, 618–624. [CrossRef] [PubMed]

137. Wang, Y.; Wu, N.; Pang, B.; Tong, D.; Sun, D.; Sun, H.; Zhang, C.; Sun, W.; Meng, X.; Bai, J.; et al. TRIB1 promotes colorectal cancer cell migration and invasion through activation MMP-2 via FAK/Src and ERK pathways. *Oncotarget* **2017**, *8*, 47931–47942. [CrossRef] [PubMed]

138. Lu, X.X.; Hu, J.J.; Fang, Y.; Wang, Z.T.; Xie, J.J.; Zhan, Q.; Deng, X.X.; Chen, H.; Jin, J.B.; Peng, C.H.; et al. A case-control study indicates that the TRIB1 gene is associated with pancreatic cancer. *Genet. Mol. Res.* **2014**, *13*, 6142–6147. [CrossRef] [PubMed]

139. Lee, H.S.; Park, S.B.; Kim, S.A.; Kwon, S.K.; Cha, H.; Lee, D.Y.; Ro, S.; Cho, J.M.; Song, S.Y. A novel HDAC inhibitor, CG200745, inhibits pancreatic cancer cell growth and overcomes gemcitabine resistance. *Sci. Rep.* **2017**, *7*, 41615. [CrossRef] [PubMed]

140. Ouaissi, M.; Sielezneff, I.; Silvestre, R.; Sastre, B.; Bernard, J.P.; Lafontaine, J.S.; Payan, M.J.; Dahan, L.; Pirro, N.; Seitz, J.F.; et al. High histone deacetylase 7 (HDAC7) expression is significantly associated with adenocarcinomas of the pancreas. *Ann. Surg. Oncol.* **2008**, *15*, 2318–2328. [CrossRef] [PubMed]

141. Jung, D.E.; Park, S.B.; Kim, K.; Kim, C.; Song, S.Y. CG200745, an HDAC inhibitor, induces anti-tumour effects in cholangiocarcinoma cell lines via miRNAs targeting the Hippo pathway. *Sci. Rep.* **2017**, *7*, 10921. [CrossRef] [PubMed]

142. Wang, J.H.; Lee, E.J.; Ji, M.; Park, S.M. HDAC inhibitors, trichostatin A and valproic acid, increase Ecadherin and vimentin expression but inhibit migration and invasion of cholangiocarcinoma cells. *Oncol. Rep.* **2018**, *40*, 346–354.

143. Lopez, G.; Braggio, D.; Zewdu, A.; Casadei, L.; Batte, K.; Bid, H.K.; Koller, D.; Yu, P.; Iwenofu, O.H.; Strohecker, A.; et al. Mocetinostat combined with gemcitabine for the treatment of leiomyosarcoma: Preclinical correlates. *PLoS ONE* **2017**, *12*, e0188859. [CrossRef]

144. Matulonis, U.; Berlin, S.; Lee, H.; Whalen, C.; Obermayer, E.; Penson, R.; Liu, J.; Campos, S.; Krasner, C.; Horowitz, N. Phase I study of combination of vorinostat, carboplatin, and gemcitabine in women with recurrent, platinum-sensitive epithelial ovarian, fallopian tube, or peritoneal cancer. *Cancer Chemother. Pharmacol.* **2015**, *76*, 417–423. [CrossRef]

145. Bruzzese, F.; Leone, A.; Rocco, M.; Carbone, C.; Piro, G.; Caraglia, M.; Di Gennaro, E.; Budillon, A. HDAC inhibitor vorinostat enhances the antitumor effect of gefitinib in squamous cell carcinoma of head and neck by modulating ErbB receptor expression and reverting EMT. *J. Cell Physiol.* **2011**, *226*, 2378–2390. [CrossRef]

146. Leone, A.; Roca, M.S.; Ciardiello, C.; Terranova-Barberio, M.; Vitagliano, C.; Ciliberto, G.; Mancini, R.; Di Gennaro, E.; Bruzzese, F.; Budillon, A. Vorinostat synergizes with EGFR inhibitors in NSCLC cells by increasing ROS via up-regulation of the major mitochondrial porin VDAC1 and modulation of the c-Myc-NRF2-KEAP1 pathway. *Free Radic. Biol. Med.* **2015**, *89*, 287–299. [CrossRef] [PubMed]

147. Wang, Z.; Tang, F.; Hu, P.; Wang, Y.; Gong, J.; Sun, S.; Xie, C. HDAC6 promotes cell proliferation and confers resistance to gefitinib in lung adenocarcinoma. *Oncol. Rep.* **2016**, *36*, 589–597. [CrossRef] [PubMed]

148. Wang, L.; Li, H.; Ren, Y.; Zou, S.; Fang, W.; Jiang, X.; Jia, L.; Li, M.; Liu, X.; Yuan, X.; et al. Targeting HDAC with a novel inhibitor effectively reverses paclitaxel resistance in non-small cell lung cancer via multiple mechanisms. *Cell Death Dis.* **2016**, *7*, e2063. [CrossRef] [PubMed]

149. Galloway, T.J.; Wirth, L.J.; Colevas, A.D.; Gilbert, J.; Bauman, J.E.; Saba, N.F.; Raben, D.; Mehra, R.; Ma, A.W.; Atoyan, R.; et al. A phase I study of CUDC-101, a multitarget inhibitor of HDACs, EGFR, and HER2, in combination with chemoradiation in patients with head and neck squamous cell carcinoma. *Clin. Cancer Res.* **2015**, *21*, 1566–1573. [CrossRef] [PubMed]

150. Vonderheide, R.H. The immune revolution: A case for priming, not checkpoint. *Cancer Cell* **2018**, *33*, 563–569. [CrossRef] [PubMed]

151. Terranova-Barberio, M.; Thomas, S.; Munster, P.N. Epigenetic modifiers in immunotherapy: A focus on checkpoint inhibitors. *Immunotherapy* **2016**, *8*, 705–719. [CrossRef] [PubMed]

152. Terranova-Barberio, M.; Thomas, S.; Ali, N.; Pawlowska, N.; Park, J.; Krings, G.; Rosenblum, M.D.; Budillon, A.; Munster, P.N. HDAC inhibition potentiates immunotherapy in triple negative breast cancer. *Oncotarget* **2017**, *8*, 114156–114172. [CrossRef]

153. Miyashita, T.; Miki, K.; Kamigaki, T.; Makino, I.; Tajima, H.; Nakanuma, S.; Hayashi, H.; Takamura, H.; Fushida, S.; Ahmed, A.K.; et al. Low-dose valproic acid with low-dose gemcitabine augments MHC class I-related chain A/B expression without inducing the release of soluble MHC class I-related chain A/B. *Oncol. Lett.* **2017**, *14*, 5918–5926. [CrossRef]

154. Krug, L.M.; Kindler, H.L.; Calvert, H.; Manegold, C.; Tsao, A.S.; Fennell, D.; Ohman, R.; Plummer, R.; Eberhardt, W.E.; Fukuoka, K.; et al. Vorinostat in patients with advanced malignant pleural mesothelioma who have progressed on previous chemotherapy (VANTAGE-014): A phase 3, double-blind, randomised, placebo-controlled trial. *Lancet Oncol.* **2015**, *16*, 447–456. [CrossRef]

155. Suraweera, A.; O'Byrne, K.J.; Richard, D.J. Combination therapy with histone deacetylase inhibitors (HDACi) for the treatment of cancer: Achieving the full therapeutic potential of HDACi. *Front. Oncol.* **2018**, *8*, 92. [CrossRef]

156. Ramalingam, S.S.; Maitland, M.L.; Frankel, P.; Argiris, A.E.; Koczywas, M.; Gitlitz, B.; Thomas, S.; Espinoza-Delgado, I.; Vokes, E.E.; Gandara, D.R.; et al. Carboplatin and Paclitaxel in combination with either vorinostat or placebo for first-line therapy of advanced non-small-cell lung cancer. *J. Clin. Oncol.* **2010**, *28*, 56–62. [CrossRef] [PubMed]

157. Wilting, R.H.; Yanover, E.; Heideman, M.R.; Jacobs, H.; Horner, J.; van der Torre, J.; DePinho, R.A.; Dannenberg, J.H. Overlapping functions of Hdac1 and Hdac2 in cell cycle regulation and haematopoiesis. *EMBO J.* **2010**, *29*, 2586–2597. [CrossRef] [PubMed]

MDPI

St. Alban-Anlage 66

4052 Basel

Switzerland

Tel. +41 61 683 77 34

Fax +41 61 302 89 18

www.mdpi.com

Journal of Clinical Medicine Editorial Office

E-mail: jcm@mdpi.com

www.mdpi.com/journal/jcm

9 783039 434060